D0949937

SOURCE BOOK IN ASTRONOMY
1900–1950

The Radio telescope of the U. S. Naval Research Laboratory, Washington.

SOURCE BOOK
IN ASTRONOMY
1900–1950

HARLOW SHAPLEY
EDITOR

HARVARD UNIVERSITY PRESS

CAMBRIDGE, MASSACHUSETTS

1960

GENERAL EDITOR'S PREFACE

In his original preface, the founder of this series of Source Books, Professor Gregory D. Walcott, wrote, "The general plan is for each volume to present a treatment of a particular science with as much finality of scholarship as possible from the Renaissance to the end of the nineteenth century." Under his guidance, the concept was extended both backward into the medieval and ancient worlds and forward into the first half of the present century. This enlargement of the plan will appear in two ways, namely, in a more extended historical coverage in Source Books in fields not hitherto represented, and in special volumes devoted to scientific contributions in periods not originally covered, most particularly materials that appeared between 1900 and 1950.

The series was made possible by a grant in 1927 of $10,000 by the Carnegie Corporation to the American Philosophical Association, which set up a revolving fund to help defray expenses incurred by editors of individual volumes. This fund still exists and, as a matter of fact, has increased modestly in amount. The American Association for the Advancement of Science and the History of Science Society have approved the project and are represented on the Editorial Advisory Board.

As of June 1960, this Board was composed of the following members:

Marshall Clagett	History of Science	University of Wisconsin
I. Bernard Cohen	History of Science	Harvard University
C. J. Ducasse	Philosophy	Brown University
Ernest Nagel	Philosophy	Columbia University
Harlow Shapley	Astronomy	Harvard University
Edmund W. Sinnott	Botany	Yale University
Harry Woolf	History of Science	University of Washington

The General Editor wishes to acknowledge the indispensable help he has received from the Advisory Board and the whole-hearted cooperation of the Harvard University Press and particularly of Mr. Joseph D. Elder, its Science Editor. He wishes to express his gratitude, and that of many others interested in the history of science, to his predecessor for his devoted labors in starting this series and carrying it on successfully through several crises. It is his conviction that, in this age of science, an increased understanding

of the history of the sciences can broaden both the nonscientist's acquaint-
ance with scientific thought and the scientist's cultural outlook.

EVERETT W. HALL
General Editor

Department of Philosophy
The University of North Carolina
Chapel Hill

Professor Hall died suddenly on June 17, 1960. His successor as General
Editor of the series will be announced in the next volume.

PREFACE

An attempt has been made to collect in this volume sixty-nine contributions that satisfactorily illustrate the vigorous march of astronomical science during the interval from 1900 to 1950—contributions which are at the same time informative to the general reader. To serve the nonscientific reader it has seemed advisable to avoid as much as possible mathematical equations, spectroscopic terminology, and chemical formulas. Nevertheless, some fairly technical contributions, such as those by Einstein, Hertzsprung, Ambartsumian, Edlén, Chandrasekhar, Reber, Lyot, and van de Hulst have been included, since they represent high points in our understanding of the astronomical world. (Incidentally, these eight scientists are from eight different countries—an illustration of the wide spread of astronomical inquiry.)

Among the several important astronomical contributions that do not find place in this volume because they are too technical, too fragmentary, or because the manner of presentation is unsuitable, are several papers by Karl Schwarzschild, who not only touched effectively the theory of relativity and derived the Schwarzschild line element, but also pioneered in photographic photometry and made lasting contributions in the fields of radiative equilibrium and stellar statistics.

Examples of other technical or too tabular contributions that would have given professional astronomers a fuller historical picture of the half-century are S. W. Burnham's and R. G. Aitken's work on double stars; the important early discussion of the relative abundances of the elements by Cecelia Payne (Gaposchkin) in her "Stellar Atmospheres" (1925); the discovery of interstellar gas clouds by W. S. Adams; Bengt Strömgren's fundamental analysis of interstellar hydrogen, and his pioneer work on stellar evolution; E. A. Milne's report on Sir James Jeans' mathematical treatment of rotating gaseous spheres and his own novel work on relativity theory; A. Unsöld's early astrophysical explorations in his "Physik der Sternatmosphären"; the stellar parallax work at the Allegheny, McCormick, Yale, van Vleck, Mount Wilson, Yerkes, and Cape observatories; the work by J. S. Plaskett, R. M. Petrie, and their colleagues at Victoria on the spectra and orbits of close binary stars; and S. Chandrasekhar's work on many mathematical phases of theoretical astrophysics.

The valuable series of eighteen articles on "Physical Processes in Gaseous Nebulae" by D. H. Menzel and his associates: L. H. Aller, J. G. Baker, M. H. Webb, Leo Goldberg, and George Shortley, are too specialized and technical for inclusion.

This quality of not being quotable in a book that will probably be used not so much by practicing astronomers as by historians of science affects much of pre-1950 theoretical astrophysics. The outstanding work, for example, of G. Abetti, A. A. Belopolski, L. Biermann, A. Danjon, R. Emden, W. Grotrian, K. O. Kiepenheuer, E. A. Milne, H. Mineur, M. G. J. Minnaert, J. H. Moore, A. Pannekoek, H. H. Plaskett, P. Swings, M. Waldmeier, and C. F. von Weizsäcker must be examined in the original publications by those desirous of acquainting themselves with the technical methods and hypotheses of astrophysics.

It is noteworthy and pleasing, however, that many major advances in the photometry of celestial bodies, in the astronomy of position, in stellar structure and evolution, and even in celestial mechanics and cosmogony, have been presented in a manner very suitable for inclusion in this general survey of astronomical progress. For example, the account of the founding of the Palomar Observatory by G. E. Hale and of the discovery of Pluto by C. W. Tombaugh provide very readable contributions, as do also the approach to cosmogony by Henri Poincaré, the story of the Tunguska Meteorite by L. A. Kulik and E. L. Krinov, W. de Sitter's introduction to relativity, the survey of cosmic chemistry by R. Wildt and of astrophysics by Otto Struve and S. Rosseland, and, of course, almost anything that A. S. Eddington has written.

I have not excluded various speculations that are now less credible than when first put forward, provided the hypotheses have had a considerable effect on subsequent work—for example, J. C. Kapteyn's Two Star Streams (which have been largely explained by galactic rotation), and some of the adventures in cosmogony by Eddington, G. Lemaître, and W. de Sitter.

The assignment of papers to specific parts is arbitrary, but in general obvious. The order within a part is nearly always chronological. The square-bracketed notes throughout the volume, as well as the introductions, have been written by the editor. The 1950 terminal date must be kept in mind when one looks for a report on recent developments. Almost the whole of radio astronomy and much of the writing on stellar and planetary origins and evolutions are post-1950. The same is true of the synthesis of the elements in stellar interiors.

Seventy scientists are directly quoted in the volume, but more than 375 names of astronomical investigators are listed in the index. Although this Source Book, limited in size as it must be, can scarcely claim to be a history of modern astronomy, it should serve as a general collateral guide to historical study; for it includes contributions from all astronomical fields.

It should be specifically noted that other editors might have made other selections and might have leaned to greater or lesser technicality. The present selection undoubtedly reflects to some extent the bias of the compiler. But several astronomers have aided in the selection of the items that make up the volume. In particular, Owen Gingerich, Otto Struve, and D. H. Menzel have made useful suggestions. In library work and in the preparation of the illustrations Mrs. Jaqueline S. Kloss has been of great assistance. Mrs. Shapley has assisted in all phases of the preparation of the manuscript.

The sources of the material for each chapter are specified in the Appendix, where also the investigators are identified and the original titles of their contributions given. The editor has frequently chosen titles that indicate more clearly the nature of the selection.

To the editors and publishers of the many books and journals that contain the selections I have used, I am indebted for reprinting permission. I am also indebted to Mount Wilson-Palomar, Lick, Yerkes, and Harvard Observatories for illustrative materials that were not in the original papers. Figures 3:2, 5:1, 7:1, and 38:1 have been substituted for the originals. The frontispiece and Figures 1:1, 1:2, 6:1, 14:1, 18:3, 24:1, 24:2, 50:1, 51:1, 58:1, and 69:1 have been added by the editor. The living authors have all consented to the use I have made of their articles. *The Astrophysical Journal* has been an important source. A surprising number of the most quotable discussions have first appeared in the *Leaflets* of the *Astronomical Society of the Pacific*, a circumstance in part due to the large number of contributors from the California observatories.

HARLOW SHAPLEY

January 11, 1960

CONTENTS

SOURCE BOOK IN ASTRONOMY
1900–1950

I

INSTRUMENTATION

The mechanical engineer, the electronic physicist, the optician, and the laboratory spectroscopist have made profound contributions in many astronomical fields, even though generally quite innocent of knowledge about stellar matters. Good instrumentation is a large part of our armor for the attack on the astronomical unknowns. Inspired gadgetry, sometimes taken fully developed from another area of investigation, sometimes thought out and built in a time of necessity for the astronomer's use only, has guided a large part of modern astronomy.

Among these accessories, most of them developed in the past half-century, are the quarter-wave plate in the solar observatory; the selenium cell that Joel Stebbins turned on the moon and stars; the photoelectric cells of many sorts and uses, much developed from the early tubes used by Paul Guthnick, Hans Rosenberg, J. Stebbins, A. E. Whitford, and others; the coronagraph of Bernard Lyot's designing; the Schmidt telescope and its adaptations by E. H. Linfoot and especially by J. G. Baker; [1] radio telescopes in great variety; balloons, rockets, and satellites; the polarimeter as used by J. S. Hall, W. A. Hiltner, and Elske Smith for the study of interstellar particles (see part IX); the Ross lenses and the Maksutov reflector; red-sensitive photographic emulsions; the F. E. Ross field flatteners; J. F. Hartmann's simple apparatus for testing lenses and mirrors; [2] A. Lallemand's photoelectric image converter; and the great electronic calculators, of various forms and potentialities, used widely for the solution of obstinate problems of astrophysical theory, celestial mechanics, and stellar positions and motions.

In addition to this parade of ingenuity we have accessories that count photons, split images to increase spectrographic speed, reduce measures automatically, and so on, but in this section of the Source Book we are limited by space considerations to only four astronomical instruments which stand out as exceedingly important in the recent advances of astronomy.

The contribution by Meghnad Saha concerns what could be done in spectrum analysis of sun and stars from an appropriately equipped strato-

---

[1] "Optical systems for astronomical photography," *Amateur Telescope Making*, Book Three (Scientific American, New York, 1953), pp. 1–34.

[2] "Objectivuntersuchungen," *Zeitschrift für Instrumentenkunde 24*, 1, 33, 97 (1904).

spheric observatory, but not how to build one. Professor Saha's most deservedly famous contribution to astrophysics was his pioneer and epochal work "On a Physical Theory of Stellar Spectra," *Proceedings of the Royal Society* (London), A, *99,* 135–153 (1921). It is too specialized for inclusion in the Source Book. He used both atomic physics and stellar spectroscopy to show that atomic ionization at various temperatures could account for the great variety of stellar spectral types.

We should note that almost all of the developments in radio astronomy and in the building of great radio telescopes have occurred since 1950, our terminal date.

1. THE 200–INCH REFLECTOR ON MOUNT PALOMAR

By George Ellery Hale

A single decade includes two highly significant events in the endowment of research in America. On October 1, 1892, the University of
Chicago, founded by John D. Rockefeller, opened its doors to students.
About ten years later, on January 29, 1902, the trustees of the Carnegie
Institution of Washington, established by Andrew Carnegie, held their
first meeting in Washington. The first of these events marked the incep-
tion of the long series of national and international gifts for research
which we owe to the generosity of the Rockefellers. The initiation of
the Carnegie Institution also meant, not merely one of several large en-
dowments by Carnegie, but the inauguration of an extensive plan for
the development of research, subsequently strengthened by his establishment of the Carnegie Corporation of New York.

The Carnegie Institution of Washington, through its organization of
widely distributed laboratories and observatories and the substantial
aid given to its research associates in many educational institutions,
opened extraordinary opportunities for scientific investigation. The
Rockefeller Foundation, the General Education Board, the International Education Board, and the Rockefeller Institute for Medical Research, with their broad and liberal policy, both national and international, proved admirably adapted to meet a wide variety of needs. The
present undertaking, involving the cooperation of the Rockefeller and
Carnegie groups, is but one of the countless results of their great gifts
for the increase in knowledge. . . .

Such considerations doubtless determined the Rockefeller boards,
which had previously made large grants to the California Institute, to
offer it funds in 1928 for the construction of a 200-inch reflecting telescope, together with all the buildings and equipment necessary to render
this instrument as efficient as possible. Two conditions were made by
the donors: the assurance of the active cooperation with the California
Institute in this project of the Mount Wilson Observatory of the Carnegie Institution of Washington and the provision by the California
Institute of an adequate endowment for the new Astrophysical Ob-

servatory. The president and Executive Committee of the Carnegie Institution, as well as the director of the Mount Wilson Observatory, quickly and cordially assented to the first of these conditions, while the trustees of the California Institute were no less prompt in agreeing to obtain an endowment.

Prior to the suggestion of President Wickliffe Rose of the International Education Board, the California Institute had not planned to establish a department of astronomy or astrophysics. The recent rapid development of astrophysics, however, and the great part in this development played by physics and chemistry, clearly made the most intimate possible cooperation between observatories and laboratories essential. A good beginning had already been made, but there was room for a still more effective union of effort. It was therefore decided to erect the Astrophysical Laboratory of the new Observatory next to the Norman Bridge Laboratory of Physics, in close proximity to the Gates Laboratory of Chemistry and not far from the Pasadena headquarters of the Mount Wilson Observatory. . . .

In harmony with the plan previously drawn up by the writer at the request of Dr. Rose, it was decided that the new Astrophysical Observatory thus provided for should include (1) a 200-inch reflecting telescope and other instruments, mounted at the most favorable high-altitude site to be found within effective working distance of Pasadena; (2) an Astrophysical Laboratory, on the campus of the California Institute, to serve as the Pasadena headquarters of the resident and visiting members of the research staff, the reduction of photographic and visual observations made with the 200-inch telescope and other instruments, the performance of experiments for the interpretation of these observations with the aid of apparatus supplementing that of the other laboratories of the California Institute and the Mount Wilson Observatory, the development of new instruments and methods of research, and the work of fellows and students in astrophysics; (3) Instrument and Optical shops, also on the campus of the Institute, for the construction of special apparatus and the figuring and testing of the mirrors and lenses required for the 200-inch telescope and other instruments. . . .

The 200-inch Telescope

In designing the instrumental equipment of the new Observatory the chief points to be borne in mind are (1) our purpose to supplement, rather than to duplicate, the existing apparatus of the Mount Wilson Observatory and the California Institute; (2) to multiply the efficiency

Fig. 1. The dome of the 200-inch Hale reflector on Mount Palomar.

of all our instruments not merely by increase in size but by every possible improvement in methods of design and construction and especially
by the development of new auxiliary apparatus; and (3) to render the
results of these investigations available at once to other observatories
and research laboratories, where their immediate utilization might lead
to new advances, long before the completion of the 200-inch telescope.
As the annual reports of the Mount Wilson Observatory sufficiently indicate, this hope has been amply justified.

Our experience at Mount Wilson clearly demonstrates that the new
telescope should be an equatorial reflector of large angular aperture, so
mounted as to permit observation at the principal focus of the 200-inch
mirror, at the Cassegrain focus below the large mirror, and at the coudé
focus within a constant temperature laboratory south of the polar axis.
These possibilities exist in the case of the 100-inch Hooker telescope,
but this instrument cannot reach the region surrounding the north pole

(its farthest range in north declination is 63°58′), observation at its principal focus is difficult, and the time required to change the "cages" carrying the Newtonian and the two Cassegrain mirrors affects materially the observing program.

As for the optical parts, it was evident that in passing from a 100-inch to a 200-inch mirror a new type of disk would be required. The 100-inch mirror disk, made at St. Gobain of the best glass then available (ordinary plate glass), gives excellent results under almost all conditions. But even with the improved support system recently designed by Pease, it still shows at times sufficient change of figure with temperature to preclude the use of similar glass for a mirror of 200-inch aperture. The first step in the new undertaking therefore involved a comparative study of different means of making a suitable 200-inch disk.

Mirror disks.—After excluding one or two types of mirror disks which we and our advisers unanimously agreed to be unsuitable, the following possibilities remained: fused silica; some form of Pyrex glass; stainless steel or some other metallic alloy; and a special type of metallic disk, with a thin layer of glass of the same coefficient of expansion securely fused to its surface, proposed by the Philips Lamp Works of Eindhoven, Holland. Of these various materials, fused silica has the lowest coefficient of temperature expansion, and as Dr. Elihu Thomson and his associates had already made excellent small mirror disks of fused silica for use in telescopes, we arranged with the General Electric Company to have this investigation continued by Dr. Thomson and Mr. Ellis at the Thomson Research Laboratory at Lynn, Massachusetts. I trust they will publish a full account of the methods they devised and the results they accomplished. Only a brief statement regarding this long research is therefore included here.

In its earlier stages the process consisted in fusing a mass of nearly pure quartz sand in a circular electric furnace comprising the mold. The disk thus obtained, which contained a great number of bubbles, was then ground to the approximate curvature desired and coated to a sufficient thickness with a layer of perfectly transparent fused quartz, free from bubbles. This layer was produced by spraying pure crystalline quartz, in finely granular form, onto the surface of the hot disk by means of an oxyhydrogen flame. Excellent mirror disks up to a diameter of 25 inches were produced in this way. Unfortunately, however, it finally became evident that the much larger disks required could not be made by this process. . . .

Fortunately, our next line of attack was so strong that we could rely upon it with confidence. For many years the staff of the Geophysical Laboratory of the Carnegie Institution of Washington, under the direction of Dr. Arthur L. Day, had been engaged in developing new types of glasses, of which Pyrex is the best known. Under the leadership of Dr. Day, these glasses have been manufactured by the Corning Glass Works at their factory in Corning, New York. Every housewife is familiar with the remarkable qualities of the Pyrex glass used for cooking utensils, as its very low coefficient of expansion permits it to be subjected to rapid change of temperature without cracking. Glasses of this general type thus offer qualities for telescope disks far surpassing those of the glass used for the 100-inch and all other large telescope mirrors.

During the last ten years we have become familiar with the good performance of ordinary Pyrex in the solar telescopes on Mount Wilson and my laboratory in Pasadena. In the case of the 200-inch mirror, however, it was necessary to know whether any danger of devitrification might arise during the period of annealing and also to solve various other problems involved in its manufacture. A long series of new investigations was accordingly undertaken at the Research Laboratory of the Corning Glass Works, under the personal supervision of Dr. Hostetter and Dr. McCauley. These studies led to the development of special methods of casting, preliminary cooling, and annealing, which proved very successful in the case of a 60-inch disk made at Corning in July, 1932. This disk, after careful consideration, had been designed by Dr. Pease and the Corning research staff with a ribbed back, which affords the necessary stiffness.[1] It also permits the use of a greatly improved support system, besides very materially diminishing the weight and the annealing time, as contrasted with a solid disk of the old type. Since the casting of this 60-inch disk a new Pyrex glass of still lower coefficient of expansion has been thoroughly tested at Corning, with highly satisfactory results. The many superior qualities of this glass include reduced annealing time and complete freedom from any danger of devitrification. An annealing oven for the 200-inch disk and the special apparatus for casting and handling disks of large dimensions were also built. The 120-inch disk, required for testing the 200-inch mirror during the process of figuring, was then cast and annealed with great success. It is now being figured in our Optical Shop in Pasadena.

After the completion of the 200-inch mold and the casting of several large telescope disks for other observatories, a 200-inch disk was cast for us on March 25, 1934. The glass was poured from steel buckets

---

[1] A smaller ribbed coelostat mirror, made at Corning of Pyrex glass, has successfully passed very severe tests in our Optical Shop in Pasadena.

containing 750 pounds each, and many hours were required to fill the mold, as only about one-half of the glass actually entered the mold from each bucket. The extremely high temperature of the glass caused some of the supports of the mold material to break down near the end of the pouring process. Thus some portions of the mold rose into the molten disk. After a subsequent heating these were scooped out and the disk recooled. Although considerable work would be necessary to restore the symmetry of the rib system on the back of the mirror, this disk is being retained for possible use in case of any accident to its successor.

The second 200-inch disk was poured with perfect success on December 2, 1934, into a greatly improved mold designed by Dr. McCauley, of the Corning Glass Works, who also designed the electric control system for gradually lowering the temperature of the disk during the annealing process, which will be completed before the end of the present year.

The 200-inch telescope mounting.—Telescope mountings vary widely in their requirements. Visual and spectroscopic observations do not suffer greatly from flexure in the mounting, but the demands of direct photography are much more stringent. We have studied several possible designs for the mounting of the 200-inch telescope, and our investigation of the problem is still in progress. The engineers whom we have consulted call for two points of support, on opposite sides of the tube, and then endeavor to attain the greatest possible degree of rigidity conformable with the requirements of the optical design and the necessary range of motion in right ascension and declination.

As we wish to reach all parts of the sky visible in our latitude and to provide for the various optical combinations already mentioned, we have selected two possible mountings for rigorous comparison: the well-known fork type, first used on a large scale by Lassell for his 4-foot Malta reflector; and the English yoke type, illustrated in one form by the 100-inch Mount Wilson reflector. A yoke mounting, in another form suggested by Anderson, Edgar, and Serrurier, was recently embodied in a model built in our shop. The advantage of this modified yoke type, some of which were also included later in a modified fork type, are as follows:

The telescope tube, instead of being hung near its lower end between the arms of the yoke or fork, has its declination axis not far below its center. This increases its stability, and allows the use of a dome of much smaller relative diameter than the domes of the 60-inch or 100-inch reflectors on Mount Wilson.

The increased stability of the tube will allow the observer to be carried in a cartridge-shaped house at its principal focus, thus permitting the elimination of the large Newtonian mirror.

The troublesome and time-consuming task of changing the huge "cages" carrying the Cassegrain and Newtonian mirrors will be eliminated. There will be two Cassegrain mirrors, mounted in a tube extending a short distance below the principal focus of the 200-inch mirror. Each of these mirrors can be instantly turned into position by means of a worm gear, driven by an electric motor.

One of the convex mirrors will give at the secondary focus, just below the hole in the 200-inch mirror, a large and sharply defined star field, with a focal ratio of $f/16$ (equivalent focal length $= 3200$ in.). A stellar spectrograph with one, two, or three prisms will be mounted here, or a double-slide plate-carrier can be quickly substituted for the spectrograph when desired.

The same convex mirror, used in conjunction with a plane mirror standing at an angle of 45° with the axis of the telescope tube, will send the converging beam through the hollow declination axis to a totally reflecting prism, mounted before the slit of a long-focus grating or prism spectrograph. This spectrograph will be hung within a large hollow cylinder, supported parallel to the polar axis and so geared that the spectrograph slit will always remain vertical. A second cylinder on the opposite side of the polar axis will be provided to carry a radiometer or any other auxiliary instrument which must remain in a vertical plane. This ingenious scheme is a valuable feature of this form of mounting.

Another Cassegrain mirror, aided by the plane mirror suitably orientated, will send the converging beam toward the south pole through the hollow polar axis into a constant-temperature chamber, as in the case of the 100-inch telescope. This will permit the use of a fixed spectrograph or other auxiliary apparatus of any focal length desired. . . .

Site

The question has sometimes been raised whether the 200-inch telescope would not be more useful in the southern hemisphere. As all astronomers will recognize, a very large telescope, having a small field, should be rarely employed for scouting purposes, and never for work within the ready range of smaller instruments. It must be devoted almost exclusively to the study of selected objects. Thus the vastly greater knowledge of stars and nebulae acquired during centuries of

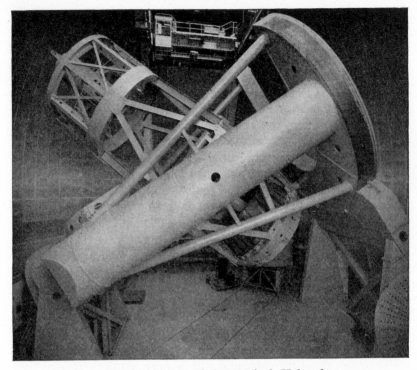

Fig. 2. An interior view of the 200-inch Hale telescope.

observation north of the equator is precisely what is needed in making adequate selections for more intensive study. Moreover, nearly two-thirds of the nearest extra-galactic nebulae [galaxies], offering the greatest possibilities of research, can be observed only from a northern station. The importance of a detailed investigation of such unique spirals as the Great Nebula in Andromeda and Messier 33, with the highest instrumental power, cannot be overstated. As for the brightest stars, the only ones available for spectroscopic examination with very high dispersion, three-quarters of them are accessible from Mount Wilson. Then, too, our existing knowledge of tens of thousands of faint stars and millions of very remote nebulae, derived from many years of work with the largest existing telescopes, provides the necessary means of selecting suitable objects for further investigation with the 200-inch telescope. Finally, the southern hemisphere can at present furnish no co-operating research institutions comparable in staff and equipment with the California Institute and the Mount Wilson Observatory.

It is, however, very desirable that much work be done south of the equator with large telescopes, thus gradually preparing the way, as in the north, for another instrument as large as the 200-inch telescope.

A long study of meteorological, topographical, and astronomical conditions, reinforced by more than thirty years of experience at Mount Wilson, leaves no room for doubt regarding the advantages of the southwestern corner of the United States as the site for a 200-inch telescope. Its latitude gives access to the whole of the northern heavens, as well as a broad zone south of the celestial equator. Its mountain ranges afford easy means of escape from the dense air of the lower atmosphere, which not only produces poor images but absorbs and scatters much light. On some of the mountains of southern California the daily range in temperature is remarkably low, while the annual range is moderate. A large daily range is very objectionable, because of the difficulty of preventing distortion of the telescope mirrors and its effect on the telescope mounting and auxiliary instruments. A large annual variation is also to be avoided, as this means low winter temperatures and greatly reduced efficiency of the observer. The average wind velocities, as shown by the Mount Wilson and other mountain records, are also very low, while the comparative freedom from clouds permits observations to be made on a large percentage of nights during the year. All these favorable conditions are readily explained when the paths of the principal storms in the United States are taken into account. . . .

In his valuable report to the Carnegie Institution of Washington on possible sites for an astrophysical observatory, the late Professor W. J. Hussey, then of the Lick Observatory, gave an excellent survey of the atmospheric conditions in southern California. Equipped with a portable 9-inch Clark refractor, he examined several mountain sites lying within the region extending from the Mexican border to the ranges lying north and east of Los Angeles, and added many useful comments on the topographical and climatic conditions in this portion of the Southwest. After following up his telescopic tests over a much longer interval of time, I fully agreed with his tentative selection of Mount Wilson, where I began work in the winter of 1903–1904.

Among other sites inspected by Professor Hussey was Mount Palomar (latitude 33°21'20'', altitude about 5600 ft.). However, after long consideration, he thought Mount Wilson would be preferable to Palomar, partly because of its easy accessibility from Los Angeles and Pasadena. At that time, in the absence of motor cars and roads worthy of the name, Palomar was very difficult to reach. Moreover, Hussey had ascended the mountain from a station on Cuyamaca, about 30 miles toward the southeast, where the conditions were poor and the desert near. Impressed by his unfavorable tests at Cuyamaca, he feared similar atmospheric disturbances at Palomar, and did not take his telescope there, though he regarded it as the most promising site in San Diego County.

We have no reason to regret Hussey's choice, as Mount Wilson has proved to be an admirable site, and even since the great increase of population in the San Gabriel Valley and the consequent multiplication of electric lights, we would not wish to move the Observatory. The choice of a site for a telescope of the large linear and angular aperture of the 200-inch is, however, a different problem under present conditions. The darkest possible sky is required for long-exposure direct photography at the principal focus of the large mirror, in addition to such low wind velocity and other good qualities as Mount Wilson enjoys.

Through the courtesy of Dr. Marvin, lately chief of the United States Weather Bureau, several sets of recording instruments were loaned to us for a considerable period. One of these sets was kept in operation on Palomar for five years, so that we are now well informed regarding the meteorological conditions. Many tests of seeing, made by several skilled observers of the Mount Wilson staff, have also been made there. These show that the average seeing differs very little from that at Mount Wilson, though Ellerman frequently found the Palomar seeing superior. The average wind velocity and rainfall are also similar at the two sites. As for accessibility, we are promised by the San Diego County authorities a broad and well-surfaced road, of about half the grade of the Mount Wilson Toll Road, and the motor trip from Pasadena can now be quickly made over perfect concrete routes. If other necessary conditions are fulfilled, and if we can be certain of permanent protection from all appreciable sources of disturbance, the 200-inch telescope will presumably be erected on Palomar Mountain.[2]

[2] [The 200-in. telescope has subsequently been officially named the Hale Telescope.]

2. A NEW TYPE OF MIRROR SYSTEM

By Bernhard Schmidt

If losses of light of a mirror and of a lens system are compared with each other, then for the same aperture ratio the mirror shows a smaller loss of light than the lens system. A freshly silvered mirror reflects at least 90 percent of the incident light, while a two-lens system transmits at most 80 percent, and a three-lens system 70 percent of the incident light. In the case of large lenses, the situation is still more unfavorable because of the stronger absorption of short wavelengths by the glass.

In large telescopes, the parabolic mirror thus would be more advantageous, in general, than a lens system, but unfortunately with large aperture ratios the usable field of view is very limited by coma. For an aperture ratio of one to three, the spreading due to coma amounts to 37 seconds of arc for a field diameter of only one degree; moreover, the spreading due to astigmatism becomes five seconds of arc. Coma increases in direct proportion to the field diameter, astigmatism quadratically. As a result, astigmatism in the vicinity of the axis is negligibly small and almost pure coma is present, while at greater distances from the axis it is modified by astigmatism.

Nevertheless, a parabolic mirror of aperture ratio 1 to 8 or 1 to 10 surpasses the ordinary two-lens objective as regards image sharpness, which is due to the fact that chromatic aberration is entirely absent in the mirror. But it is a disadvantage that the light-distribution in the aberration disk of the mirror image is one-sided, for this condition can produce systematic radial displacements in measurements of such images.

But it will be shown below that even a purely spherical mirror of aperture ratio 1 to 8 or 1 to 10 is still quite usable. If the aperture stop were brought directly in front of the mirror, then there would be no advantage over the parabolic mirror, since the spherical mirror has exactly the same aberrations; besides, spherical aberration would be present, which increases the existing aberrations over the whole field of view. However, if the aperture stop is brought to the center of curvature, the spherical mirror no longer has any but longitudinal aberrations, for coma and astigmatism are zero. The image surface lies on a sphere whose radius is the focal length and which is concentric with the mirror surface, so that the image surface is turned with the convex side to the mirror.

The aberration of a spherical mirror of 1 to 8 or 1 to 10 ratio amounts to 12.5 or 6.4 seconds of arc at the paraxial image point, and the smallest possible aberrations are only one fourth of that, 3.1 or 1.6 seconds of arc.[1] In practice, even sharper images can be obtained if the focus is set between these two positions. Under normal conditions these aberrations are smaller than the spreading inherent in the photographic layer. Therefore, even with the use of flat plates, the image quality at the edge of the field is better than with a parabolic mirror of corresponding aperture ratio; the star images are round everywhere, with a symmetrical light-distribution.

Moreover, if a round, flat film is curved by pressing it with a ring over a spherical surface corresponding to the image surface, which is easily possible without wrinkling, then the confusion disks are the same size over the entire field. The same thing can be accomplished with a sharp-edged plano-convex condenser lens in front of a flat photographic plate (plane side of the lens toward the plate).

If the aperture ratio is greatly increased, however, then the spherical aberration becomes very large, since it increases with the third power of the aperture ratio. For 1 to 3, or 1 to 2, the aberrations at the paraxial image point are 240 or 800 seconds of arc. The smallest possible confusion disk has a diameter of 60 or 200 seconds of arc. With a focal length of 1 meter (39.4 inches), the paraxial rays then would be 1.2 or 4 mm (0.047 or 0.157 inches), or the smallest possible ones 0.3 or 1 mm (0.012 or 0.038 inches). In this case, therefore, the spherical mirror no longer would be useful.

I shall now show how completely sharp images can be obtained with a spherical mirror of large aperture ratio.

In order to produce a parabolic mirror from a spherical mirror, the latter's edge must be flattened, that is, be given a greater radius of curvature. However, a concentric curved glass plate (of the same thickness everywhere) can be placed on the spherical mirror, and one of its surfaces deformed. But now the curvatures must be reversed, and its edge must be more strongly curved than its center. Also, the amount of the deformation must be twice as great, because now the deviation

---

[1] *Translator's note.* The pair of larger figures refers to the size of the aberration disk at the focus for paraxial rays, the pair of smaller ones to the size of disk at the focus for rays from the outermost zone.

results from refraction. In general, in order to obtain the same deviation by refraction as by reflection, about four times as great inclinations have to be given, but in this case, since the rays go through the glass surfaces twice, only twice as large deformations are necessary.

This plate can be optically sagged to such an extent that one side becomes plane again, while the other has a pure deformation curve. That is to say, a plane-parallel plate, instead of a zero-power meniscus, can be deformed just as well from the beginning. Almost the same effect is thus obtained optically with this correction plate as with a parabolic mirror.

A suitably shaped cover plate of this kind for a spherical mirror [2] also has the practical advantage that the silver coat of the mirror is well protected. It is a disadvantage in that, owing to the passage of light twice through the glass plate, the loss of light reaches about 20 percent.

The correcting plate can also be placed in another position in the optical path. If it is located beyond the focal surface, then the light goes through the plate only once. The plate then obviously must have twice as much deformation as in the first case. The loss of light is then only 10 percent.

If the correcting plate is now brought to the center of curvature of the mirror, then there result the same relationships as before in the case of the spherical mirror with aperture stop in the center of curvature, but with the difference that now the spherical aberration is abolished, even over the whole field. Thus it is possible to use aperture ratios of 1 to 3 or 1 to 2, and to obtain freedom from coma, astigmatism, and spherical aberration.

If the inclination of the incident rays is very large, then the correcting plate is projected as an ellipse, and the deformation is not projected on the correct places of the mirror, so that the correction is variable and even introduces an overcorrection in the radial direction. However, large inclinations do not need to be considered at all, since the photographic plate soon would become greater than the clear aperture. In practice, photographic plates greater than one-fourth to one-third the aperture can hardly be used, the inclination aberrations then being negligibly small.

The case is somewhat different for the chromatic aberrations of the correcting plate. In order to keep these as small as possible, the cor-

[2] *Translator's note.* One form of this optical system is known as the Mangin mirror; it is described in Czapski-Eppenstein, *Grundzüge der Theorie der Optischen Instruments* (ed. 3; pp. 110–11 [1924]).

recting plate is so shaped that the central part acts like a weak condensing lens and the outer parts have a divergent effect. If the neutral zone is placed at 0.866 of the diameter, then the chromatic aberration is a minimum. If the point of inflection of the curve is at 0.707, then the thickness of the edge is equal to the central thickness. The remaining difference in thickness between the thickest and thinnest parts of the plate is very small, only several hundredths of a millimeter, so that a disturbing color effect does not occur; in any case, the effect usually is much smaller than the secondary spectrum of a corresponding objective.

This chromatism is identical with the so-called "chromatic difference of the spherical aberration."

If the mirror has the same diameter as the correcting plate, then the incident cylinder of rays for outer images falls eccentrically on the mirror, and there is a part of it left out, so that the outer portions of the plate obtain somewhat less light. If this is to be avoided, the mirror must have a greater diameter than the clear aperture, and of course it must be greater by about twice the plate diameter. In a mirror of 50 cm (19.7 inches) clear aperture (diameter of the correcting plate) and of 1 m (39.4 inches) focal length, the photographic plate for a field of 6° has a diameter of 10.5 cm (4.1 inches), and accordingly the mirror must have a diameter of 71 cm (28 inches).

The rapid coma-free mirror system described here offers, according to the preceding considerations, great advantages in regard to the light-gathering power and aberration-free imagery. There is assumed, however, a technically complete understanding of the correcting plate.

3. INVENTION OF THE CORONAGRAPH

By Bernard Lyot

The rareness of total eclipses of the Sun, their short duration, and the distances one has to travel to observe them, have, for more than a half century, led astronomers and physicists to seek for a method which enables them to study the corona at any time.

The first attempts made to reveal the corona without an eclipse date from 1878. It would take too long to enumerate all those that have since been made. Most of the observers simply photographed that part of the sky where the Sun was, either direct or through coloured filters. An opaque disc, somewhat larger than the image of the Sun, was placed at the focus of the instrument to screen off the direct rays of the Sun and so avoid the photographic halo. At first, only blue, violet, and ultra-violet rays were used; but, later on, from 1904, the progress of sensitizing allowed the use of red and infra-red radiations, which are not so scattered by our atmosphere. A great number of plates have thus been obtained, generally from high places, sometimes at an altitude of more than 4000 m, and in a very pure atmosphere. They show halos, often irregular, surrounding the Sun, and resembling the corona. Unfortunately, these halos have been obtained with exposures much too short to register the corona itself; they do not show the prominences which are more brilliant than the corona, and, during partial eclipses of the Sun, they even spread on the disc of the Moon without showing the limb of our satellite, which should have been outlined by the light of the corona. These halos were due principally to the light of the Sun scattered by the lens or the mirror of the telescope.

Other observers applied indirect means. In 1893, Deslandres, and after him Hale and Ricco on Mount Etna, tried, but without success, to photograph the corona by isolating the dark K line of the solar spectrum with the spectroheliograph. The inner corona, which has no absorption lines, thus should have appeared relatively more brilliant.

I might mention also: (1) The thermo-electrical procedures tried in 1900 by Hale, then by Deslandres, to detect the infra-red radiations of the corona; (2) the polariscope of Savart, used in vain by Wood, to recognise the polarized light of the corona; (3) the spectroscope used by Millochau, in 1906, to look for the green line of the corona close to the limb of the Sun. In consequence of these numerous failures, the problem had long been considered impossible.

In October 1929 I had the pleasure of speaking with the late professor Henry Osmund Barnard, former Director of the Observatory at Colombo, Ceylon, and giving him the results I had obtained on the planets with my polarimeter, sensitive to a proportion of polarised light of one-thousandth. Professor Barnard advised me to renew the attempts of Wood with this polarimeter.

The measurements made during eclipses showed that the brightness of the corona, at 2′ from the Sun's edge, is about a million times less strong than that of the solar disc. To succeed, it therefore appeared necessary, above all, considerably to reduce the diffusion of the light by the telescope. . . .

Coronagraph.—The principal defect, that is to say, the light diffracted by the edge of the lens, still had to be overcome. With this object I have made several coronagraphs, of which the plan is seen in Fig. 1.

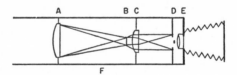

Fig. 1. Plan of the coronagraph.

The single lens, plano-convex, is shown at A; it forms an image of the Sun on a blackened brass disc at B projecting beyond the Sun by 15″ to 20″. A field lens C, placed behind the disc, produces an image of the lens A on the diaphragm D in the shadow. The edge of the diaphragm cuts off the light diffracted by the edge of the first lens. A small screen placed in the centre of the diaphragm cuts off the light of the solar image produced by reflection on the surfaces of the lens A. Behind the diaphragm and the screen, protected from the diffused light, a carefully corrected objective E forms an achromatic image of the corona on the plate. The whole apparatus is contained in a tube F, open only during the observations, and coated with thick oil in the interior, to collect the particles of dust. The first lens must be wiped frequently, and with particular care.

Atmospheric diffusion.—To give good results, the coronagraph must be in a high place, because the dust, which is always present in the

lower layers of the atmosphere, produces a whitish aureole of diffracted light around the Sun. This varying aureole is generally more
than a hundred times brighter than the corona. The distance to which
this aureole spreads varies in inverse ratio to the diameter of the particles producing it; a few degrees for dust, 5′ to 10′ for the ice crystals
of cirrus. When the sky is perfectly clear, the aureole completely disappears, and the brightness of the atmosphere, in red light, diminishes
to half a millionth of that of the Sun—that is to say, becomes fainter
than that of the corona.

From observations made in the mountains, it appears that the particles of dust are carried by rising currents of air due to the heating of
the ground by the Sun's rays; a layer of air, relatively warmer, suffices
to stop them, and they then spread out widely, forming a horizontal
sheet of a brownish colour. The height to which the dust rises increases during the period of fine weather and decreases when it rains or
snows; it is greater in summer than in winter, greater towards the
equator than towards the pole. In France, during the summer, the
highest level of dust is often between 2500 m and 3000 m. This fact
has led me to set up my apparatus in the Observatory of the Pic du
Midi, situated in the Pyrenees at an altitude of 2870 m above sea level.

This observatory offers advantages not found anywhere else for
studying the corona: on the one hand, a sky, the pureness of which is
often perfect, chiefly at the end of spring and the beginning of summer; and, on the other hand, a dome containing a very firm equatorial,
6 m in length, on which heavy and cumbersome apparatus can be
fixed. . . .

My work on the corona so far has been made in two stages: the first
one from 1930 to 1934, during which I sought to test the new method
to obtain the principal results found during eclipses; the second stage,
since 1934, during which I tried to obtain new results.

Observations made in 1930.—The first coronagraph was built in the
Observatory of the Pic du Midi, and fixed on the equatorial on 1930
July 25. This very primitive apparatus contained the lens 8 cm in
diameter and 2 m in focal length. . . . This lens was placed in a tube
built with planks found on the spot. An eye-piece, a polarimeter or a
direct vision spectrograph could be fixed at the end of the tube. After
reducing the aperture to 3 cm to hide the principal defects of the lens,
prominences with a violet-pink tint were seen. The use of a red glass
greatly increased the contrasts and allowed of good observations.

The Sun was surrounded by a slight halo without details. The polarimeter showed that this halo was polarised in a radial plane like the

corona, but more faintly. The polarisation appeared at 6' from the limb; it increased rapidly towards the Sun and then remained almost constant under 3'. The more transparent the sky, the stronger it was.

Diagrams drawn from the observations [1] show the proportion of polarised light found in different directions at 80'' from the edge of the Sun, on 1930 July 29 and July 31. The axis of the Sun was inclined 10° to N.E.; all the diagrams show minima at the poles and maxima above the zones of the sunspots. The proportion of polarised light attained, in some parts, 20- to 25-thousandths, about one-fifth of that of the corona. We can conclude that, in these parts, one-fifth of the light of the halo belonged to the corona, the rest being due to diffusion. On July 30 and the following days the spectroscope showed, in the light of the halo, among the solar lines due to diffused light, an emission line, with a wave-length equal to that of the green line of the corona within 1 angstrom.

The first photograph of the spectrum of the corona without an eclipse was taken on 1930 August 8, with the slit of the spectrograph tangential to the eastern edge of the Sun. When the slit of the spectrograph was moved, the green line underwent great changes in intensity, agreeing with the indications of the polarimeter. By covering the head with a black cloth the red line was seen, but fainter, and with variations of intensity different from those of the green line.

In 1932, the solar activity was approaching its minimum and the coronal lines were faint. I preferred to cease observations for a time and study new apparatus in view of the forthcoming increase in solar activity. This apparatus consisted chiefly of a more powerful coronagraph than the earlier one, connected with a spectrograph easier to manage, very luminous in the infra-red, and with very slight flexure.

The final coronagraph.—The chief lens of the corona is planoconvex, with a diameter of 20 cm, and a focal length of 4 m. It was ground at the Paris Observatory, by M. Couder, from a disc of excellent borosilicate crown glass, kindly presented by the firm of Parra-Mantois. The tube is no longer of wood, but of duralumin. It is 6 m long and weighs only 22 kg. It consists of four parts, which can be put one inside the other to facilitate transport. The eyepiece holder is provided with a joint allowing the observer to bring any point of the field to the centre of the eyepiece without moving the coronagraph. The latter is carried on two large ball-bearing pedestals and can be turned on its axis. To the coronagraph is joined a single spectrograph with a

---

[1] *L'Astronomie 45,* 248 (1931).

plane Rowland grating of 4 inches, reflecting, in the second order, 10 per cent. of the visible and next infrared rays, and, in the first order, 35 per cent. beyond 9000 A. . . .

Figure 2 shows the whole of the installation, the tube of the corona-graph A and the ball-bearing pedestals B and C. The chief lens is at

Fig. 2. The coronagraph on the equatorial telescope of the Pic du Midi.

D; it is held in its cell by a groove and can be taken out to be cleaned, by means of a handle. Also shown are the spectrograph E, the frame-holder F for the objective and the camera G. This apparatus was set up in 1935 and has been used each year since. . . .

Direct photography of the corona.—When the sky is very pure, by adapting an orange filter and a weak eyepiece to the coronagraph, the strongest parts of the inner corona are sometimes seen. More often, on the contrary, the coronal forms have slight contrasts and the diffused light makes them disappear in the eyepiece. They can, however, still well be photographed if a red filter and highly contrasting panchromatic plates are used. The photographs . . . have been obtained through a filter transmitting wavelengths of more than 6200 A., on slow panchromatic Guilleminot plates multiplying the contrasts by 3½, for these radiations.

The first direct photograph of the corona without an eclipse was obtained at 16^h on 1931 July 21.[2] Twelve exposures were taken; after each one, the coronagraph was turned through 30° on its axis. This proceeding enables us to distinguish with certainty the coronal features from the defects due to the instrument, chiefly the spots produced by particles of dust deposited on the second lens of the coronagraph. One positive has been made by printing successively, on the same plate, the six best negatives placed in such a manner as to obtain the coincidence of the images of the prominences; this diminishes the spots which do not coincide, as well as the irregularity of the grain of the negative.

[2] *L'Astronomie 46,* 272 (1932).

4. A STRATOSPHERIC SOLAR OBSERVATORY

By Meghnad Saha

It is well known that our observations on the spectra of the Sun and the stars are limited to the redward side of $\lambda 2900$, the ultraviolet part being absorbed in the upper atmosphere, at a height of between twenty and fifty kilometers, by a layer of ozone (equivalent to 3 mm of gas at N.T.P.) now known to arise from the photochemical action of the ultraviolet rays of the sun on oxygen molecules. This amount of ozone, tiny as it is, is sufficient however to cut off the spectrum between $\lambda\lambda 2900$ and 2200 almost completely, though absorption begins to be perceptible from $\lambda 3200$. Below $\lambda 2060$, the extinction of the spectrum is due to absorption by molecular oxygen and nitrogen. According to some investigators, there is a so-called window between $\lambda\lambda 2300$ and 2100, but evidence on this point is divergent.

The abrupt termination of solar and stellar spectra below $\lambda 2900$ has been a great handicap to the advancement of our knowledge of the heavenly bodies, because the information gained from study of the spectrum beyond $\lambda 2900$ is not sufficient to explain the problems of stellar mechanisms operative there. To take one example: the great intensity of the Balmer series and the associated continuous spectrum in the chromosphere has given rise to a large number of speculative theories which have again and again been obliged to fall back upon certain plausible hypotheses regarding the strength of the Lyman lines. If these lines could have been observed, the problem of hydrogen excitation in the Sun and stars would probably have received complete elucidation, and the problems of stellar atmospheres would have been nearer solution.

It is therefore not surprising that when some years ago Cario made the suggestion that the North Polar region, being free from illumination by the Sun during the winter, might not contain any ozone, and therefore observations carried out there might extend stellar spectra much further on the violet side of $\lambda 2900$, the suggestion was greatly welcomed. It was a disappointment for the astronomical world when observations by Rosseland did not confirm Cario's hypothesis. The reason for this failure is now well understood, for Dobson and Götz, in their survey of the ozone content of the atmosphere at different latitudes, have shown that the amount of ozone in the atmosphere fluctuates with the season, rising at Abisko (latitude 68° N) from 2.40 mm in the middle of September to 3.6 mm in the middle of March. There is thus actually an increase in the ozone content during winter. This fact, apparently at variance with the theory of the photochemical origin of O_3, has been satisfactorily explained by S. Chapman. The explanation is roughly as follows: The solar rays not only form ozone, but also destroy ozone. Every quantum of light between $\lambda\lambda 1750$ and 2060 produces, on being absorbed, two molecules of O_3 out of atmospheric oxygen. But this ozone absorbs strongly the light between $\lambda\lambda 2300$ and 3000, and every quantum absorbed converts two O_3 molecules into three O_2 molecules after a number of subsidiary reactions. The actual number of ozone molecules existing at any time depends upon the equilibrium between these two groups of opposing reactions. It appears that during a polar winter, when sunlight no longer illuminates the upper atmosphere, the ozone molecules already formed continue to exist, the destructive agency having been withdrawn. One always incurs a risk in extrapolating, but, as far as evidence goes, it appears certain that during winter the atmosphere of regions a few degrees removed from the North Pole also retains its ozone screen, so that observations of stellar spectra (the Sun does not come into view, as it is below the horizon) will have no chance of taking us beyond the limit attainable in more hospitable climates.

Regener's work.—The recent discovery of Götz, Meetham, and Dobson that the ozone screen does not lie between fifty and one hundred kilometers as was formerly thought, but is confined between twenty and forty kilometers, affords a definite opportunity of extending stellar spectra beyond $\lambda 2900$, as has actually been demonstrated by E. and V. Regener. Professor Regener has developed a fine technique of sending into the upper atmosphere balloons carrying automatic recording apparatus for measurement of the intensity of cosmic rays. His highest record has been thirty-one kilometers where, according to the estimates of Dobson and Götz and confirmed independently by the Regeners, two-thirds of the total ozone remains below. In the course of his last reported work in 1934, he sent along with his cosmic ray apparatus a quartz-spectrograph provided with automatic shutters and pointed toward a matt surface below, which was illuminated by sunlight. The time of exposure was short, and the reflecting power of the matt surface for the ultraviolet rather feeble. In spite of these disadvantages they were able to show that with increasing altitude the spectrum extended further into the ultraviolet, and that at the greatest height reached by their apparatus the limit was extended by about a hundred units beyond the limit reached by the same apparatus for the same exposure on the ground. If the exposures had' been longer, and the surface had had a better reflecting power for λ2800, it is clear that the spectrum might have extended much further than the lowest limit attained so far. They also confirmed the findings of Dobson and Götz that most of the ozone is to be found between twenty and thirty-five kilometers, that above forty kilometers the total amount is one-twelfth of the whole, and that above fifty kilometers it is barely two per cent of the whole.

The pioneering work of Regener has shown the practical possibility of having a "Stratosphere Solar Observatory." It can now be confidently expected that if a regular program can be organized for sending balloons to a height of thirty-five to forty kilometers, provided with quartz, fluorite, and vacuum spectrographs of sufficient light-gathering power, our knowledge of the solar spectrum beyond λ2900 will receive a great impetus. The Russian worker Moltchanoff claims that he has reached a height of forty kilometers with a balloon provided with Radio-Sonde signaling apparatus; it is therefore to be hoped that within the near future the problem of photographing the solar spectrum at a height of forty kilometers will be definitely solved. . . .

We conclude from the above discussion that a spectrophotogram of the sun, taken at a height of 40 km, will extend the spectrum to λ2000, and probably no atmospheric bands will appear between λλ2900 and 2000. Between λλ2000 and 1700 the Runge-Schumann bands of O_2 may appear in absorption. The region λλ1700—1250 will probably be completely cut off. A strip between λλ1250 and 1000 may be expected to be transmitted. Below λ1000 no prediction can be made, as laboratory data are not available.

But access even to these limited regions will result in invaluable additions to our knowledge, for they will afford information about the behavior of the resonance lines of most of the elements which occur in the Fraunhofer spectrum and thus ease our way for the final solution of the mysteries of solar physics; e.g., we expect to get information (a) about L_α λ1216 of H; (b) about

$$\lambda 1640 \left(= 4R \left(\frac{1}{2^2} - \frac{1}{3^2} \right) \right), \quad \lambda 1215 \quad \left(= 4 R \left(\frac{1}{2^2} - \frac{1}{4^2} \right) \right) \text{of He}^+ ;$$

(c) about the existence or otherwise of the Li-continuum at about λ2300; (d) about the resonance lines of elements from Be to O (4→8);

we shall not probably obtain any information about F and Ne, but we may obtain the Na-continuum. (e) As regards Mg, we shall obtain much desired information about the resonance lines of Mg and Mg^+ which are just beyond $\lambda 2900$; (f) the same is true of the resonance lines of the elements Al to S. (g) We hope also to obtain very valuable information regarding transitional elements, particularly Fe^+.

The above short account will indicate how much we should gain from a Stratosphere Observatory.

5. THE BEGINNING OF RADIO ASTRONOMY

By Grote Reber

Several papers [1] have been published which indicate that an electromagnetic disturbance in the frequency range 10–20 megacycles arrives approximately from the direction of the Milky Way. It has been shown [2] that black-body radiation from interstellar dust particles is not the source of this energy.

The antenna system shown in Figure 1 was constructed for the investigation of this phenomenon.[3] The receiver can be set at the desired declination by rotating along the meridian on the circular tracks at each side. Readings at a fixed declination are taken over an interval of several hours, the rotation of the earth providing the change in right ascension.

The drum at the focal point is an artificial black body described elsewhere.[4] The entire receiving system has an effective cone of acceptance approximately 3° in diameter.

The output is indicated by a microammeter so connected that any intercepted energy will cause the readings to decrease. A few typical records are shown in Figure 2, in which the individual points are omitted because they lie too close together. The magnitude of the dip in the curve gives a measure of the intensity of the received energy. Over a long period, as the apparatus warms up, the zero level will gradually rise. The dotted line indicates the run which would have been obtained had no radiation been captured.

The results of preliminary measures of the variation of the static disturbance as a function of galactic longitude are shown in Figure 3, in which each point represents the central intensity of the dip on one night's record. The magnitude of the systematic error may be ±50 per

[1] K. G. Jansky, *Proc. I. R. E. 20,* 1920, December, 1932; *21,* 1387, October, 1933; *23,* 1185, October, 1935; *25,* 1517, December, 1937. H. T. Friis and C. B. Feldman, *Bell System Tech. J. 16,* 337, July, 1937.

[2] Whipple and Greenstein, *Proc. Nat. Acad. Sci. 23,* 177, 1937.

[3] For details of the instrumental design and the method of reduction of the data see G. Reber, *Proc. I. R. E. 28,* 68, 1940.

[4] G. Reber, *Communications 18,* 5, December, 1938.

Fig. 1. The antenna system for Reber's radio telescope.

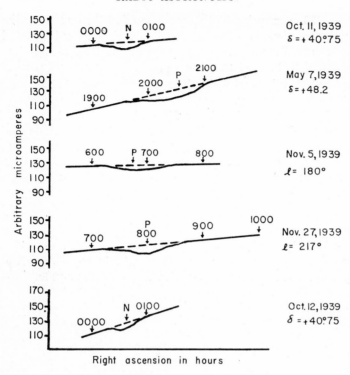

Fig. 2. Typical records of radio signals.

cent, but the general order of the disturbance is correct. The plane from which this energy arrives is tilted about 5° south of the plane of the galaxy in the vicinity of longitude 150°.

Since the theory of black-body radiation predicts an intensity proportional to the square of the frequency in this range, the first tests were

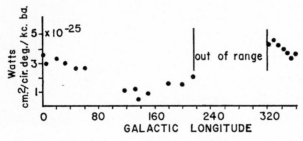

Fig. 3. Preliminary measures of the variation of the static disturbance as a function of galactic longitude.

made at 3300 megacycles. Nothing was found at the sensitivity limit of 10^{-20} watts per square centimeter per circular degree per kilocycle band width. Improved equipment for the frequency of 900 megacycles gave no results at the limit of 10^{-22} watts per square centimeter per circular degree per kilocycle band width. The data of Figures 2 and 3 were obtained at 162 megacycles.

A few bright stars, such as Vega, Sirius, Antares, Deneb, and the Sun, gave negative results. Mars and the Orion nebula also gave no readable indication. If radiation is present from any of these objects, the intensity is below 10^{-25} watts per square centimeter per circular degree per kilocycle band width at 162 megacycles. The only other positive results are from the great nebula in Andromeda, with a mean of four readings giving a maximum intensity of 8×10^{-26} watts per square centimeter per circular degree per kilocycle band width.

The foregoing observations confirm previous evidence that radiation in the radio spectrum is apparently coming from the direction of the Milky Way. The intensity is a function of galactic longitude.

II

THE SUN

The subject of the sun was pretty well in hand at the beginning of this century. Sunspots and sunspot cycles had been systematically investigated for many decades. The solar eclipse had provided for us an introduction to the corona, to its streamers, arches, and especially its peculiar bright line spectrum. The spectroscopes in Italy, America, England, and France had provided abundant information on solar prominences. But now, halfway through the century, we come into a new era of solar discoveries. In the next decade we shall go deeper than we once imagined possible into the surface and subsurface activity of this convenient star.

Instrumentation and astrophysical theory have opened new approaches during the past half-century. A half-dozen European solar observatories have been created. G. E. Hale and Bernard Lyot found new ways of looking at the sunspots and the corona, respectively. Bengt Edlén, following the lead of Walter Grotrian and others, solved the mystery of coronal radiation, but provided other puzzles. W. O. Roberts and colleagues with their coronagraphs on a Colorado mountain top, Orren Mohler with eclipse photographs, and S. L. Lippincott with Richard Dunn's Sacramento Peak chromosphere-camera films, have started the unraveling of the evanescent solar spicules. Pettit has put order into the varied phenomena of solar prominences through providing a systematic classification of their forms. His is probably a preliminary step, as was that of Angelo Secchi ninety years ago; D. H. Menzel and J. W. Evans have in 1953 proposed other criteria for classifying these solar uprisings, and Menzel and Shirley Jones have extensively used the new classification.

With the current building of new equipment for solar probing, including rockets, balloons, satellites, cosmic ray counters, and radio telescopes, the sun should soon reveal some very important new information concerning stellar structure and behavior.

6. THE MAGNETIC FIELD IN SUNSPOTS

By George Ellery Hale

We shall consider a little later the nature of sun spots, but for the present we may regard them simply as solar storms. When spots are numerous the entire sun is disturbed, and eruptive phenomena, far transcending our most violent volcanic outbursts, are frequently visible. In the atmosphere of the sun, gaseous prominences rise to great heights. . . . But such eruptions as the one of March 25, 1895, photographed with the spectroheliograph of the Kenwood Observatory, are clearly of an explosive nature. As these photographs show, it shot upward through a distance of 146,000 miles in 24 minutes, after which it faded away.

When great and rapidly changing spots, usually accompanied by eruptive prominences, are observed on the sun, brilliant displays of the aurora and violent magnetic storms are often reported. The magnetic needle, which would record a smooth straight line on the photographic film if it were at rest, trembles and vibrates, drawing a broken and irregular curve. Simultaneously, the aurora flashes and pulsates, sometimes lighting up the northern sky with the most brilliant display of red and green discharges. . . .

In certain regions of the sun we have strong evidence of the existence of free electrons. This leads us to the question of solar magnetism and suggests a comparison of the very different conditions in the sun and earth. Much alike in chemical composition, these bodies differ principally in size, in density, and in temperature. The diameter of the sun is more than 100 times that of the earth, while its density is only one-quarter as great. But the most striking point of difference is the high temperature of the sun, which is much more than sufficient to vaporize all known substances. This means that no permanent magnetism, such as is exhibited by a steel magnet or a lodestone, can exist in the sun. For if we bring this steel magnet to a red heat it loses its magnetism and drops the iron bar which it previously supported. Hence, while some theories attribute terrestrial magnetism to the presence within the earth of permanent magnets, no such theory can apply to the sun.

If magnetic phenomena are to be found there they must result from other causes.

The familiar case of the helix illustrates how a magnetic field is produced by an electric current flowing through a coil of wire. But according to the modern theory, an electric current is a stream of electrons. Thus a stream of electrons in the sun should give rise to a magnetic field. If the electrons were whirled in a powerful vortex, resembling our tornadoes or waterspouts, the analogy with the wire helix would be exact, and the magnetic field might be sufficiently intense to be detected by spectroscopic observations.

A sun spot, as seen with a telescope or photographed in the ordinary way, does not appear to be a vortex. If we examine the solar atmosphere above and about the spots, we find extensive clouds of luminous calcium vapor, invisible to the eye, but easily photographed with the spectroheliograph by admitting no light to the sensitive plate except that radiated by calcium vapor. These calcium flocculi, like the cumulus clouds of the earth's atmosphere, exhibit no well-defined linear structure. But if we photograph the sun with the red light of hydrogen, we find a very different condition of affairs. In this higher region of the solar atmosphere, first photographed on Mount Wilson in 1908, cyclonic whirls, centering in sun spots, are clearly shown.

The idea that sun spots may be solar tornadoes, which was strongly suggested by such photographs, soon received striking confirmation. A great cloud of hydrogen, which had hung for several days on the edge of one of these vortex structures, was suddenly swept into the spot at a velocity of about 60 miles per second. More recently Slocum has photographed at the Yerkes Observatory a prominence at the edge of the sun, flowing into a spot with a somewhat lower velocity.

Thus we were led to the hypothesis that sun spots are closely analogous to tornadoes or waterspouts in the earth's atmosphere. If this were true, electrons caught and whirled in the spot vortex should produce a magnetic field. Fortunately, this could be put to a conclusive test through the well-known influence of magnetism on light discovered by Zeeman in 1896.

In Zeeman's experiment a flame containing sodium vapor was placed between the poles of a powerful electromagnet. The two yellow sodium lines, observed with a spectroscope of high dispersion, were seen to widen the instant a magnetic field was produced by passing a current through the coils of the magnet. It was subsequently found that most of the lines of the spectrum, which are single under ordinary conditions, are split into three components when the radiating source is in a suffi-

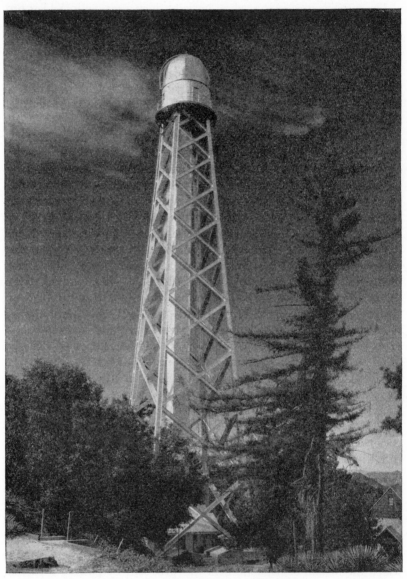

Fig. 1. Tower telescope on Mount Wilson.

ciently intense magnetic field. This is the case when the observation is made at right angles to the lines of force. When looking along the lines of force, the central line of such a triplet disappears, and the light of the two side components is found to be circularly polarized in opposite directions. With suitable polarizing apparatus, either component of such a line can be cut off at will, leaving the other unchanged. Furthermore, a double line having these characteristic properties can be produced only by a magnetic field. Thus it becomes a simple matter to detect a magnetic field at any distance by observing its effect on light emitted within the field. If a sun spot is an electric vortex, and the observer is supposed to look along the axis of the whirling vapor, which would correspond with the direction of the lines of force, he should find the spectrum lines double, and be able to cut off either component with the polarizing attachment of his spectroscope.

I applied this test to sun spots on Mount Wilson in June, 1908, with the 60-foot tower telescope, and at once found all the characteristic features of the Zeeman effect. Most of the lines of the sun spot spectrum are merely widened by the magnetic field, but others are split into separate components, which can be cut off at will by the observer. Moreover, the opportune formation of two large spots, which appeared on the spectroheliograph plates to be rotating in opposite directions, permitted a still more exacting experiment to be tried. In the laboratory, where the polarizing apparatus is so adjusted as to transmit one component of a line doubled by a magnetic field, this disappears and is replaced by the other component when the direction of the current is reversed. In other words, one component is visible alone when the observer looks toward the north pole of the magnet, while the other appears alone when he looks toward the south pole. If electrons of the same kind are rotating in opposite directions in two sun-spot vortexes, the observer should be looking toward a north pole in one spot, and toward a south pole in the other. Hence the opposite components of a magnetic double line should appear in two such spots. The result of the test was in harmony with my anticipation. . . .

The intensity of the magnetic field in sun spots is sometimes as high as 4,500 gausses, or 9,000 times the intensity of the earth's field. In passing upward from the sun's surface the magnetic intensity decreases very rapidly—so rapidly, in fact, as to suggest the existence of an opposing field. It is probable that the vortex which produces the observed field is not the one that appears on our photograph, but lies at a lower level. In fact, the vortex structure shown on spectroheliograph plates may represent the effect rather than the cause of the sun-spot field.

7. A CLASSIFICATION OF SOLAR PROMINENCES

By Edison Pettit

When at a solar eclipse the moon shuts out the intense light of the sun, the pearly green corona with its delicate arches and streamers is revealed. Close to the black moon's edge a reddish ring encircles the sun. From it rise flame-like clouds of luminous gas which sometimes extend well up into the lower regions of the corona. The clouds are the solar prominences and the colored layer of the sun's atmosphere from which they seem to rise is the chromosphere. Until the invention of the spectroscope, over seventy years ago, prominences and the chromosphere could be seen only during a total eclipse of the sun, but now with the aid of the spectroscope and its offspring, the spectroheliograph and the spectrohelioscope, they may be photographed and viewed whenever the sun is visible.

More recently, a new type of [instrument, Lyot's polarizing monochromator (see selection 3 above)] has been perfected in which the scattered light has been greatly reduced. With it prominences may be viewed directly if the observations are made on high mountains above the dense dust-filled air which makes the sky so bright at lower elevations. Most of our information about prominences, however, has come from photographs made with the spectroheliograph.

If we observe the diameter of the sun's photosphere with the spectrohelioscope using the light from the spectrum between the absorption lines we find that it is essentially the same for all colors. If, however, we use light emitted by hydrogen, helium, or ionized calcium, we find that the diameter is somewhat larger than before, as if a layer four or five thousand miles deep had been placed upon the sun. This layer is the chromosphere. It is more irregular than the photosphere and above it the prominences rise to enormous heights like clouds in an invisible atmosphere.

Unlike clouds on the earth, prominences consist of great volumes of incandescent gases whose density is of the order of one-millionth that of our atmosphere at sea level; about that of a fairly good vacuum. They extend far into the corona, but it is difficult to believe that the corona supports them in any way, such as that by which terrestrial clouds are supported by the earth's atmosphere.

Although an observer's first impression is that prominences are entirely irregular in form, a more careful study indicates that they may be arranged in five main types according to their shape and behavior. The

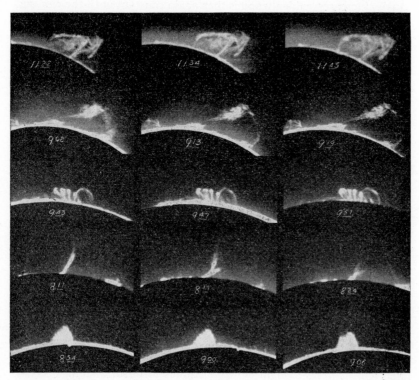

Fig. 1. Types of solar prominences. a. Active; b. Eruptive; c. Spot; d. Tornado; e. Quiescent. The three views of each type show changes in half an hour or less.

following names are used to characterize the different types of prominences: (1) active, (2) eruptive, (3) spot, (4) tornado, (5) quiescent.

Active prominences are by far the most numerous. Their three-dimensional form is somewhat like a thin sheet of interlacing filaments standing on edge and connecting with the chromosphere in only a few points. A thickness of 6,000, height of 30,000 and length of 50,000 miles would be representative dimensions. Knots and streamers continually pour from the prominences into one or more centers of attraction in the nearby chromosphere. These centers are not visible and probably they attract a prominence by electrical forces. As the material is pulled into the chromosphere, it moves along curved lines with velocities sometimes exceeding 100 miles per second, but more commonly with speeds of about half the amount. The speed is constant over great distances, but usually increases suddenly and is then again constant until the next sudden change which may take place within a few minutes. Eventually the prominence, thus pulled to pieces, thins out and disappears.

If the intensity of the center of attraction increases considerably, its electrical energy reaches out into space beyond the prominence and brings in any chromospheric matter that may be there. We did not expect that chromospheric matter existed in coronal space until these prominences were studied. The evidence now is that a center of attraction excites some of the previously invisible matter and pulls it into the sun from the coronal regions in the form of long streamers which appear out of blank space and descend with uniform velocities of about 120 miles per second. One almost expects them to splash as they hit the sun, but nothing of the kind has been observed.

When the intensity of attraction is further enhanced, the whole body of the prominence rises, and great ribbons are torn from it which traverse a large arc, and disappear into the center of attraction.

Eruptive prominences, so far as we know, begin as active or spot prominences; none have erupted directly from the chromosphere. When the center of attraction becomes very intense, the whole prominence rises and continues to move upward. During the ascent streamers continue to pour into the original center of attraction and into others that form in the vicinity. Their velocities are 60 miles or more per second.

Ordinary eruptive prominences rise to heights of 200,000 or 300,000 miles and the record height of 963,000 miles (1.11 solar diameters) was observed at Mount Wilson two years ago. Small ones which rise only a few thousand miles are also seen. The time required to attain maximum height is usually only a few hours and sometimes even less than an hour, for they move with velocities ranging from a few miles per second to the record speed of ascent, 452 miles per second, observed at Lake Angelus in 1937. Velocities of 50 or 60 miles per second are fairly common. When a prominence begins to rise, it is necessary to start a program of rapid exposures with the spectroheliograph immediately, if results of scientific value are to be obtained. . . .

Eruptive prominences fade away rather suddenly in the last stage of the eruption. Expansion partly accounts for their disappearance and there is some loss of matter by returning streamers, but a reduction of the proportion of excited atoms seems to be the chief cause of disappearance. Whether this prominence matter eventually returns to the sun we do not know, but the prominences originating in the corona suggest that it does.

The prominences thus far described are not necessarily associated with sunspots and are found anywhere on the solar surface from the poles to the equator. *Sunspot prominences* are found only over spots, or in regions where spots have just vanished or are about to form. They are typical and so well marked that frequently, before we see it, a spot-group heralds its coming over the east limb by the prominence that appears above it.

There are several sub-groups of this type. One consists simply of broken filaments, arranged in the shape of a fan, which move into the spot along converging lines. In another, complete loops or arches form over the spot and, curiously, the motion is down both branches of the arch into the spot. In both types material is often seen to form like balls of cotton in the space above the spot and lengthen out into short ribbons as it falls. Sometimes so many form at once that the effect is that of a snow storm in slow motion. Here again we must admit that the corona contains nonluminescent chromospheric gases which become excited by the attracting center within the spot.

Active prominences are often associated with spots. The motions of such prominences are probably caused by centers of attraction within the spot-group.

Out of a spot-group a solid sheet or ribbon of incandescent gases is often projected which rises to heights varying from a few thousand to 50,000 miles and then subsides without becoming detached from the chromosphere. The velocity may be so high (more than a hundred miles per second) that the prominence breaks into a spray. Such prominences are called surges and often they rise and subside over the same area at short intervals. The life of the smaller ones is only about 10 or 20 minutes, but the largest may last about an hour. In one spot area as many as 16 surges have been seen in five hours.

Occasionally small faint rings of chromospheric gases, 3,000 or 4,000 miles in diameter, are ejected from sunspot areas. They appear brighter at the periphery, but we cannot be sure whether they are actually rings or hollow globes. When they leave the sun they move in straight lines, rise a few thousand miles and fade out within a few minutes.

Low bright prominences are sometimes seen as a bump in the chromosphere over a spot-group. These are 5,000 to 10,000 miles high and 20,000 to 25,000 miles long, like a low bank or desert mesa; they may be nearly hemispherical though often they are irregular in outline. They seem to be the focus for incoming filaments that form above them.

Tornado prominences resemble twisted ropes or waterspouts. Fine atmospheric conditions and considerable magnification are required to reveal their true character. Their diameters are usually less than 10,000 miles and their heights less than 50,000 miles. Only a few good photographs of such prominences are on record; that shown in the illustration on page 37 is the only one for which the final stages of dissolution were observed. Apparently they increase in rotational velocity until they blow to pieces like the small dust storms of the plains.

Although there probably is no such thing as an absolutely motionless prominence, we do find many which change very slowly in general outline, and these are called *quiescent prominences*. They sometimes remain unchanged, as far as we can judge, for several days or even longer. As seen on the disk, they look like long ribbons standing on edge, occasionally reaching a third of the way around the sun. Within them, knots and streamers are seen to be in continual motion with velocities of 3 to 6 miles per second. Eventually centers of attraction form and they become active and disappear.

We may reasonably inquire what relation bright chromospheric eruptions, which are associated with short-wave radio fade-outs, have to prominences in general. Like the sunspot prominences, they are nearly always associated with sunspots. Their lifetime is usually only 10 to 25 minutes and they show no marked radial velocities, as rapidly rising matter should. In fact, there is little evidence that they ordinarily rise much above the chromosphere and commonly they appear to be simply a brightening of the chromospheric structure already existing in a spot area. It would seem that "eruption" is the wrong word to apply to them. "Flare" and "disturbance" have been suggested, but possibly "coruscation" (sudden flash or play of light) would be more descriptive.

Prominences appear the same, when viewed in either neutral hydrogen (the Balmer series) or ionized calcium (H and K). This is surprising since either radiation pressure or electrical energy should separate the two kinds of atoms.

It is not certain that prominences in general come from the chromosphere. The surges and ejections certainly do, but it is possible that the others may be formed, at least partially, by accumulations from the chromospheric atoms in the corona. We have witnessed the destruction of prominences many times, but of their origin we know little. Their observed uniform motion over immense distances is equally puzzling because their motion should be accelerated if they move subject to light pressure, electrical forces, or gravitation.

8. THE MYSTERY OF CORONIUM SOLVED

By Bengt Edlén

Ever since the first emission lines in the solar corona were discovered some 70 years ago, many suggestions of the most different kind have been put forward to explain their origin. So far, however, none of these attempts has led to an acceptable result. Even a critical discussion of the matter by Swings finished with the conclusion that the origin of the coronal lines remains as mysterious as ever.

In spite of the well-known difficulties of coronal observations, there are at present somewhat more than 20 relatively well-measured lines, which can be considered as certainly belonging to the specific coronal spectrum (table 1). None of these lines has ever been observed either in a laboratory light-source or in any other astronomical object, except for a short appearance at a nova outburst of RS Ophiuchi. In the present article an identification is proposed which accounts for the main part of the coronal lines and gives the explanation for the unique character of the coronal spectrum. At the same time the reason for the failure of all previous attempts to connect it with any known atomic or molecular spectrum becomes evident.

Grotrian's recent observation that the wave-numbers of the coronal lines 6374 and 7892 coincide with the separations $^2P_{3/2} - {}^2P_{1/2}$ and $^3P_2 - {}^3P_1$ in the ground terms of Fe X and Fe XI, gave the impulse to a systematic search for analogous term separations. Hereby two other coronal lines were found to coincide with the separations $^2P_{3/2} - {}^2P_{1/2}$ and $^3P_2 - {}^3P_1$ of Ca XII and Ca XIII as determined by unpublished measurements of the author. This fact increased the probability that the coincidences did not happen by chance and that the whole coronal spectrum might consist of "forbidden lines" analogous to those of the gaseous nebulae but from atoms much more highly ionised than had previously been considered at all.

As the absence of any more coincidences might well be attributed to the impossibility of analyzing the corresponding extreme ultraviolet spectra, an attempt has been made to predict by extrapolation such term separations as could be expected in the corona according to those

Table I. The emission lines in the spectrum of the solar corona.

λ^1 A	cm^{-1}	Intensity[2]			Identification	I. P.
3328.1	30039	8	1.0		Ca XII $^2P_{3/2}$—$^2P_{1/2}$	589
3388.10	29506.6	20	16		Fe XIII $^3P_2$—$^1D_2$	325
3454.13	28942.6	8	2.3			
3600.97	27762.4	12	2.1		Ni XVI $^2P_{1/2}$—$^2P_{3/2}$	455
3642.87	27443.1	—	—		Ni XIII $^3P_1$—$^1D_2$	350
3800.77	26303.0	1	—			
3986.88	25075.2	5	0.7		Fe XI $^3P_1$—$^1D_2$	261
4086.29	24465.2	6	1.0		Ca XIII $^3P_2$—$^3P_1$	655
4231.4	23626.2	8	2.6		Ni XII $^2P_{3/2}$—$^2P_{1/2}$	318
4311.5	23187.3	1	—			
4359	22935	1	—			
4567	21890	3	1.1			
5116.03	19541.0	5	4.3	2.6	Ni XIII $^3P_2$—$^3P_1$	350
5302.86	18852.5	20	100	120	Fe XIV $^2P_{1/2}$—$^2P_{3/2}$	355
5694.42	17556.2	—	—	1.5		
6374.51	15683.2	6	8.1	28	Fe X $^2P_{3/2}$—$^2P_{1/2}$	233
6701.83	14917.2	4	5.4	3.3	Ni XV $^3P_0$—$^3P_1$	422
7059.62	14161.2			4		
7891.94	12667.7			29	Fe XI $^3P_2$—$^3P_1$	261
8024.21	12458.9			1.3	Ni XV $^3P_1$—$^3P_2$	422
10746.80	9302.5			240	Fe XIII $^3P_0$—$^3P_1$	325
10797.95	9258.5			150	Fe XIII $^3P_1$—$^3P_2$	325

[1] Wavelengths from S. A. Mitchell, *Hb. der Astroph.* Vol. VII, p. 400, 1935 and B. Lyot, *M. N. 99*, 580, 1939.

[2] First column: estimated by Grotrian, *Zs. f. Astroph., 2*, 106, 1931; second column: measured by Grotrian, *Zs. f. Astroph., 7*, 26, 1933; third column: measured by Lyot, l.c.

already found. We have then to consider the configurations s^{2}p, s^{2}p^2, s^{2}p^4, and s^{2}p^5 of cosmically abundant atoms in appropriate ionisation stages. There were good reasons to expect intense lines especially of Fe XIII and Fe XIV because of (1) the rich abundance of iron as judged from the intensity of Fe X and Fe XI, (2) the very high degree of ionisation indicated by the nonappearance of the well-known forbidden lines of Fe VII, and the presence of Ca XII and Ca XIII with much higher ionisation potentials than Fe XIII and XIV. It is thus of great significance indeed that the extrapolated separation in the ground term of Fe XIV definitely points to the strongest line of the coronal spectrum 5303, and that the three next strongest coronal lines 10747, 10798 and 3388 permit a spontaneous identification with the transitions in Fe XIII as given in table 1. Besides, the fainter 3987 can be attributed to a transition in Fe XI. Together with 6374 and 7892 these identifications of Fe X, XI, XIII, and XIV comprise more than $\frac{9}{10}$ of the total intensity of the coronal line emission. The fact that no other ionisation stages of iron appear is quite naturally explained, as the extrapolated Fe XII-transitions are unobservable and the neighbouring Fe VIII, IX, XV, XVI and XVII have no deep metastable levels. The absence of $^3P_2 - {}^3P_0$ of Fe XI and Fe XIII as well as the comparatively very low intensity of Fe XI $^3P_1 - {}^1D_2$ is in agreement with the theoretical transition probabilities recently calculated by Pasternack.

If now the iron identifications mentioned above are accepted, one can predict with considerable accuracy the position of the corresponding transitions in the elements nearby. It then appears that within the observable range the complete set of transitions extrapolated to Ni is found among the remaining coronal lines with roughly the same relative intensities as for iron. In this way six more coronal lines have been identified. The average intensity ratio of Fe to Ni lines is approximately 10 : 1, in accordance with the usual cosmical abundance of these elements. The proposed identifications for both Fe and Ni seem furthermore to compare well with all reliable physical classifications of the coronal lines as shown, for instance, by Lyot's "groups" in table 1.

Comparing the intensities of lines from different ions, one finds that for Fe the two higher stages of ionisation are the more frequent, while for Ni the opposite is true. This indicates a maximum abundance for ions with ionisation potentials of about 400 volts. In this connection we may remind that Lyot, assuming the profiles observed for some coronal lines as due to a thermal Doppler effect in oxygen atoms, found an equivalent temperature of 660,000° which becomes 2,300,000° when recalculated to iron atoms.

With the present identifications we have reached a fairly complete explanation of the corona spectrum, leaving unidentified only some very faint lines forming less than 3% of the total line emission. So far, only the three elements: calcium, iron, and nickel, have been documented. It is quite possible that still others will be found through the identification of the remaining faint lines. It seems fairly certain, however, that the elements K, Cr, Mn, and Co, for which possible transitions can be accurately predicted, do not account for any of the coronal lines as yet observed. This gives an important information about the composition of the coronal matter, for if present in the same ratios as in the solar atmosphere, these elements should give quite observable lines. On the other hand the absence of K, Cr, Mn, and Co and the proportion of Fe to Ni agrees well with the average composition of meteorites, which sug-

gests a rather plausible origin for the coronal matter. It may be remarked that all the other elements which are abundant in meteorites:
O, Mg, Al, Si, and S, have at the expected ionisation stages all their
strong lines outside the observable range.

The more detailed argumentation underlying the explanation of the
coronal spectrum which has been proposed in this preliminary note will
be given in a later report.

INTERPRETATION

According to the explanation now described the essential part of the
coronal lines is due to iron, nickel, calcium, and argon atoms deprived
of 10 to 15 electrons, *i.e.*, about half of their normal electron envelope.
The discovery of this enormously high stage of ionization has obviously
introduced a new argument in the discussion of solar phenomena. Several attempts to give the established facts a physical explanation have
already been made.

Let us first recall some of the more obvious arguments for the existence of a very high temperature in the corona:

1. The high mean stage of ionization as revealed by the emission lines.
2. The breadth of the emission lines, if due to thermal Doppler effect. The
broadening of the lines might also be caused by macroscopic irregular motions
(turbulence) or radial motions of the matter.
3. The blurring out of the Fraunhofer lines in the continuous spectrum of the
inner corona, assumed to be an effect of the velocities of the scattering electrons.
4. The absence of the Balmer lines in the emission line spectrum of the corona,
explained by the electrons being too fast to be captured by the protons.
5. Dynamical considerations showing that great thermal velocities are necessary to balance the gravitational forces in order to explain the observed density
gradient of the corona.

All these observations point to temperatures higher than a quarter of a
million degrees.

Independently of the identification of the coronal lines, Alfvén came
to the conclusion that the corona might consist altogether of particles
with very high energy and derived from the density function a temperature of about one million degrees. On certain assumptions Alfvén finds
that the energy necessary to maintain this high temperature of the corona would be about 10^{-5} of the total energy radiated by the Sun, and
that the total energy contained in the corona would be produced in
about two hours. The "heating" mechanism conceived by Alfvén is described in the following way:

"Motion of solar matter in magnetic fields on the Sun, especially the vortical motion in a sunspot, must bring about potential differences between different points of the solar surface, and it was shown [in a previous paper] that under certain conditions this gives rise to discharges above the surface of the Sun. Calculations indicate that the electromotive force can be as high as 10^7 volts, so that even if charged particles are usually accelerated only by a small fraction of this potential, they attain rather high energies.

"The process is most conspicuous in the prominences, where consequently we can expect a very intense production of high energy particles. As the mechanism is of a very general character the same process is likely to take place very frequently on a smaller scale. If—as many authors mean—we can regard the chromosphere as a multitude of small prominences, it is likely that a production of high energy particles takes place almost everywhere on the solar surface or in some layer above it."

A different view on the origin of the highly ionized particles in the corona was taken by Menzel by the suggestion that the coronal matter was ejected from the hot interior of the Sun through holes and cracks in the solar surface. A somewhat similar opinion is expressed by Vegard in a recent paper. According to his theory, "the highly ionized heavy ions present in the corona come from the Sun's deeper layers and are driven away from the Sun at great speed through the electric fields resulting from photo-electric effect produced by soft X-rays."

Finally, a quite different explanation has been put forward by Saha, who suggests that the highly ionized atoms emitting the coronal lines are the fragments of a kind of nuclear fission, similar to the uranium fission, occurring somewhere near the solar surface.

Before the various suggestions have been more thoroughly examined it would be unwise to judge in favour of the one or the other. In that respect the physical explanation of the solar corona still remains a problem.

9. SOLAR SPICULES

By Walter Orr Roberts

For many years solar observers have been aware of the irregular appearance of the chromosphere of the sun when viewed under the best observing conditions. Frequent reference has been made to Secchi's observations of the "vertical flames" of the chromosphere in polar regions of the sun. In discussing Lick Observatory eclipse photographs Menzel has called attention to the "spike" prominences of the chromosphere and to the fact that "the difference between the chromosphere and the prominences is merely one of degree." The irregularities of the chromosphere are evident in the eclipse photographs reproduced by Menzel and, still more strikingly, in the photographs taken by Marriott, of Swarthmore College, at the eclipse of October, 1930. In the latter, small vertical polar prominences are clearly visible in some sections of the solar limb.

During the fall of 1943 I noticed that small chromospheric spike prominences were clearly discernible in photographs of the polar regions of the sun which I was taking with the Lyot-type coronagraph of the Harvard College Observatory located at Climax, Colorado.[1] I was amazed at the extremely brief lifetimes and the great frequency of occurrence which visual observations of these spicules indicated. Consequently, I decided, toward the end of the year, to undertake a further investigation of this interesting phenomenon. On December 12, 1943, under the very best conditions of atmospheric steadiness and purity, I obtained a motion-picture film centered near the position angle of the north pole of solar rotation. The 35-mm film was exposed at a rate of one picture per minute through an interference polarization monochromator designed and built by Dr. John W. Evans and similar to that used by Pettit for prominence observations. Exposures were about 2 seconds on a 103 H sensitization film produced especially by Eastman Kodak Company. With a red-glass filter and the special film, the over-all

[1] [For a report on "The Numbers and Motions of Solar Spicules," see *Miscellaneous Reprint* No. 75 (1957) of the High Altitude Observatory—an article by R. G. Athay and R. N. Thomas.]

transmission band utilized was about 4 *A* wide, centered on 6563 *A* of *H*α.

From this first film and from subsequent visual and photographic observations, I have ascertained, at least roughly, the behavior of the typical polar chromospheric spicule. The typical spicule shows first as a barely detectable lump on the solar limb. The lump rapidly enlarges and brightens to a maximum intensity that is still relatively faint for prominences in general. Within a minute or two of first appearance the lump elongates to maximum height, after which there is no detectable motion but simply a relatively gradual fading-out. Average proportions for spicules at maximum extension are approximately 3″ or 4″ × 10″, though strong observational selection undoubtedly favors the recording of the large spicules and the overlooking of the small ones. Average total lifetimes are about 4–5 minutes. Spicules much smaller than the typical one exist, and perhaps the frequency of occurrence for the smaller ones is even greater than for those I have called typical.

At a given moment of time I have observed visually more than 25 of the spicules within a 60° arc centered on the solar pole. Only the larger ones show on even the best photographs. On no occasion of satisfactory seeing and atmosphere purity have I failed to find the spicules present in considerable numbers in both polar regions. The intensity of the spikes is very low, on the average, and a great number of them exists at the very threshold of visibility. It is doubtful that they would be visible in apparatus not possessing relatively high resolution and high image-to-background contrast. Thus it is not surprising that the spicules have not attracted greater attention previously. Also it is unlikely that observers would be attracted to them in preference to the more striking prominences usually in evidence in other portions of the sun's disk. To study the spicules in detail from drawings is almost impossible, in spite of the advantages of visual observation, because of their intricacy and their rapid changes and because they are frequently obliterated momentarily by atmospheric unsteadiness. . . .

I have partially analyzed the film obtained on December 12, 1943; but for more complete studies I shall need improved measuring equipment. Preliminary results from the first fifty frames of this motion picture, taken at the rate of one picture per minute, disclosed the formation of just over 50 distinct spicules. The 48 which could be measured with improvised apparatus displayed the frequency distribution of lifetimes given in Figure 1. These tabulated lifetimes represent simply the difference in number between the frame on which the spicule was first seen and the one on which it was last seen. The results are rough be-

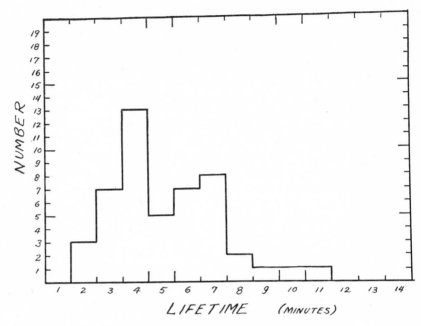

Fig. 1. Frequency distribution of spicule life-times.

cause of the difficulty of ascertaining the frame of first and last appearances with variable picture quality and too long an interval between pictures. The double maximum in the frequency distribution need not be regarded as significant in view of the grossness of the measurements and the small number of spicules involved. The drop in frequency for the shorter lifetimes may possibly be simply the result of observational selection. I have secured other films of the spicules at the more suitable rate of 6 frames per minute and these should reveal more accurately the lifetime and behavior of the small spikes when adequate measuring apparatus is devised and constructed.

The diagram in Figure 2 exhibits the distribution along the limb of the spicules detected in the first 50 frames of the December 12 film. No significant clustering exists; no large number of spicules came from any particular point of the arc; nor was any significant part of the arc devoid of the small spikes. No significant relationship to the position angle of the pole appears to exist. . . .

Figure 2 also displays to some extent the variation in behavior of the spikelets, one from another. Considerable range in size and shape exists. The size and length of life seem from rough comparison to be corre-

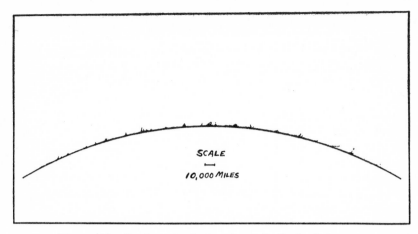

Fig. 2. Distribution of spicules along the limb of the sun.

lated strongly. The largest spicule in Figure 2 possessed the 11-minute lifetime, and some of the smallest as short as 2-minute lifetimes. This correlation serves further to accentuate the observational selection favoring observation of the larger and longer-lived spicules.

I have also made numerous visual observations of the small-scale chromospheric structure in the low latitudes of the sun. In most cases the spicules, if present, appear to have a more complex structure at low latitudes than at high. The inclinations to the vertical seem systematically greater, and the curvatures of the spikes themselves seem to be greater. Typical spicules definitely appear in low latitudes at times and apparently most conspicuously when the activity of larger prominences is at a minimum. I have taken motion-picture films to check this observation, but these will be too difficult to examine without improved measuring apparatus. At present the data are too meager to say more than that the typical spicule structure is generally most pronounced in polar latitudes.

An interesting feature of the spicules—as, indeed, of all prominences —is the extreme sharpness of the boundaries of the emitting gases. In all cases the edges of prominences, from the faintest spicules to the brightest surges and the largest eruptives, appear perfectly sharp with no evident diffusion or haziness at the borders. This is definitely not an effect of the use of high-contrast photographic records, since not only are many of my films developed to a gamma well under 1 but also the effect is most noticeable by visual observation. To the very limit of resolution of the coronagraph (about 1″) the boundaries appear clear and distinct, and even the internal structure of the larger prominences seems invariably to be a complex interlacing of perfectly sharp filaments and knots of prominence gas.

There appears to be no mention elsewhere of measurements of the lifetimes or frequencies of spicules. Hale has referred to "spike" prominences, but the phenomenon to which he was referring is evidently of a somewhat different nature, being very bright and related closely to a sunspot group. Hale's observation probably relates to a prominence of the "sunspot" classification (Class III) according to Pettit. The small chromospheric spicules are not mentioned by Pettit in his prominence classification scheme, nor does reference to them appear in the works of McMath and his collaborators or in the publications available here of Lyot or Waldmeier.

Measures of the dimensions of the spicules and the number visible at a given moment were made, previous to my observations, by Mohler, of the McMath-Hulbert Observatory, from the excellent eclipse plates of Marriott exposed at the Swarthmore eclipse expedition of October, 1930. Mohler found from the eclipse plates average dimensions of about $2'' \times 11''$ for the spicules, in close agreement with my independent measures. Mohler estimated from the number visible along the limb that the disk of the sun should possess approximately 2×10^6 such spicules.

Of course, it is difficult to decide when a given prominence is to be classified as a spicule. Especially is this true at lower solar latitudes where the extreme simplicity of behavior of the spicules is not preserved uniformly. Undoubtedly all gradations of intermediate behavior between spicules and other prominences are possible. Still, the spicules seem to fall well enough into a pattern of behavior sufficiently typed and sufficiently distinguishable to be regarded as a characteristic phenomenon of the solar poles, at least at this phase of the solar cycle, much as the well-known coronal plumes are considered a phenomenon characteristic of the corona in the polar latitudes of the sun near minimum activity. And it is inviting to consider possible physical associations between the two.

The spicules and their constant rapid changes provide additional evidence of large-scale turbulence in the chromosphere. The velocity of motion of a typical spicule is apparently of the order of 30 km/sec radially outward from the sun's center. It must be remembered, however, that this "typical" spicule is possibly one of the less frequent and larger spicules because of observational selection. Spicules possess characteristics very close to those now ascribed to solar granulations, namely, diameters of 1″–2″ and lifetimes of about 3 min. The recent interest in granulations stimulated by the work of Plaskett and ten Bruggencate impresses one with the importance of the possibility of associating spicules with the granulations. However, the expected velocity of outflow from granulations is about 1 km/sec, only one-thirtieth that estimated for the spicules. Further measures of the spicules and attempts to measure Doppler displacements in the granules should be made in order to determine whether this discrepancy is real. Still further, if direct photography of the polar coronal plumes and spicules can establish a link between the two, it may be reasonable to consider that the spicules associate granulations with the polar corona.

It is also reasonable to hope that a more detailed investigation of spicules may contribute to an understanding of the process of supply of material to the corona and regular prominences. The spicules are the only solar feature thus far discerned which seems observationally to indicate a continuous outflow of material through the chromosphere to the coronal regions. Indeed, it may prove that the spicules and the somewhat similar small-scale, low-latitude prominence structure are all that there is to the chromosphere and that actually the only difference between the chromosphere and the prominences is, as Menzel suggests, one of degree. In this case, the meaning of the "height of the chromosphere" becomes vague and the comparison of different measures extremely uncertain. In any event, the spicules are certainly further evidence of large-scale turbulence at the sun's surface; and, as such, they may provide, after careful study, valuable observational information regarding problems of disk-prominence-coronal interactions in general.

III

THE PLANETARY SYSTEM

The past half century has shown that, notwithstanding the greater glamour of stars, galaxies, and interstellar space, much has been done, and there is still much more to do, in the local family of planets. The emphasis on planets during the nineteenth century did not exhaust that field of inquiry. In fact, of all the bodies of the solar system, only Venus, Mars, and the comets fail to appear in this book. The rotation period and the atmosphere of Venus have in our 1900–1950 interval received some serious attention, but no entirely decisive results have been attained except for the recording by W. S. Adams and T. D. Dunham of no trace of oxygen or water vapor lines in the atmospheric spectrum; and the comets, ably treated by C. P. Olivier, F. L. Whipple, L. Biermann, J. H. Oort, R. A. Lyttleton, and others, have still not convincingly yielded the secrets of their origin and nature.

The work of twelve astronomers is presented in this report on the planetary system. Proceeding outward from the sun, we note that Mercury, which is discussed in W. de Sitter's lucid article on relativity in part XVII, had to wait for Einstein before its non-Newtonian behavior was fully explained. The planet Earth receives attention by way of H. Kimura's report on the wobbling polar axis. The peculiar irregularities in the length of the day (earth's rotation) have been exhaustively analyzed by H. Spencer-Jones (*Monthly Notices of the Royal Astronomical Society 99*, 541ff [1939]); he confirms that the slight steady lengthening of the day is due to tidal friction in the earth's shallow seas, and the irregularities in rotation are due to changes in the moment of inertia of the earth; but the details of observation and theory are too complicated for this survey.

The Earth's atmosphere is explored by F. L. Whipple by way of meteors in selection 16. The spectacular 1908 meteorite in Siberia is described by L. A. Kulik, with supplementary comments by E. L. Krinov; and the remarkable phenomenon of persistent meteor trains is interpreted by C. C. Trowbridge and C. P. Olivier.

To continue the comments on the contents of this part: the moon's temperatures and reflectivity have been accurately determined by Pettit and Nicholson at Mount Wilson, as reported in selection 13. A less technical discussion of the work is given by Pettit in the *Astronomical Society of the Pacific*, Leaflet 35 (March 1931).

In the 1930's the asteroid Eros provided opportunity for one of our most important international cooperative researches: the measurement of the solar parallax, which is in fact the base line of our planetary system; it is reported here by the former Astronomer Royal, Sir Harold Spencer-Jones. The subject is not closed, and already (1960) further work by new methods is under way.

In an impressive contribution, G. M. Clemence, D. Brouwer, and W. J. Eckert have accurately located the positions and determined the motions of the five outer planets throughout the four-century interval from about 1660 to 2060—a calculational enterprise involving many millions of numerical operations. The work would have been totally impractical without modern computing machinery. For this enterprise they used the IBM Selective Sequence Electronic Calculator which was made available without cost to the project by Mr. Thomas J. Watson. Only the five outer planets entered this major work because "the effects of the remaining principal planets, Mercury, Venus, Earth, and Mars, are hardly appreciable."

Pluto, the outermost planet, was discovered during this half-century. In selection 14 the discoverer describes step by step his procedure; providing a remarkable exhibition of the scientific method. We still know very little about that small cold body; nor all that asteroids, meteors, and comets can tell. We are not completely satisfied with any of the many theories of the origin of the planets. The work of C. F. von Weizsäcker, Harold Urey, and G. P. Kuiper appear, however, to be leading us to an acceptance of a modified Kantian-Laplacian theory of the origin. Obviously we have not yet finished asking deep questions and making significant discoveries and interpretations in our little planetary system.

10. THEORY OF THE MOON

By Ernest W. Brown

The completion of a laborious piece of work which has occupied many years for its execution furnishes a suitable opportunity for giving a general account of the object for which it was undertaken and of the methods by which the results have been obtained. The problem under consideration was that of the motion of the Moon as deduced solely from the Newtonian law of gravitation. It is limited, in the first instance, to the solution of an ideal problem in which the bodies are considered as particles, and two of them move in fixed elliptic orbits around one another. This constitutes the "main problem." The history of the attempts to obtain a solution with sufficient accuracy is well known, and I shall only touch on that portion of it which is directly connected with the new theory.

The original idea of the method here adopted to obtain a complete solution—as, indeed, of nearly all the methods of those who followed—is due to Euler. The pioneer work done by him on the lunar problem has, in my opinion, never received the full credit which it deserves. This may, perhaps, be partly due to the way in which he set forth his ideas; but it is, I think, mainly owing to the fact that his work was immediately followed by that of Laplace, whose justly great reputation in every department of mathematics and especially in celestial mechanics, has overshadowed the claims of his predecessor. However this may be, Euler fully recognized the importance of the special method under consideration. In the introduction to a paper published in 1768, "Reflexions sur la Variation de la Lune," he states the problem to be considered in the paper as follows: "Déterminer le mouvement d'une Lune qui feroit ses révolutions autour de la Terre dans le plan de l'écliptique et dont l'excentricité seroit nulle, pendant que le Soleil se mouvroit uniformément dans un cercle autour de la Terre." After some general remarks he writes: "Quelque chimérique cette question j'ose assurer que, si l'on réussissoit à en trouver une solution parfaite on ne trouveroit plus de difficulté pour déterminer le vrai mouvement de la Lune réelle. Cette question est donc de la dernière importance et il sera toujours bon d'en approfondir toutes les difficultés, avant qu'on en puisse espérer une solution complète." He then proceeds to find the solution, now known as the "variation orbit," as far as the fourth power of the only small parameter present. One may almost see in the few lines just quoted a germ of the magnificent work done by Poincaré on periodic orbits within the last twenty years.

The development of this idea of Euler is mainly due to G. W. Hill, who put the earlier steps into such a form that high accuracy could be obtained without excessive labour. J. C. Adams had also taken it up and worked at it in a somewhat similar manner. Hill determined the variation orbit and the principal part of the mean motion of the perigee, while Adams also found the variation orbit, but by a less powerful method, and the principal part of the mean motion of the node.

Before taking up a complete treatment from this stage it was necessary to consider as carefully as possible the amount of labour which would be demanded. The working value of a method of treatment is not really tested by the closeness with which the first or second approximation will make the further approximations converge quickly to the desired degree of accuracy; the real test is, perhaps, the ease with which the final approximation can be obtained. Here we have the essential difference between the present method and all other methods. The approximations of the latter proceed along powers of the disturbing force. Euler's idea was to approximate along powers of the other small constants present. This gives a more rapid convergence and a degree of certainty in knowing the limits of error of the final results which no other method approaches.

With this in view it was necessary to cast the equations of motion into such a form that the degree of accuracy demanded should be capable of being obtained with a reasonable amount of labour, and it must be made clear that this degree of accuracy had actually been attained when the work was completed. Precautions against errors of computation must be taken, and the results should, if possible, be expressed in such a form that comparison with those of previous theories is possible. These and other points are considered in the following paragraphs:—

First, every coefficient in longitude, latitude, and parallax which is as great as one-hundredth of a second of arc has been computed, and is accurate—apart from possible errors of calculation—to at least this amount. Hansen, indeed, gives his results to thousandths of a second, but certain of them are in error by some tenths of a second; indeed, it was not possible to obtain all of them more accurately without much increasing the extent of the calculations. Some of Delaunay's coefficients, owing to slow convergence, are not accurate to one second of arc. As a matter of fact my results are obtained correctly to one-thousandth of a second, and there are comparatively few coefficients greater than this quantity which have not been obtained.

Second, the theory is expanded algebraically in powers of four of the five parameters, the fifth (the ratio of the mean motions of the Sun and Moon) having its numerical value substituted at the outset. The last is known with a degree of certainty which satisfies all the possible needs of the theory, and the effect of any possible change which may be made in its observed value can be easily deduced from Delaunay's purely literal theory. The chief advantage gained is due to the fact that slow convergence (perhaps divergence) occurs only along powers of this ratio, while there is little loss of theoretical interest in using its numerical value. Moreover it is not difficult to find out how many places of decimals are necessary at the outset in order to secure a given number of places in the results.

Third, exceptional precautions have been taken to avoid the occurrence of errors during the course of the work. Equations of verification have been computed, not only at the end of large masses of calculations, but at practically every step in the process; in fact each manuscript page of work has, on the average, not less than two test equations computed. The most dangerous sources of error—the omission of a whole set of terms or the use of a wrong set—was partly guarded against by a property of the method itself. A comparatively small error of this kind produces in the final results large systematic errors, which a rough comparison with the values of Delaunay and certain properties of the solution will detect without difficulty. Very searching final tests, eleven in number, are furnished by the remarkable relations known to exist between the expressions for the mean motions of the perigee and node and the constant term of the parallax. These were all completely satisfied. Finally, the work was so arranged that computers could be engaged to do considerable portions of it. Only one, Mr. I. I. Sterner, has actually been employed, but this disadvantage was counterbalanced by the very high accuracy of his work. A rough calculation of the chances of an error slipping by both of us in the work turned over to him and verified by me, and through the special test equations, gave two or three possible errors in the whole. Two such errors were actually detected by the numerous final tests, and these were, of course, traced down and corrected.

Fourth, the results originally computed for the rectangular coordinates of the Moon have been transformed to polar coordinates, and thus a direct comparison with those of Delaunay has been rendered possible. Newcomb had previously transformed Hansen's results to the same system, so that these are also available for comparison. This comparison will appear in a following number of the *Monthly Notices*. Nearly all the differences Delaunay—Brown can be explained by slow convergence of the Delaunay series, and in most of the remaining cases the differences Hansen—Brown are very small. Unexplained disagreements between the new results and those of both the earlier theories only occurred in the cases of coefficients difficult to determine accurately by the latter methods, owing to the occurrence of very small divisors and the slowness of approximation proceeding along powers of the disturbing force.

Fifth, comparison of the new coefficients with those deduced from observation has at present only been possible to a limited extent, but in two cases—the mean movements of the perigee and node—it has been completed. The net result is very satisfactory. The differences in the annual motions of these two lines are less than three-tenths of a second of arc, and these are capable of being explained by reasonable suppositions concerning the figures of the Earth and Moon, the constants connected with these bodies not being yet known with sufficient accuracy. One of the most important coefficients—that of the principal parallactic inequality in longitude—appears to furnish a value for the solar parallax very near the mean of all the values obtained by other methods.

A few brief details about the amount of time and labour expended may not be uninteresting. From 1890 to 1895 certain classes of inequalities were calculated, but the work was only begun on a systematic plan, which involved a fresh computation of all inequalities previously found, at the beginning of 1896. Mr. Sterner began work for me in the autumn of 1897 and finished it in the spring of the present year, though neither of us was by any means continuously engaged in calculation during that period. He spent on it, according to a carefully kept record, nearly three thousand hours, and I estimate my share as some five or six thousand hours, so that the calculations have probably occupied altogether about eight or nine thousand hours. There were about 13,000 multiplications of series made, containing some 400,000 separate products; the whole of the work required the writing of between four and five millions of digits and *plus* and *minus* signs.

Although the problem now completed constitutes by far the longer part of the whole, much remains to be done before it is advisable to proceed to the construction of tables. The problem solved is that of the Moon under the attractions of the Earth and Sun, the centre of mass of the Earth and Moon being supposed to move in a fixed elliptic orbit. There remains to be found the effect of the figures of the bodies—mainly that of the Earth—the effects of the differences of the actual motion of the centre of mass of the Earth and Moon from fixed elliptic motion, due to the attractions of the planets; and, the most difficult of all, the effect of the direct attractions of the planets on the Moon. There are many periodic coefficients, due to the last, larger than one-tenth of a second of arc in magnitude, and the whole subject needs a careful and extended investigation. An attempt to complete the problem by considering anew these remaining sources of disturbance has been already started. The difficulties presented appear to arise much less from intricate calculations than in the construction of a satisfactory method which will give the assurance that no sensible terms have been neglected. If a moderate degree of success attends these efforts it should be possible to discover whether, within the limits set by the observations, the motion of the Moon shows effects which cannot be traced to the direct operation of the Newtonian law of gravitation.

11. TROWBRIDGE'S CLASSICAL WORK
ON METEOR TRAINS

By Charles P. Olivier

Even the most casual observer has noticed that some meteors leave a train or luminous streak behind them, while others seem to leave nothing. In most cases this train is as ephemeral as the meteor itself, but a few hours of observation generally will show one or more that last several seconds, and years of such work will furnish a few that last several minutes. We have on excellent authority that, if a telescope is at hand and is turned upon a meteor train, its visibility is greatly prolonged, so that one which to the unaided eye would disappear in ten or twenty seconds might thus be seen two or three minutes. In extreme cases even to the naked eye they have remained visible over half an hour. The smoke cloud left by the explosion of a bolide in daylight sometimes remains visible even longer.

Fairly accurate observations, made simultaneously at two or more places upon remarkable trains, have proved that they were several miles in length and certainly occasionally one or more miles in diameter. We thus know that they seem to fill several cubic miles of space with their faint glow. As again we know positively that the bodies which produced them are all very small, no matter whose estimate we accept, and usually could not on their passage have filled one millionth part of the space so illuminated, it long has been a mystery how such trains were produced. Or, if produced, how they could remain luminous in the (supposed) intensely cold and tenuous atmosphere at that altitude. The chance of irradiation playing an important part in their apparent size, as it does for the fireball or meteor itself, is wholly untenable because telescopic observations, made at leisure, confirm their dimensions. Also the trains are not very brilliant per unit area hence irradiation would not be a large factor.

The explanation that appeared simplest at first sight and that often has been given was that the trains consisted of the incandescent material and glowing gas left by the passing body; in the case of a meteor most of the material going thus to form the train, and for a fireball at least that [material] on the surface. That this explanation is false is obvious; first because the tiny particles which might be assumed to be left, as well as the heated gas, would cool in a second of time or less; second because not enough of them possibly could be thrown off nor enough gas be heated to fill so great a space. It might be added that certain meteors, usually with slow angular motion, do leave a train of sparks behind them, but this is a different phenomenon and each tiny spark is a unit, which quickly ceases to glow,—goes out as would a white hot coal of its size. No observer need mix this phenomenon with the usual trains, particularly the long enduring ones.

It so happens that the labors of one man, the late C. C. Trowbridge, of Columbia University, have done more for the study of meteor trains than all others combined, hence his work will be reviewed here. He was not himself an observer, hence depended upon the careful data of astronomers, among whom we may mention Barnard, Denning, A. Herschel, and Newton, as well as many others. Nevertheless to Trowbridge's zeal in collecting and studying the records obtained by these men is due the best theory of what produces this mysterious train.

These data were tolerably extensive, for in 1907 [1] he had collected reports of thirty-seven trains which remained visible from over five to forty minutes each; fifty-three in all over one minute with an average duration of 14.8 minutes. His conclusions are as follows:

1. The meteor trains are self luminous gas clouds combined with very minute meteoric dust particles, the latter in daylight reflecting light like any ordinary clouds.

2. The height of meteor trains seen at night appears to be at a definite altitude indicating that the phosphorescence is dependent on the gas pressure where the trains are found.

3. The diffusion of the trains is gas diffusion and its rate depends on the temperature of the atmosphere, and probably on the initial intensity of the train.

4. Many meteor trains appear to be tubular in form, that is, the luminosity is greater near the border.

5. Experiments have been made by the writer which give the law for the rate of decay of the phosphorescence of the air at very low pressure, and these experiments explain the long duration of the meteor trains, on the hypothesis that it is a phosphorescence which decays according to the same law.

6. Statistics on the color of trains show that, excluding those illuminated by sunlight, trains are as a rule green or yellow fading to white, colors which are typical of the phosphorescence of the air.

The average mean altitude of 54 miles $= 87$ km seems excellently well determined, and the upper and lower limits of 60 and 50 miles

---

[1] *Astroph. Jour.*, 26, 95 (1907).

TABLE I. Altitudes of trains.

Number	Altitude limits of visible track of nucleus—miles		Altitude limits of train—miles		Mean altitude of train	Altitude computations made by
10					50	A. S. Herschel
12	100	53	59	53	56	A. & J. Thomson
26					54	A. S. Herschel
29 *	120	60	65	60	63	H. A. Newton
44	68	49	59	49	54	H. A. Newton
46					51	H. A. Newton
47	65	52			58	H. A. Newton
52 †					59	H. A. Newton
80	90	30	58	50	54	W. F. Denning
120	78	47	59	47	53	W. F. Denning
121	65	37	57	45	51	W. F. Denning
125 ‡	65	28			45	W. F. Denning
129	90	41			54	A. S. Herschel and Greg
Mean	82.8	44.1	59.5	50.7	54.0	

* Over 60 miles—train five miles long, carefully calculated.
† Altitude of lower part of train.
‡ Short train ½° long at 45 miles.

scarcely less so. It further is seen that the altitude at which the nucleus itself became visible seems to have no effect upon the position of the train, in some cases 40 to 50 miles having been travelled before its production. Also in some cases the nucleus continued to move on some distance after the train had ceased to form. That a train is often visible for only a part of a meteor's course is a phenomenon well known to every observer, as well as the fact that the nucleus sometimes goes farther without the train following to the very end. This fact was stated in 1866–1867 by Greg, Glover, and Goulier independently.

As for meteor trains seen in daylight, according to Trowbridge [1] they seldom occur above 40 miles and sometimes as low as 25 miles. They appear as if composed of thin smoke. This, in connection with the much higher altitudes of the night trains, seems to indicate that in the upper levels of the atmosphere the glow does not arise mainly from light reflected from fine meteoric dust, but is a luminosity of the gas in the meteor's train. Attention is called to the Leonids leaving greenish trains,

[1] *Pop. Sci. Monthly 79,* 191 (1911).

the Perseids yellowish, and most daylight trains (i.e., smoke) appearing reddish—this last merely reflected sunlight. His general idea is that the night trains, which for long durations always expand, are due to gas diffusion but are tubular in form. Hence when seen from the outside the two sides of the relatively empty tube would appear brighter than the middle, which is actually confirmed by observations. The conclusion, in connection with laboratory experiments, is that the barometric pressure at these heights, i.e., 50 to 60 miles altitude, is not far from 0.2 mm.

In his second paper mentioned he elaborates what occurs on the passage of a meteor as follows:

It has been shown . . . that when a body is very hot an immense number of negatively charged corpuscles or ions are given forth by the body. Air containing free ions becomes a conductor of electricity, hence we have in a meteor rushing through the atmosphere a condition extremely like a very long electrical discharge tube containing gas at low pressure. . . . The burning meteor . . . must form a column of highly ionized air. . . . Moreover at a certain altitude, corresponding to a pressure 0.2 mm . . . or about one two-thousandth to about one four-thousandth of one atmosphere pressure, the conditions are precisely right for the formation of phosphorescence in the meteor track. . . . When the meteor nucleus has been consumed, all that remains visible in the dark sky is the body of phosphorescent gas in the part of the track where the gas pressure conditions were correct for the formation of the persistent glow.

He adds that it is not certain that electrical discharges take place in the meteor's track, and they may not even be essential for the formation of the phosphorescence. The air heated and ionized by the rapidly moving, burning meteor may readily suffer chemical or physical changes in its composition which on gradually reverting to its original state give out such a glow. Both papers quoted, as well as others by the author, contain much valuable detail and excellent drawings of long enduring meteor trains, as well as a fuller explanation of the theory than can be given in this brief review. To date they may be considered classical in this subject.

12: AN INTERIM REPORT ON AN INTERNATIONAL RESEARCH PROJECT: THE WANDERING OF THE NORTH POLE [1]

By H. Kimura

The present state of the work on latitude variation:

(1) International Latitude Service on the north parallel $+39°$ $8'$.— The number of international stations was six in the beginning of their cooperation (namely, 1900.0), which number was reduced to four because of several unavoidable circumstances about the years 1914–1915, and then again one was lost about the year 1919. Thus now only three stations in America, Italy, and Japan have remained. These three independent stations are satisfactory for the finding of the two coordinates of the polar motion (x and y) and the unknown declination error, together with the non-polar variation (z) common to all stations. But certainly the observations at these stations must contain local systematic errors and also accidental errors. As a consequence, the final results are not accurate enough.

Fortunately, through the great effort of Professor Subbotin, the Director of Taschkent Observatory, the establishment of a new station at Kitab near Samarkand in Uzbekistan was decided upon, and the observations will probably be commenced in the autumn of this year.

(2) Free cooperations.—The following observatories have been making observations of latitude variation for many years: Washington, Greenwich, Pulkowa, and only one in the southern hemisphere, Rio de Janeiro. Besides these observatories, a new station at the Lembang Observatory, Java, will be established by the kindness of its Director, Dr. Voûte. This may also be commenced within this year. All persons who are interested in the work would welcome the results from this observatory as having a particular advantage, lying near the Equator.

(3) A new enterprise of the international cooperation in the southern hemisphere, namely two latitude stations in the observatories of La Plata and Adelaide, will probably be realized within a few years. Those observatories lie fortunately on nearly the same parallel, within

[1] [The Editor has made a number of verbal changes in Dr. Kimura's translation.]

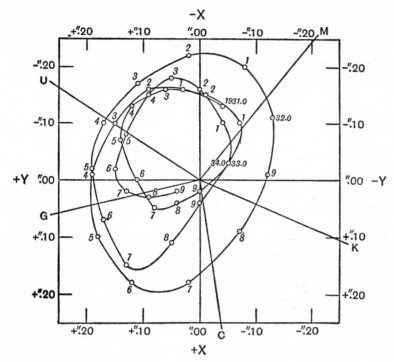

Fig. 1. A sample record of the wandering of the North Pole.

one minute of latitude, and on nearly the same but opposite meridians, like the old stations at Oncativo and Bayswater.

Next I have to add a few words about the results of the recent observations made at the international stations, Greenwich, and Washington. Generally speaking, those of the two observatories agree very well with that of the international service, provided that the former two are corrected by their own z. It is of importance that the instruments used at Greenwich and Washington are not only different from each other but also different from the visual Zenith telescope used in all international stations. However, in the results of the last years at both observatories some abrupt decreases in their mean latitudes of a pretty considerable amount, say $0''.07$, have occurred, while the same phenomena did not appear in the international result for the corresponding epochs. But it must be understood that such phenomena have happened in almost all observations of the latitude variation.

Now we pass to the z-phenomenon (Non-polar variation).

(a) *International service.* In my annual reports for the latitude variation the part of the non-polar variation was given in another form than z. However, if we calculate it in the same manner as that made usually for the chain method, nearly the same curve of z as that in the previous years can be obtained.

(b) *Greenwich.* At this observatory z consists principally of a semi-annual form, the annual one being very small. It is noteworthy that the observations during the period 1890–1900, which were carried out in the latitude of about $+52°$, for instance, at Prague and Potsdam, gave also z of such double period annually. But there is one exception, with the observations made at Leiden, which contain an annual period of a considerable amount.

(c) *Washington.* In this observatory z is very similar to that of the international service. And the old observations made at New York gave also nearly the same z. . . .

Since a few years ago, the 14-month motion contained in the polar motion has become considerably smaller and its period has shortened. As Chandler had noticed already, the similar status had once occurred in the year about 1840. And I think the same state of smallness of the 14-month motion will continue a few more years.

13. ABSTRACT: LUNAR RADIATION AND TEMPERATURES

By Edison Pettit and Seth B. Nicholson

Methods of Observing.—The vacuum thermocouple attached to the 100-inch telescope was used as in measuring stellar radiation. A microscope cover-glass was used to separate the reflected light from the planetary heat. A fluorite screen and water-cell were used for special purposes. . . .

Limiting measurable temperature.—Temperatures below 100°K are measured with difficulty, since the value of [the radiometric magnitude] for that temperature is 12.47 mag.; and it is probably not feasible to detect planetary radiation from a celestial body whose temperature is below 70°K.

Reflection of solar radiation between 8 and 14µ.—Even if the moon reflected all the solar radiation in this region, it would add only 1½ percent to the planetary heat and affect the computed temperature by only 1°. With the fluorite screen the emissivity between 8 and 10µ is shown to be the same as between 8 and 14µ, which indicates that the silicates, if present in the lunar crust, cannot be detected by the selective emissivity between 8 and 10µ as in laboratory tests, probably because they are in a finely divided or porous state for which the emissivity would be more like that of a black body.

Distribution of radiation over the disk.—From drift-curves it is found that the distribution of planetary heat over the disk at full moon does not follow the formula

$$E = a \cos \theta,$$

derived from the Lommel-Seeliger law, but more nearly the formula

$$E = a \cos^{3/2} \theta.$$

This is explained by the rough surface. From the water-cell drift-curve it appears that the general trend of reflected light at full moon is uniform, but that at the limb in the E–W direction the intensity is 60 percent greater than that of the neighboring *maria*.

Temperature of the subsolar point.—The measures indicate that at full moon, when the subsolar point is nearly central on the disk, its temperature is 407°K [273°F], but at quarter phase, when it is on the limb, they show a temperature of 358°K [185°F]. The directive effect of the rough surface on the planetary heat probably accounts for this difference.

Temperatures during a lunar eclipse.—Measurements made near the south limb on June 14, 1927, show that the temperature fell from 342°K [+156°F] to 175°K [−143°F] during the first partial phase, continued to drop to 156°K [−179°F] during totality, and rose again abruptly to nearly the original temperature during the last partial phase. From these data it is estimated that less than 0.1 cal cm^{-2}min^{-1} is conducted into the moon from the surface. . . .

The theoretical lunar temperature.—From the water-cell measurements and the distribution-curve, the light reflected from the subsolar point is estimated to be 0.2 cal cm^{-2}min^{-1}; and combining this with the amount conducted and with the solar constant, we find the mean spherical rate of lunar emission to be 1.61 cal cm^{-2}min^{-1} and the computed black body temperature to be 374°K. From the distribution-curve of planetary heat about the subsolar point the mean spherical rate of emission is found to be 1.93 cal cm^{-2}min^{-1} and the corresponding temperature 391°K, which is to be compared with the theoretical temperature of 374°K. . . .

Reflection of solar radiation from the lunar surface.—From the distribution-curve of reflected light about the subsolar point, and the drift-curve made with the water-cell in the beam, the formula

$$E_r = K \frac{0.46 \sec^2 i/2}{0.46 \cos \theta + \sin \theta}$$

was derived, where i and θ are the angles of incidence and reflection, respectively. The laws of Lambert, Lommel and Seeliger, and Euler cannot be made to fit the observational material.

Radiometric albedo.—The radiometric magnitude of the reflected light is found to be −13.3, and of the whole lunar radiation −14.8. The radiometric albedo of the reflected light is 0.093.

Temperature of the dark side.—This was found to be 120°K [−243°F]. The temperature is so low that considerable observing will be necessary to establish a good value.

14. THE DISCOVERY OF PLUTO

By Clyde W. Tombaugh

At the beginning of the 20th century, a few investigators, particularly Dr. Percival Lowell, became interested in the possible existence of an unseen planet beyond Neptune. Certain small residual perturbations of Uranus were interpreted by Lowell as indications of the presence of another, more distant planet. Dr. Lowell started a search program in 1906, at Flagstaff, using a camera 5 inches in aperture to record photographically the positions of the many thousands of faint stars, for purposes of comparison. About 200 plates were obtained by E. C. Slipher and 50 by K. P. Williams. The series of 3-hour exposures was centered along the invariable plane of the solar system, at every 5° of longitude. Dr. Lowell examined these plates with a hand magnifier by laying one plate of a pair over the other. The new planet could not be found, for, as is now known, the inclination of its orbit placed its images just off the south edge of the plates.

Next, a series of plates was made with the 42-inch reflecting telescope. Its small field rendered this instrument unsuitable for searching. The comparison of plates was greatly facilitated, however, by the use of a Zeiss blink comparator.

The search was resumed in 1914, as a result of the loan of a 9-inch telescope by the Sproul Observatory. From 1914 to 1916, a great many plates were taken with this instrument. The plates were examined under the blink comparator by Dr. Lowell and C. O. Lampland. Nothing was seen of the elusive "Planet X." This fact was a keen disappointment to Dr. Lowell, as he felt that theory had lent all the aid it could.

The search work from 1906 to 1916 was at some disadvantage, because Pluto was moving slowly on the more remote part of its orbit and was consequently about half a magnitude fainter than it was at the time of discovery.

In 1915, Dr. Lowell published his 125-page "Memoir on a Trans-Neptunian Planet." His solutions indicated that "Planet X" had a mass 1/50,000 that of the Sun, or 7 times that of the Earth, and a mean distance of 43 astronomical units from the Sun. Dr. Lowell assumed the unseen planet to have a low density and a high albedo, like the four giant planets, and hence a disk subtending an angle of one second of arc with a stellar magnitude between 12 and 13. On November 16, 1916, Dr. Lowell died suddenly and the search was discontinued for 13 years.

In 1919, W. H. Pickering, from discrepancies in the motion of Neptune, found indications of a planet in the same region of the sky and predicted that the new planet would be of the 15th magnitude. Four photographs were taken with a 10-inch telescope on Mount Wilson in December, 1919, of the region in which the unseen planet was supposed to be. Portions of these plates were pretty carefully examined, but nothing was found, for Pluto's images were outside the area of close scrutiny. Moreover, some of the images of Pluto were nearly obliterated by accidental defects in the emulsion. It was natural, then, for the astronomical world to doubt the existence of a trans-Neptunian planet.

The Lowell Observatory still courageously persevered with the search. In 1925, glass disks were procured for a 13-inch photographic telescope with a wide field. Percival Lowell's brother, Dr. A. Lawrence Lowell, generously provided the funds to complete the instrument. Great credit is due the skillful optician, Mr. C. A. R. Lundin, who succeeded in making a research instrument of superior quality. In January, 1929, the Director of the Lowell Observatory, Dr. V. M. Slipher, invited me to join the Observatory staff. In a few weeks, the 13-inch objective arrived from the East, and was promptly installed in its tube. In the next few weeks, tests were performed to ascertain the properties of the instrument.

The 13-inch Lawrence Lowell telescope has a focal length of 66 inches and a plate scale of 30 millimeters per degree; it gives good images over almost the entire area of a plate, 14 by 17 inches, covering an area of the sky nearly 12° by 14°. Success in using so large a plate effectively is attained by the use of special plate-holders which bend the plates into a slightly concave form by a controlled amount. Every plate was tested and made to conform to a standard curvature, so to insure a homogeneously focussed series of plates. The images of the faintest stars were consistently less than 1/30th of a millimeter in diameter.

Every precaution was taken to match perfectly the plates of each pair, so as to facilitate thorough and rapid examination under the blink comparator—a laborious work at best. Other precautions included: selecting plates of like age and sensitivity, careful judging of sky transparency and sky light, equal exposure times, same guide star, similar steadiness of seeing, same hour angle, careful guiding, avoiding moonlight, and uniform development. A 7-inch refractor was used as a guide telescope, so that the guide star occupied the center of the plate. A 5-inch camera was attached to check the brighter planet suspects. The standard exposure time was one hour, recording stars to the 17th magnitude, except at the extreme ends and corners of the plates. Guide stars, seldom fainter than the 7th magnitude, were carefully chosen on or very near the ecliptic. . . .

Since planets move slowly in the sky with reference to the stars, it should be evident that the basic technique in searching for new planets is to photograph a given region of the sky, then take a second photograph of the same region at a later time for the purpose of comparison. The scale of the 13-inch plates is such that, for a planet beyond the orbit of Neptune, the shift of the image is suitable for detection after an interval of two or three days.

By the end of March, 1929, final adjustments having been made, I embarked upon the long, hard, tedious task of taking many plates at the telescope each month and of examining them under the blink comparator. The favored region in Gemini was sinking in the western evening sky—soon it would be behind the Sun. Accordingly, these regions were promptly photographed, then regions to the east along the ecliptic, as rapidly as possible. Various improvements in the technique suggested themselves, as the work progressed. Two proved to be very important. (1) Much time, effort and expense may be lost in running down planet-suspects that turn out to be only asteroids near their apparent stationary points where they imitate the slow motion of a more distant planet. The simple expedient was to photograph each region near its "opposition point" (180° from the Sun), where the apparent retrograde motion is a maximum for all planets outside the Earth's orbit, and the daily shift in position against the star background is roughly inversely proportional to the distance of the object. As a consequence, the asteroids, on the average, moved about 7 millimeters per day on the plates, and exhibited short trails during the hour's time of exposure, whereas Pluto moved only $\frac{1}{2}$ millimeter per day. This criterion was useful in estimating at sight the distance of any suspicious object, and extremely convenient in computing a rough ephemeris when it was necessary to re-photograph a region later in running down a promising planet suspect. The known asteroids number a few thousand and are widely scattered between the orbits of Mars and Jupiter. The angular distances of the stationary points from opposition for Mars and Jupiter are, respectively, 36° and 64°. Accordingly, the areas of the sky between these limits were scrupulously avoided. On account of the revolution of the Earth around the Sun, the opposition longitude moves eastward through the constellations at the rate of 30° per month. From September, 1929, to the end of the search, the practice of photographing each region of the heavens at opposition was strictly followed. There are two other advantages to photographing at opposition. The opposition point is on the local meridian at midnight, and so allows the observer enough observing hours to obtain the necessary number of plates, without running into unfavorably low altitudes in the sky. Moreover, the Earth is

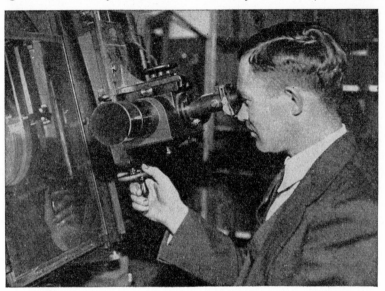

Fig. 1. The plate comparator used in the discovery of Pluto.

nearest the planet then, although for a planet as distant as Pluto, the difference in brightness between opposition and conjunction is only 0.1 of a magnitude. (2) Experience in examining pairs of plates under the blink comparator early revealed that the greatest number of planet suspects were very faint and beyond the reach of the 5-inch camera plates for checking. Almost every pair of plates had a few faint planet suspects, which were only plate defects. Some pairs had over a hundred of them. If the search program was to be thorough and worth anything, this problem had to be met in a practical way. Accordingly, the second important improvement in the technique consisted in taking three plates of each region within a total time of one week. This procedure increased the photographic work by 50 per cent, but it was well worth it. The planet suspects were marked as the "blinking" progressed. At the end of a panel, the unmarked plate was removed and the third plate slipped into place. Only a few out of several thousand suspects survived the crucial test of the third plate; thus, only a few regions had to be re-photographed. The practice of taking 3 plates of each region was advantageous in that the best matched 2 of the 3 plates were selected for blink examination.

I entertained the possibility that a new planet might be found in any part of the zodiac. After catching up with the opposition point in September, 1929, the blink examination began in real earnest on plates taken in Aquarius. The search progressed eastward through Pisces and Aries. On each of the plates the images of about 50,000 stars appeared. A pair could be thoroughly examined in 3 days. The number of stars increased in Taurus as the Milky Way was approached. The work of examining each pair became more tedious and more prolonged. The plates taken in western Gemini contained over 300,000 star images each! Since the time spent in the blink examination of plates is proportional to the number of star images, this work fell behind that of taking the plates.

Three plates centered on Delta Geminorum were taken on 1930 January 21, 23, and 29, respectively. After laboriously finishing the easternmost pair in Taurus, I happily decided to postpone the two very rich regions in Gemini and started blinking the Delta Geminorum pair. When one-fourth of the way through this pair of plates, on the afternoon of February 18, 1930, I suddenly came upon the images of Pluto! The experience was an intense thrill, because the nature of the object was apparent at first sight. The shift in position between January 23 and 29 was about right for an object a billion miles beyond Neptune's orbit. Since the plates had been taken in opposition, I was confident that the object was not an asteroid. The images were sharp, not diffuse; hence, there was no suggestion that the object was a comet. In all of the two million stars examined thus far, nothing had been found that was as promising as this object. It was about two magnitudes brighter than the faintest stars recorded on the plates. At once the 5-inch plates, taken simultaneously, were inspected with a hand magnifier. There too were the images of Pluto, clearly confirmed, exactly in the same positions! The images were quickly found on both plates of January 21, the positions conforming with the motion indicated by the plates of January 23 and 29. Thereupon, I informed the Director and other members of the staff, who came to take a look.

The night of February 18 was cloudy, but the following night was clear, and the Delta Geminorum region was photographed again with the 13-inch telescope. The new plate was put under the comparator. Within a minute, the image of the planet was found, after 3 weeks' motion, just where it was expected. On February 20, with the aid of a film print of the immediate star field from this negative, the planet was soon picked out among the faint stars in the field of view of the 24-inch visual refractor. It was disappointing that no disk was visible—merely a faint star-like object of the 15th magnitude was seen, which had moved a small distance from the photographic position of the night before. Dr. C. O. Lampland started photographing the new planet with the 42-inch reflector. The rate of motion continued to be satisfactory, and proved beyond doubt that the object was a trans-Neptunian planet.

News of the discovery was telegraphed to the Harvard College Observatory on March 13, 1930; from there it was announced to the world. A few months later the planet was named Pluto.

Pluto was discovered at the time it was crossing the ecliptic. Had it been far from its node, it would have been outside the first photographic belt and might not have been found until 1932, when two more belts, one on each side of, and parallel to, the first were completed.

After the discovery of Pluto, I was urged to continue this thorough and systematic search over a wide area of the sky for other possible distant planets. This immense and tedious task continued at Flagstaff until three-fourths of the entire heavens had been explored, and 90 million star images had passed in review in the eye-piece of the blink comparator!

Some doubt has arisen as to whether Pluto is really Dr. Lowell's predicted "Planet X" because of the apparent smallness of its mass. On the other hand Pluto was found within 6° of his predicted place, and the predicted and actual elements of the planet's orbit are in remarkably good agreement.

Thus, Pluto, a planet barely visible in a 12-inch telescope, was singled out from 20 million other equally bright, or brighter, objects in the sky.

15. THE TUNGUSKA METEORITE

By L. A. Kulik

At 7:00 on June 30, 1908, in fair weather a large bolide or fireball was observed flying from south to north through the cloudless sky over the Yenissei River basin in Central Siberia (1)[1]. After the fall of the mass, a column of fire rose over the taiga; it was observed in Kirensk, situated at a distance of 400 km. The appearance of the "column of fire" was followed by three or four powerful claps and a crashing sound which were recorded over radii of more than 1,000 kilometers, and by powerful air waves (2). The blast caused the radial windfall of trees (crowns outward) to a distance of several dozen kilometers from the point of the fall (3). The air waves were so powerful that they set in motion microbarographs in North America, as well as in Western Europe; they circled the world and were recorded a second time at Potsdam, Germany. Throughout the world, seismographs at stations as far from each other as Irkutsk and Tashkent, Tbilisi and Jena, and Washington and Java, recorded the powerful earthquake wave caused by the explosion of the meteorite as it struck the Earth's crust (4). The epicenter of this disturbance was in the vicinity of the Podkamennaya Tunguska, near the Vanovara trading station (5). The large masses of fine particles of matter sprayed in the atmosphere during the flight of the meteorite and by the explosion when it struck the Earth's crust at cosmic speed, created a thick dust layer in the upper strata of the atmosphere, formed "silvery clouds" (luminous clouds) at an altitude of 83–85 kilometers, and dust screens in the ceiling and lower layers of the stratosphere. This produced the remarkable and incomparably beautiful phenomenon known as white nights. They were observed over the territory stretching from the region of the fall to Spain and from Fenno-Scandia to the Black Sea. During the days following the night of June 30–July 1, the "lights" gradually faded (6).

Information concerning the fall of the meteorite was first obtained

---

[1] [For a useful commentary on Kulik's report, see the twelve notes at the end of this article by E. L. Krinov, who has continued the Tunguska researches after Kulik's death. Krinov also provided the translation of Kulik's report.]

in 1921 by an expedition to Siberia headed by the author, and equipped on instructions from A. V. Lunacharsky, then People's Commissar of Education. At that time certain scientists did not consider this phenomenon worthy of investigation. In 1925–26, new data confirmed the author's opinion; firstly there were the data on the seismic waves obtained by A. V. Voznesensky, and, secondly, there was information provided by S. V. Obruchev, a geologist, and I. M. Suslov, an ethnographer. Finally, in 1927, the author made a trip to the Podkamennaya Tunguska River to study the region of the fall. The investigation revealed that there had been a continuous, eccentric, radial windfall of trees over an area with a radius of about 30 kilometers and in some places it extended to hills 60 and more kilometers away. Eye-witnesses related that individual trees were felled on the hills in the vicinity of Vanovara. Further evidence of the unusual force of these air blasts is that in Kirensk, 400 kilometers away, fences were torn up; in Kezhma, grain loaders were thrown from their feet; in Kansk, 600 kilometers away, rafters were cast into the river, while south of Kansk, at a distance of 700 kilometers, horses could not stand up.

Due to the great number of hills surrounding the center of the fall, a plan of the windfall area resembles the meteorological chart known as a wind rose. Naturally, it is now difficult to estimate the windfall area without a survey. On the outer fringe of the windfall area (about 20 kilometers from the center), everything on the surface of the earth was charred. Moreover, as a rule the branches of the trees which remained standing as well as those that were felled, are broken. A typical feature is that every break surface has been charred and there is "not a single break without a burn." The central part of the windfall area lies among hilly, permanently frozen peat bogs alternating with marshes in the watershed; it is surrounded by charred trees which have remained standing but have been completely stripped of their branches. The area where every single tree has been felled begins at a distance of one-half to one kilometer from the center.

The first expedition did not discover any clearly defined meteoritic craters in the region. The author revisited the region in 1928, accompanied by a cinema cameraman, so that by filming the locality a considerably greater part of the windfall area could be studied; with a minute theodolite the direction of the tree felling and the center area were determined; rock-flour usually found in meteoritic craters, i.e., fine, sharp-edged, crushed material of surrounding rocks, was discovered in some depressions (7). At the same time the question was raised as to the origin of certain round depressions, 50 meters in diameter and

4–6 meters deep, which were discovered among the mounds of the peat bogs (8). Magnetic investigation of these depressions failed because of the presence of magnetic rocks (Siberian traps) in the area.

The remarkable film and the need for a more thorough investigation prompted the author to undertake an expedition in 1929. He was accompanied by L. V. Shumilova, a geo-botanist, and E. L. Krinov, a research worker. For eighteen months small teams of two to seven workers surveyed, excavated, dug trenches in the frozen peat bogs, bored, etc., to make a comparative study of the permanently frozen hilly peat bogs, to obtain climatological data, to ascertain the extent of the permafrost area, and to establish the optimal conditions for aerial survey. A base of three cabins and a small meteorological station were established in the center of the platform. The data obtained showed that at the beginning of January there was a complete coincidence of winter frost and permafrost throughout the area. This refuted the belief that the southern limit of permafrost was located here and that underground water bubbles and crater-like gaps could be found here. It was also established that the best time for aerial survey of the windfall area was the end of May, when the snow had melted but no leaves had as yet appeared on the trees.

In 1929, at the request of the members of the expedition, S. Y. Belykh, an astronomer, determined the geographical coordinates of the astro-radio points located on Mount Farrington (9), Mount Shakrama (25 kilometers to the south), and at the Vanovara trading station (about 90 kilometers to the southeast). That same year a round depression known as the Suslov Crater, located 200 meters west of the foot of Mount Stoikovich, was studied thoroughly. A trench excavated at the southern edge to drain off water disclosed large folds of up to 1.5 meters in the peat and underlying blue clay; they had evidently been produced by the enormous lateral force of the explosion gases emanating from the meteorite. At the northern edge of this asymmetrical crater a blue, semi-transparent silica-glass containing traces of nickel, was detected in the protruding clayey soil (10). In the center and on the northern and southern borders of the crater three 31.5-meter borings were made with a four-inch rotary shock borer of the Empire type. They terminated in a sandy aquifer. These holes revealed that there was permafrost at a depth of 25 meters. Boring was not undertaken elsewhere.

Every part of the valley in the central windfall area bears traces of a big flood. Furthermore, in the Southern Marsh the concentric peat banks are a sign of the rapid ebb of the water (11). These facts and the configuration of the site prove the veracity of the statements of local inhabitants who claimed that where "it" fell, water sprang from the Earth for some time. In all probability individual masses of the meteorite pierced the 25 meter layer of permanently frozen soil and released artesian waters which partially destroyed the permanently frozen mounds in the peat bogs.

Data confirmed by the direction of the felled trees indicate that the center of the fall of several individual meteoritic masses is on the northern border of Southern Marsh. The expedition headed by the author in 1929–1930 studied a round crater amid the frozen peat bog mounds. Samples of the silt in the bog were taken from a depth of five meters. Under the microscope it was found that they contained minute globules of nickeliferous iron, and fused clusters of quartz grains. Lastly, in 1930, Evenkis, who visited the expedition, said that immediately after the fall of the meteorite pieces of brilliant native iron had been found in the vicinity of the center of the fall!

An account of the facts pertaining to the fall of this meteorite should be completed by theoretical conclusions of vital importance: A huge iron meteorite fell on June 30, 1908, near the Podkamennaya Tunguska River. We may suppose that this body broke up first in the air, and then further when it struck the Earth's crust, which it penetrated as discrete fragments and was broken into still smaller fragments by the action of incandescent gases produced at the time. At a depth of slightly less than 25 meters crushed masses of nickeliferous iron, some pieces weighing one or two hundred metric tons, should be encountered. We believe that the entire mass of the original iron meteorite before it invaded the Earth's atmosphere weighed several thousand metric tons.

At present the author is attempting to obtain an aerial survey plan of the site of the fall of this meteorite.

[The following is an excerpt from an article by L. A. Kulik, "Data on the Tunguska meteorite obtained by 1939," published in *Dokladii A. N. S.S.S.R.* (Transactions of the USSR Academy of Sciences).]

The expedition did not succeed in taking photographs from the air in 1930. It was not before 1938 that such photographs were taken by the Chief Administraton of the Northern Sea Route. It aimed, while showing the position of the windfallen forest, to determine the loci of the fall of separate meteorite masses, and to obtain objective evidence as to the fall.

The photographs were taken between June 26 and July 18, 1938, when the taiga already bore a rather developed foliage, which obviously reduced the visibility of the windfallen trees on the photographs whose scale is 1 :4700 with a geodetic foundation of the V class. The photocuts and then the mosaic field photo-scheme, showing the centripetal direction of the windfallen trees, facilitated establishment of the starting points of the air waves. Naturally enough, this center coincided with the center which I determined in 1928 by direct theodolitic measurement of the windfallen trees.

When the photomaterial is studied, and after a photochart more perfect than the field photoscheme is made, these indications will become more accurate (12).

COMMENTARY ON KULIK'S THE TUNGUSKA METEORITE [1]

By E. L. Krinov

(1) By means of a study of seismograms, A. V. Voznesensky, former director of the Irkutsk Magnetic and Meteorological Observatory, made calculations which showed that the Tunguska meteorite fell at 0 hrs. 17 min. 11 sec. GMT. Thus, recordings of the seismic wave caused by the fall of the meteorite made it possible to establish the exact moment of the fall with a great degree of accuracy. There is not another meteorite whose fall has been determined with such precision.

(2) E. L. Krinov has interpreted the "column of fire" as the dust train left in the path of the fireball (bolide) as it moved through the sky. Since, according to eye-witnesses, the train formed a column, it may be concluded that Kirensk was in the plane of projection of the trajectory of the bolide. This is precisely the condition under which the dust train would be situated vertically in the firmament. Since Kirensk is southeast of the point of fall of the meteorite, it follows that the movement of the bolide was from southeast to northwest. The impression of a "column of fire," caused by the dust train that was actually shaped as a gray stripe, could have been created by the flight of the fireball (bolide) which directly preceded the appearance of the dust train.

[1] [Personal letter, November 1958.]

(3) The area where the trees were felled was elliptical in shape; the direction of the long axis coincided with the projection of the trajectory of the bolide.

(4) Barographic records of the air waves registered by many Siberian meteorological stations, by stations in Petersburg, and by the microbarograph at the Slutsk Observatory in Russia were discovered in the thirties.

(5) The location of the epicenter according to calculations made by A. V. Voznesensky was the following: $\phi = 60°\ 16'$; $\lambda = 103°\ 06'$ of Greenwich.

(6) By making a study of the observations of atmospheric transparency made by Dr. C. G. Abbot in California in 1908, Academician V. G. Fesenkov established in 1949 that due to the spread of a great amount of Tunguska meteorite matter during its flight through the atmosphere at cosmic speed, the atmosphere was dim for several days following the fall of the meteorite. At the same time, the dispersion in the atmosphere of meteoric dust which reflected solar light at night, is what produced a number of anomalous light nights. As regards the luminous (silvery) clouds, their simultaneous appearance in the night of June 30–July 1 and the fall of the meteorite are merely a coincidence. It is generally known that in the moderate and northern latitudes luminous clouds are observed each year during the period of white nights. It is at present believed that the nature of the luminous clouds has no connection with meteoric bodies. However, L. A. Kulik believed that the luminous clouds were a concentration of the products of the breakup of meteoric bodies in the atmosphere; this hypothesis has now been found incorrect.

(7) The discovery of rock-flour in certain depressions in the region of the fall of the Tunguska meteorite has not been reliably established.

(8) L. A. Kulik originally believed these round depressions to be meteoritic craters, and they have entered scientific literature as "meteoritic craters" (see F. G. Watson: "Between the Planets," 1956; H. H. Nininger: "Out of the Sky," 1952; R. B. Baldwin: "The Face of the Moon," 1949, etc.). However, it was subsequently established that these craters were in reality natural formations due to the permafrost in this area. The most probable (but not authentic) point of the fall of the meteorite (its explosion) is at present believed to be what is known as "Yuzhnoye Boloto" (Southern Marsh) (see below).

(9) Named by L. A. Kulik in honor of O. Farrington, the well-known American meteoriticist. The coordinates of Mount Farrington are: $\phi = 60°\ 54'\ 58''.98$; $\lambda = 101°\ 56'\ 59''.70$; Mount Shakrama: $\phi =$

60° 44′ 18″.21; $\lambda = 101° 55′ 17″.85$; the Vanovara trading station: $\phi = 60° 20′ 18″.23$; $\lambda = 102° 17′ 6″.0$.

(10) The piece of glass found subsequently proved to be a fragment of ordinary bottle glass melted when the roof of one of the huts built by the expedition caught fire. The refuse and remaining charred parts of the roof, along with the remnants of household utensils, bottles, etc., that had been kept in the attic, had been removed to precisely the spot where the above piece of glass was later found.

(11) The original idea of concentrically located "waves," gained from ground observations, was subsequently rejected when aerial surveys were made. They showed that the waves were situated more or less parallel to each other and ran approximately in meridian direction.

(12) A conclusive photochart was obtained in 1941 on the basis of aerial surveys. This photochart shows that the center of the radial windfall of trees approximately coincides with the location of Southern Marsh.

16. METEORS IN THE MEASUREMENT
OF THE UPPER ATMOSPHERE

By Fred L. Whipple

Since February, 1936, the observing programs of two patrol cameras
at the northern stations of the Harvard College Observatory have been
synchronized for the purpose of photographing meteors simultaneously.
One camera is located at Cambridge and the other, 38 km away, at the
Oak Ridge Station. The large base line provides accurate triangulation
in the determination of the position in space of any point in a meteor
trail. At first the Oak Ridge camera was equipped and now both cameras are equipped with rotating shutters operated by synchronous mo-
tors. These shutters break the photographed trails into segments equally
spaced in time, so that the angular velocity of the meteor may be
measured directly at any point of the trail. By triangulation the distance and direction of motion are known; the linear velocity and deceleration may then be calculated. Twenty-three meteors have been
doubly photographed in about 2500 hours total exposure time.

The principal aim of the photographic meteor investigations was
originally to determine precise spatial orbits of sporadic meteors in order to make certain whether they belong to the solar system or come
from interstellar space. Preliminary results indicate that they are generally, if not always, members of the solar system. It was found that the
observed data could be used to determine accurately the densities of the
earth's upper atmosphere at heights of from 50 to 100 km. A theory for
these determinations has been presented by the author; it is based upon
the theories by Opik and Hoppe. In principle the method may be explained as follows: the integrated light of a meteor with a known velocity is a direct measure of the meteor's original mass, if we assume the
luminous efficiency from physical theory. The instantaneous brightness
is taken to be proportional to the rate of loss of meteoric matter, the
light being emitted when atoms evaporated from the body by friction
collide with air molecules away from the main body of the meteor. Thus
it is possible to calculate the mass of the meteor at any point of its
photographed trail. The observed deceleration at a given point then

measures the resistance of the atmosphere to a rapidly moving body of known mass, from which data the density of the atmosphere at that point can be calculated. Uncertainties concerning the composition and shape of the meteoroid do not affect the numerical results seriously.

The accurate reduction of the photographic meteor trails requires considerable measurement and calculation so that only ten meteors are

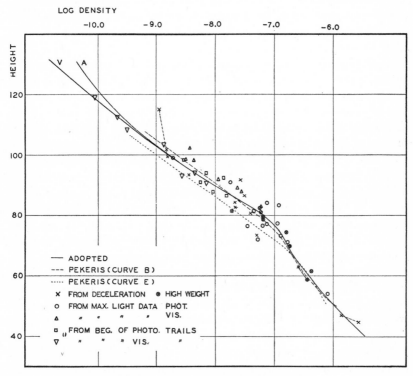

Fig. 1. Atmospheric heights and densities.

as yet completely reduced. The present results, however, give considerable information concerning densities in the earth's upper atmosphere. It is hoped that within a year the material now available will be reduced, after which a critical study of the theory and the physical constants can be made. The most reliable density determinations (circled crosses and crosses in Figure 1) depend upon the measures of deceleration by the method outlined above. When connected by dashed lines the crosses represent a range in solution caused by uncertainty in the time of appearance of the meteor. The two lower crosses at about the

100-km height represent the range of an elliptic solution for one meteor; hyperbolic solutions were also possible from the direct observations of this meteor.

Less reliable, but still consistent, density determinations (circles in Figure 1) can be obtained from the heights at which maximum luminosities occur for the various meteors. The height at which a meteor of a calculated mass and velocity will attain its maximum luminosity depends, of course, upon the density of the atmosphere. Instead of the deceleration, the height of maximum light can thus be used to determine the atmospheric densities.

A third method, less precise than these two, has also been used. In the early part of a meteor's trail, after evaporation of the surface material has begun but before appreciable mass has been lost, the rate of loss of mass and therefore the luminosity is directly proportional to the air density; the deceleration is here observed to be negligible. For a meteor of a known mass, the air density can thus be calculated directly from the luminosity. This method is useful at greater heights than the deceleration method but is, unfortunately, less reliable, because the air density depends analytically upon the first power of the meteor's luminosity, the least accurately observed datum. In the deceleration and maximum-light methods the luminosity enters only to the one-third power. In Figure 1, the hollow squares identify densities determined from the luminosity in an early part of each meteor trail.

The last two methods, which do not depend upon deceleration measures, have been applied to the visual shower meteors observed by the Arizona Expedition for the Study of Meteors and reduced by Öpik. It was assumed that the mean heights observed for the visual meteors corresponded to the points of maximum luminosity and that the meteors were first observed at visual magnitude +4.0. The luminous efficiency of the visual meteors was taken to be constant with velocity, rather than directly proportional to the velocity as for the much brighter photographic meteors. The air densities calculated from the mean heights are indicated by erect triangles in Figure 1, and those from the beginning heights as inverted triangles. Considerable uncertainty is involved in the calculation of these densities.

The density of the atmosphere at any level depends upon the temperature there and at all levels below. The total mass in a column of given cross-section is determined by the barometric pressure at sea-level, so that the density at the given level depends upon how the whole column is supported. The introduction of a higher temperature layer at a low level would tend to raise the upper part of the column and thus increase the density at all higher levels. In a region of high temperature the logarithm of the air density decreases slowly with height; in a region of low temperature it decreases rapidly. In Figure 1 a steep slope in the log-density curve indicates a higher temperature than a

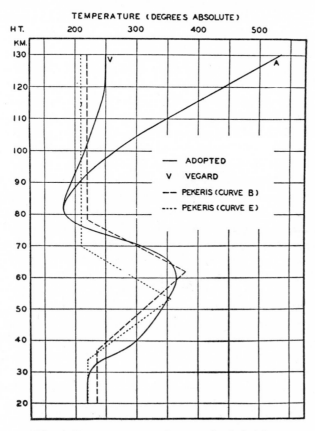

Fig. 2. Temperatures and atmospheric heights.

flat slope. The actual curves in Figure 1 were calculated from the corresponding temperature curves of Figure 2. It was assumed that the atmosphere is homogeneous in composition with a molecular weight of 18.8 (mostly molecular nitrogen) and that at 20 km the pressure is 41 mm of mercury at a temperature of $-58°C$. The two temperature curves in Figure 1 by Pekeris were based upon a theory of atmospheric oscillations. The "adopted" curve to 50 km is based on F. J. W. Whipple's determinations of temperatures by the speed of sound from gun fire. A maximum temperature has been assumed at 60 km. The minimum at 82 km is assumed because this is the mean level of the noctilucent clouds which may be due to water-vapor condensation at a low temperature. Above this level the temperature must again rise to attain the higher temperatures calculated from the nitrogen band spectrum of the aurora. The lower-temperature curve passes through Vegard's value of the auroral temperature and the upper-temperature curve through Rosseland and Steenholt's correction to his value.

The temperature maximum near 60 km is confirmed by the steep slope of the log-density curve at this height and by the numerical values of the density. The observations would suggest that perhaps the layer of maximum temperature is thicker than was assumed in the "adopted" temperature curve and that the temperature minimum occurs at a higher level than 82 km. A subsequent rise in temperature at greater heights is clearly shown by the slope of the observational curve, but the accuracy is not sufficient for an exact evaluation of the rate of rise. In general, the agreement between the adopted and observed curves is so close that the author does not wish to correct the "adopted" curve until more material has been analyzed, and until the physical theory of a meteor's luminous efficiency has been reconsidered.

The temperature inversion above 60 km is verified by independent observations of meteor light curves. From the photographic meteor trails in the Harvard plate collection Miss Hoffleit observed that the faster shower meteors attain their maximum brightness nearer the ends of their trails than do the slower meteors. Her results are quantitatively explained by the present theory of the meteor phenomenon, coupled with the temperature inversion above 60 km as represented by the "adopted" temperature and density curves of Figures 2 and 1. The explanation follows simply. When a meteor is evaporating by its passage through the atmosphere the rate of evaporation, and therefore the luminosity, depends directly upon the air density and upon the meteor's cross-section. At first the loss of mass is small and the rate of increase of brightness depends only upon the rate of increase in the air density. Eventually the meteor decreases in size until the rate of decrease in cross-section matches the rate of increase of air density. At this point maximum light is attained. Deceleration has little numerical effect in the theory until after maximum light.

For meteors of the same luminosity the faster ones will attain maximum light at greater heights than the slower ones. The faster meteors happen to occur in the low-temperature region of the atmosphere where the logarithmic density gradient is great. Consequently they continue to increase in brightness until near the ends of their trails where the rate of decrease in cross-section is great enough to match the rapid increase in air density. The slower meteors occur at lower levels where the temperature of the atmosphere is greater and the logarithmic density gradient is smaller. The decrease in cross-section dominates the light curve at an earlier stage and maximum light occurs relatively earlier in the trail, as was observed by Miss Hoffleit.

17. THE SOLAR PARALLAX: A COORDINATED INTERNATIONAL MEASURE OF A FUNDAMENTAL CONSTANT

By Harold Spencer-Jones

The reduction and discussion of the observations of Eros made at the opposition of Eros in 1931 [by the astronomers of 15 countries], have just been concluded. Ten years may seem a long time for the completion of the reduction of these observations, but it must be remembered that a number of observatories took part in this programme and that the material had to be coordinated in successive stages. The pace was, in consequence, regulated by that of the slowest participant, and, in fact, the completion of the discussion has now been possible only by the omission of the results of one important series of observations which are not yet available. This afternoon I propose to deal only with the solar parallax, leaving the mass of the Moon for a future occasion.

As most Fellows will know, Eros was discovered in 1897. The planet occasionally makes a close approach to the Earth, the closest possible approach being at a distance of 13.8 million miles. The minimum distance at the opposition in 1901 was nearly 30 million miles; in 1931 it was 16.2 million miles. The latter opposition, therefore, provides the most favorable opportunity for determining the solar parallax and the mass of the Moon since the discovery of Eros.

Before opposition, in October and November, 1930, the motion of Eros was mainly in right ascension, which was increasing. The parallax of the planet at the beginning of October was approximately 16″. By the beginning of December the motion in right ascension was decreasing and a rapidly increasing motion in declination had set in. The closest approach to the Earth took place on January 30 when the parallax was 50″; the motion in declination was then about $1\frac{1}{4}$° per day, 2″.8 per minute of time. Opposition occurred on February 17 and the planet reached a stationary point on March 16, when the parallax was 32″. After this date the motion, of course, was reversed. Though the parallax was less in March than in January or February, observations of the planet were then made under specially favorable conditions, as the slow motion enabled the same reference stars to be used over a longer period.

The first piece of work that had to be done was that of determining the positions of a number of reference stars. Prof. Kopff [of Germany] prepared a list of stars selected to lie in the belt one degree in width on either side of the computed path of Eros, the number and distribution being so chosen as to provide eight to twelve stars for the reduction of an astrographic plate centered at any point along the path. The work of determining the places of these stars by meridian observations was carried out by a number of observatories. During the southward sweep of Eros, the extreme limits of right ascension differed only by 1 hr. 20 min., and as only two seasons were available for the determination of the positions of the stars the programme proved somewhat difficult. An unfortunate circumstance arose later, when the early observations of Eros showed that it was not following the computed path, but departed from this by an increasing amount which reached 15′ at the time of closest approach. This was subsequently found to be caused by an error in the computation of the perturbations of Jupiter, arising from an error in copying a figure from the bottom of one page of computations to the top of the next page.

It was decided that exposures upon Eros should be centered on the true path of the planet and that the positions of extra stars, selected to fill the quarter degree vacant belt on one side of the path, should be determined later.

The primary stars were too few in number to serve for the reduction of plates taken with long focus instruments. A list of fainter secondary stars was prepared by Prof. Schorr for the portion of the path down to Dec. −14°; for the southernmost portion of the path, observatories were asked to select their own stars and send their lists to me, so that positions might be supplied. The positions of the secondary reference stars were obtained from plates taken with astrographic telescopes at Greenwich, the Cape, Bergedorf and Leipzig, supplemented by some wide-angle plates secured at Lick. The Greenwich and Cape plates were measured at Greenwich, and a list of nearly 6000 star positions was published by this [Royal Astronomical] Society. Of the various suggested methods of exposure upon Eros the most satisfactory proved to be that of "interrupted exposures." This was the method adopted at the Cape, and though never, as far as I know, used before, it proved to be very accurate. The telescope is guided upon any convenient star in the field, and a series of interrupted exposures given, the length of each exposure and the interval between them being chosen so that the images of Eros are stellar in appearance, well separated from one another, and comparable to the average comparison star image. The position of Eros can, in this way, be determined much more accurately than from a long trailed image, and the mean of the Eros images is comparable, so far as guiding error is concerned, to the comparison star images.

The solar parallax can be determined either from observations in right ascension or from observations in declination. To obtain the parallax from observations of right ascension, plates taken with individual instruments were used, where exposures had been made upon the planet on both sides of the meridian. This method has the advantage of largely eliminating systematic errors peculiar to the particular instrument. The great disadvantage is that it involves a considerable wastage of material when observations are not well balanced on either side of the meridian. It is desirable, therefore, to combine all material, making a determination of the parallax from the combined observations on each night; but it then becomes necessary to take into account systematic instrumental errors. These were investigated by means of extensive inter-comparisons of all available material. From the declination observations, the parallax is determined by comparison of observations in the northern and southern hemsipheres; systematic differences enter directly into these comparisons. The systematic errors in declination were therefore determined separately for the northern and southern observatories, each being referred to their respective means.

The systematic corrections are not negligible, having a range of $0^s.021$ in right ascension and $0''.23$ in declination.

When the weights came to be investigated, it soon became apparent that the weight of several plates obtained on one night with the same instrument was not proportional to the number of plates. For each instrument the internal probable error of an observation on any particular night was determined from their internal agreement; the error of the night was determined from the comparison of the results of adjacent nights, allowance being made for the change of tabular error of the ephemeris, and from the intercomparison of the results of different instruments on the same nights. The longer series were dealt with first, and a least squares adjustment made. The minor series were tied on to as many as possible of the longer series. Having determined the internal probable error and the probable error of the night for each instrument, tables were drawn up for both right ascension and declination, giving the weight of any number of observations on one night with each instrument.

It may be mentioned that, for instruments of comparable focal length, there is a surprising variation in the weight of a single plate. The unit of weight was chosen to correspond to a probable error of $\pm 0''.32$. For instruments of astrographic focal length, there is one group with an average weight for a single plate of about 15 and another group with much smaller weights, from 0.2 to 3. Though the number of exposures for the average plate differ somewhat from one instrument to another, the differences in weight cannot be accounted for in this way.

A part of the error is probably to be attributed to inaccuracy in the timing of exposures. Observatories had been duly informed of the importance of accurate timing, but that this importance was not in all cases appreciated is shown by the fact that only approximate times were given by one observatory. When it was requested that more precise times should be supplied it was learned that exposures had only been recorded to the nearest quarter of an hour! Inaccuracy in timing the exposures would mainly impair the accuracy of the declination observations, because of the rapid motion in declination. As, however, a considerable variation in weight is shown by the right ascension observations, other causes must be partly responsible. Bad guiding, improperly adjusted objectives, unsuitable methods of observation and insufficient care in measurement are doubtless contributory factors.

In the case of Tokyo, the observations of which are seen to stand out, it is known that the mounting of the instrument is weak. This may be sufficient to explain the discrepancy for this particular instrument.

Separate determinations of the solar parallax are possible from the observations with sixteen different instruments. These may be combined in various ways. The two Cape instruments and the Cordoba [Argentina] astrographic have a predominant weight. The remaining thirteen instruments agree, however, in the mean with these three. The mean result from long-focus instruments, depending upon the positions of the secondary comparison stars, is in agreement with that from short-focus instruments, depending upon the positions of the primary comparison stars. The mean result from photographic instruments is in close agreement with that from visual instruments, employed photographically with yellow filters and yellow-sensitive plates.

The results may be summarized as follows: —

From 16 instruments combined	$8''.7900 \pm 0''.0013$
From Cape 13-in., 24-in., Cordoba 13-in. . . .	$8''.7900 \pm 0''.0015$
From remaining 13 instruments	$8''.7901 \pm 0''.0026$
From primary stars (6 instruments)	$8''.7894 \pm 0''.0019$
From secondary stars (10 instruments) . . .	$8''.7907 \pm 0''.0019$
From photographic instruments (11)	$8''.7903 \pm 0''.0014$
From visual instruments (5)	$8''.7876 \pm 0''.0043$

With these results may be compared the results obtained from the whole of the material in right ascension and declination separately, in deriving which the systematic corrections required by each instrument have been applied.

From all R.A. observations . . . . 8″.7875 ± 0″.0009
From all Dec. observations . . . . 8″.7907 ± 0″.0011

However the material is analysed or subdivided, it is impossible to depart appreciably from the value 8″.790. It would seem, therefore, that unless there exists some source of systematic error common to all the observatories, the solar parallax must be very close to this figure.

18. PLANETARY MOTIONS AND THE ELECTRONIC CALCULATOR

By G. M. Clemence, Dirk Brouwer, and W. J. Eckert [1]

Measuring the motions of the planets and devising a rigorous theory to represent these motions have played a basic role not only in developing physical science, but also in solving practical problems such as those encountered in navigation. As the wealth of observational data accumulates over the years, however, the problem of adequately representing planetary movements by a consistent theory has become increasingly difficult. The recent advent of the electronic calculator has at last made it feasible to perform the theoretical analysis with accuracy far greater than that of the accumulated observations.

During the past three years a project has been underway to calculate the precise positions of the five outer planets of the solar system, Jupiter, Saturn, Uranus, Neptune, and Pluto. This work has been done as a portion of a cooperative undertaking (sponsored by the Office of Naval Research) of the U. S. Naval Observatory, the Yale University Observatory, and the Watson Scientific Computing Laboratory. The paths of the planets have been traced out for 400 years, from about 1660 to 2060, by calculating their actual positions at 40-day intervals. These five planets were chosen because of the influence they exert on the solar system. The first four are the controlling bodies (except for the sun), each being far more massive than all the other five principal planets and remaining material put together. Pluto, the outermost, although it has relatively small mass, exerts appreciable effects on the others. The effects of the remaining principal planets, Mercury, Venus, Earth, and Mars, are hardly appreciable, and can best be calculated separately from the main problem. The project described here involved many millions of arithmetic operations, and the discussion of many thousands of observations has produced results of a new order of accuracy.

[1] [Although Dr. Eckert was not listed as an actual author of this report, he was an equal participant in the research and a joint author of the monograph in the *Astronomical Papers of the American Ephemeris* entitled "Coordinates of the five outer planets, 1653–2060."]

At least two of the planets studied in this project have probably been of interest since man first became curious about celestial bodies; Jupiter and Saturn must have been among the earliest objects of intellectual curiosity. Before recorded history it was undoubtedly noticed that Jupiter appeared in the evening sky a month later each year, requiring twelve years for a complete circuit of the heavens, and that Saturn moved among the stars at less than half of Jupiter's speed. During the few centuries immediately preceding the Christian Era, the Babylonian astronomers represented the observed motions by empirical arithmetical formulae; and somewhat later the Greeks represented them by the means of hypothetical geometric systems of motions in space. The Greek theories of the planetary motions culminated in the system constructed by Ptolemy, some 1800 years ago. In this system, the motion of each planet is represented by superimposing one or more small circular motions in space upon a motion in a large circular orbit around the earth. The small circular motions account for the periodically varying speed with which the planet appears to move among the stars.

Ptolemy's system remained the accepted standard theory among astronomers for 14 centuries, and the combinations of circular motions were retained by Copernicus when, in the sixteenth century, he proposed the theory that the planets move around the sun instead of around the earth. Meanwhile, however, observations of the planets had become precise enough to show that their motions were not accurately represented by these theories. In the early seventeenth century, Kepler found that the orbits around the sun were ellipses, and formulated laws from which formulae and tables for computing the positions of the planets were prepared. But Jupiter and Saturn departed from Kepler's theory, and it was not until after Newton had formulated the theory of universal gravitation that a satisfactory explanation for the divergencies was advanced. Newton arrived at this theory in 1665 or 1666, but did not announce it until 1686. It postulates that every particle of matter in the universe attracts every other particle with a force along the line joining them, proportional to the product of their masses and inversely proportional to the square of the distance between them. If a planet were influenced only by the gravitational attraction of the sun, the orbit would be an exact ellipse; the observed departures form elliptic motion may be attributed to the disturbing effects of the other planets. Assuming the correctness of this simple law, it becomes theoretically possible to solve what may be called the fundamental problem of celestial mechanics: Given the positions and directions of motion of several bodies (say Jupiter, Saturn, and the sun), together with their speeds and masses (weights) at a particular time, it is required to find their positions at any other time.

Such a simple statement of the problem gives no idea of the formidable character of its solution. Ever since Newton's time the most eminent mathematical astronomers have devoted their lives to devising practical methods for solving it. The difficulty is illustrated by the fact that more than a century elapsed after the publication of Newton's theory before a mathematical representation of the motions of Jupiter and Saturn was obtained that came reasonably close to meeting the needs of that time.

Two methods of determining the motions of planets have proved to be successful in practice. In one, called the method of general perturbations, the coordinates of a planet in space are given by trigonometric series, in which the time remains an algebraic symbol. To find where the planet is, at any time, it is only necessary to substitute in the formulae a number expressing the years and fractions thereof since some initial epoch—let us say January 1, 1900, Greenwich Mean Noon. It is this method that has been used most for the principal planets. The latest application to Jupiter and Saturn was made about 60 years ago by G. W. Hill, an eminent mathematical astronomer then employed in the U. S. Nautical Almanac Office. The expression of his formula for the position of Saturn requires seven large pages, printed in small type; moreover, the tables that he devised for easy evaluation of the formula fill up 84 pages! Yet these complicated tables have been used for more than 50 years to calculate the positions of Saturn published annually in the navigational almanac.

The method of general perturbations suffers from two drawbacks. In the first place, the amount of calculation is considerable and it is difficult to be sure that all of the significant effects have been taken into account. Comparison of Hill's tables of Saturn with actual observations, for example, shows that the tables are in error by continually increasing amounts. The errors are small, and are now only of theoretical importance, but before many more years they will be large enough to be of practical importance as well. Similarly, Newcomb's tables of Uranus and Neptune, constructed about 50 years ago, show increasingly large errors. The errors must be due to defects in the theory, but the precise nature of the trouble is not yet clear in every case.

The second drawback of the method of general perturbations, as it has so far been applied to planetary motions, is that although any value of the time can be substituted in the formula, the result is not necessarily valid. The formula can be used only for a limited period, perhaps a few hundred years on either side of the initial epoch, without serious loss of accuracy. Consequently the method is without value for tracing the motions of the planets over hundreds of thousands of years, or even millions, as we should like to do. Since it is probable that the limitation is not inherent in the method itself, several ways have been suggested for overcoming it, but a truly frightful amount of calculation would be involved. Although electronic calculators may solve the problem eventually, even the task of setting up a machine to perform the necessary operations automatically is a staggering one.

The other method that has been used for tracing planetary motions is called the method of special perturbations—in which the position of a planet is actually calculated step by step from the initial epoch. The steps are so close together that the planet does not move far enough from one step to the next to change the attractions of the other planets upon it by very much. This method also had two drawbacks. In order to find the position of a planet at any time it is necessary to calculate it for all the intervening times since the initial epoch; and if many steps are required it is necessary to use many significant figures in the calculations, to avoid excessive accumulation of error in the end-figures.

The method of special perturbations has not been much used for the principal planets; its applications have usually been to the minor planets and to comets, where the number of steps is not very great, and where the highest precision has not in general been required. In these applications it is assumed that the positions of the disturbing planets are already well enough known, so that at each step it has been necessary only to calculate the attractions of the principal planets (and of the sun) on the body being studied. The masses of the minor planets and comets are so small that they do not appreciably disturb each other, or the principal planets.

The application to the five outer planets of the solar system is by far the most extensive that has ever been made of the method of special perturbations. It is the first application in which the actions of the planets on one another have been calculated at each step, instead of assuming that the paths of all except one are known in advance. Thus, at each step, the attraction of each of the five planets on the other four was calculated, as well as the attraction of each on the sun. Each step of the integration involved 800 multiplications of large numbers, 100 divisions, 1200 additions and subtractions, and the recording of 3200 digits. The large number of steps made it necessary to use 14 decimals in the calculations.

So large an amount of calculation could not have been accomplished in a reasonable length of time without an electronic calculator of high capacity. Fortunately such a calculator was placed in operation in January, 1948. This machine, the IBM Selective Sequence Electronic Calculator, was made available without cost to the project by Mr. Thomas J. Watson, Chairman of the Board of the International Business Machines Corporation. This machine was able, in less than three minutes, to make all the calculations, in duplicate, for a single 40-day step. At each step the machine automatically compared the two independent results, and in case of disagreement it automatically repeated the calculation. Even electronic calculators make mistakes occasionally, but it was found that the machine was able to correct most of its mistakes on the second attempt. In case of disagreement on the second trial the machine stopped, indicating need for servicing. Several estimates were made of the time that would be required to make the calculation in the ordinary way, with an electric desk calculator. It appears that the ratio is in favor of the electronic calculator, by about 1000 to one! In other words, an operator with a desk machine, working 40 hours a week, might have done the job in about 80 years, if he had made no mistakes. Experience shows, however, that in work of this kind the occurrence of mistakes may double the time required. Furthermore, a mistake occurring at any step causes every subsequent step to be wrong. Thus, it is not certain that this work could have been successfully accomplished at all by ordinary methods.

As was mentioned earlier, the initial data required for solving the problem are the position, speed, and direction of motion for each planet at some initial epoch. To specify the position, three numbers are needed while the speed and direction each require two, making a total of six numbers for each planet, or 30 in all. These 30 numbers have to be determined by observing the planets with telescopes. When great accuracy is wanted it is necessary to have many observations, extending over a long period of time, and this is one reason why the calculations were extended so far into the past. The 30 constants of the orbits were known at the start with insufficient accuracy, but were improved by successive approximations. The first calculation extended for only 30 years. The resulting positions of the planets were compared with the observed positions, and the constants of the orbits were adjusted to bring about better agreement. Then a second calculation for 180 years was commenced, and carried out; after its completion a new comparison with observations was made, and the constants adjusted a second time. Another calculation was then made for the same 180 years, and another comparison with observations showed that the constants needed no further improvement. The final step was to extend the 180-year stretch to 400 years.

About 15,000 observations of Jupiter and Saturn were used in adjusting the constants of the orbits, but for Uranus and Neptune fewer observations were available. (Uranus was not discovered until 1781, and

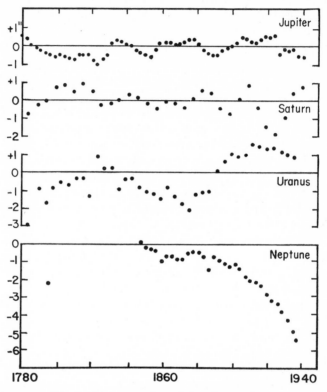

Fig. 1. Representation of the observations.

Neptune not until 1846.) For Pluto the number of observations was comparatively small, because this planet was the most recently discovered of the five. (It was first observed in 1930.) For this reason it may be expected that the calculations for Pluto will prove to be less accurate than the others, and by 2060 the error may be appreciable. It was still worthwhile to include Pluto in the work, however, on account of its influence on the other four planets; its motion is known with ample accuracy for this purpose.

The adjustment of the constants of the orbits by comparison with observations is a task of considerable magnitude. It has not yet been practicable to use electronic machines for this work, because too much discriminating judgment is needed at nearly every stage. Altogether several thousand hours of labor by expert astronomers were necessary, not counting the even larger task of making the original observations. Fortunately, however, much of the work had already been done several

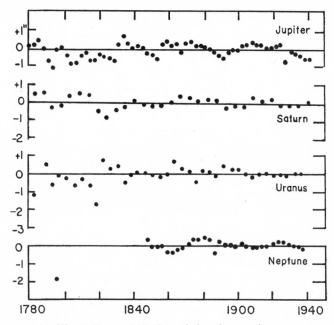

Fig. 2. Representation of the observations.

years earlier, with this very object in view; and during the course of the electronic calculations only a few hundred hours of an astronomer's time were needed. This work was done at the Yale Observatory and the Naval Observatory, principally by the [first] two authors. The electronic calculations were planned and directed by Dr. W. J. Eckert.

Figures 1 and 2 contrast the representation of the observations by the planetary tables and by the numerical integration. A positive value of a residual indicates that the observed position in the orbit is ahead of the calculation.

The plots for Jupiter and Saturn in Figure 1 are of a somewhat different type from those for Uranus and Neptune. For the former two planets, the discrepancies between the observed positions and the positions predicted by Hill's tables were read from a smooth curve drawn through the numerous residuals that represented the deviations from the tables for every year. Moreover, they were subsequently modified by a provisional adjustment of the elliptic motion which serves as the basis of the tables. In the case of Saturn this brings out very clearly the inability of the tables to represent the observed positions. For Uranus and Neptune the plotted residuals represent the deviations from the tables without any smoothing or adjustment. In the case of Uranus, such an adjustment would have produced a considerable improvement. The deviation of the path of Neptune from the tables since 1900 must be ascribed primarily to the omission of the attraction by Pluto from the tables. Pluto's existence was, of course, unknown 50 years ago, when the tables for Neptune were constructed. Finally, the observations of Neptune that were available when Newcomb constructed the tables covered only 50 years since the planet's discovery in 1846. There are, however, two precious pre-discovery observations made in 1795 when the planet was recorded as a ninth-magnitude star in the course of a series of observations of faint stars. The importance of the pre-discovery observations is that they are compatible with modern observations only if the attraction by Pluto with a mass comparable with the mass of the earth is included.

In connection with Figure 2, two points of importance should be kept in mind. First, the observations of Jupiter, primarily because of the large size of its disk, are considerably less accurate than those of the other planets. Second, the diagram shows that there was a considerable improvement in the quality of the observations about 1840 and again about 1900. In this light, the residuals for Saturn and especially Uranus look entirely normal. Large scatter is present before 1840, while improved representation is found from 1840 to 1900 and further improvement is shown since 1900. While the improvement since 1900 is also very noticeable in the case of Neptune, the residuals for this planet seem to have a systematic character. This may well be due to the fact that this planet moves very slowly among the stars—in fact at the rate of only two degrees a year. Thus for many years, the observations of the planet are based upon a comparison with the same stars in the same part of the sky. This tends to carry into the residuals the small systematic errors in the star positions.

Comparison with the observations as shown in the figures does not complete the work. It was mentioned that in addition to the constants of the orbits, the masses of the planets must be known at the outset. Fortunately, reasonably good values for the latter were available from previous investigations, but improvement in these will be possible, when a more refined analysis of the present work can be made. In this connection, about a dozen observations of Uranus during the century preceding its discovery may be very useful for the evaluation of the mass of Pluto. These observations were made under circumstances similar to the observations of Neptune in 1795. Although they are of low accuracy compared with modern observations, their early date makes

Fig. 3. A partial view of the IBM Selective Sequence Electronic Calculator.

them at least interesting. The calculations in connection with these observations have not yet been completed.

As to the value of the work, there are two points of view to be considered: (1) The basic scientific value and (2) the immediate practical value. From the scientific point of view the object of studying the motions of celestial bodies is to learn more about how the universe operates. For two hundred years after Newton's formulation of his laws of motion it was thought that they would suffice to explain all of the celestial motions. The astronomer Leverrier, however, discovered, about a hundred years ago, that the planet Mercury did not move strictly in accordance with Newton's laws. The discrepancy remained unexplained until 1915, when Einstein formulated the general theory of relativity.

Then it appeared that the discrepancy could be completely removed. But what is meant by complete removal? Only that the contradictions, if they exist, are smaller than the errors of the observations. As the observations continue to improve with the invention of new techniques of observation, it is always possible that a discrepancy will arise where none was known before. Then attempts to explain it may lead to a fundamental advance in knowledge. It is after this general pattern that *all* the sciences advance.

It is known that gravitation is not the only force acting within the solar system. For example, the presence of meteors and the zodiacal light indicate something in the nature of a resisting medium. Whether this is dense enough to affect the motions of the planets appreciably is not known. The present work may help to provide an answer, or it may lead to knowledge about still other forces in the solar system.

As to the practical value of the present work, it is anticipated that it will suffice, so far as these planets are concerned, for the foreseeable needs of navigation for the next hundred years.

IV

THE POSITIONS AND MOTIONS
OF THE STARS

At the beginning of this twentieth century we knew very little about the motions of the stars in the line of sight. A. A. Belopolski at Poulkova, W. W. Campbell and his associates at the Lick Observatory and in Chile, Frank Schlesinger and associates at the Allegheny Observatory, and the Potsdam observers were among the early workers in this exciting project of using the Doppler principle to interpret blue and red shifts of the spectral lines as indicators of speeds in the line of sight. Spectroscopically detected double stars attracted much attention then as now. But so far as stellar motions in bulk were concerned, it was the component of velocity at right angles to the line of sight, the *proper motion*, that was principally exploited.

To prepare for the future work on proper motion, great catalogues of star positions were undertaken in a fine display of international cooperation and of trust in the future. To make these catalogues, the photographic plate some 80 years ago began to displace the human eye, as Frank Schlesinger explains in selection 20.

One of the impressive catalogues of positions and proper motions is described below by R. E. Wilson who was deeply involved in its construction. But long before the "Boss" catalogues were available the star motions had been analyzed by J. C. Kapteyn, the discoverer of the "two star streams" (selection 19), and by many others. Now that the rotation of the galaxy has been deduced, chiefly by B. Lindblad and J. H. Oort, we hear little of the two star streams; but in the first two decades of the century the Kapteyn hypothesis was of much use in guiding observation and theory.

Two types of star motion other than radial velocities and proper motions (and double star orbital motion) are now under study; they are the axial rotations, which are detected and measured spectroscopically, and Ambartsumian's "expanding stellar associations." The latter represents motion by inference only, based on the present distribution of groups of stars with peculiar spectra. Both of these developments are rich in implications concerning stellar evolution.

To invade the distant realm of faint stars, the Lick Observatory, as reported in selection 24 by W. H. Wright, has undertaken the long-term project of determining very precisely the present positions of millions of stars. The stars are stored as images on photographic plates that will be duplicated many years hence—perhaps in 20 years, perhaps 100, probably both. These pairs of plates will make it possible to derive the motions of the millions of stars within their range, and thereby provide a gold mine that will one day buy good new knowledge of the structure and behavior of our galactic system.

A similar program is in progress in Russia.

19. DISCOVERY OF THE TWO STAR STREAMS

By J. C. Kapteyn

In deriving the constant of precession, and in investigating the motion of the sun through space, it is usual to start from the hypothesis that the real motions of the stars, the so-called *peculiar* motions, have no preference for any particular direction.

Of late I have found anomalies in the distribution of the *apparent* proper motions, of so strongly systematic a character that I feel convinced that we are compelled to give up this hypothesis.

It will be the aim of this paper to show the nature of these anomalies, and to explain the conclusion to which they lead us.

It is only just to mention that as early as 1895 Kobold called attention to a fact which seems incompatible with a random distribution of the direction of the motion of the stars. Had Kobold been more successful in separating the systematic motions of the stars from the displacements caused by the sun's motion, he would probably have been led to conclusions similar to those which I am now about to submit to you.

In order to show clearly the anomaly in the distribution of the proper motions here alluded to, it will be necessary to call to mind how this distribution must present itself if the hypothesis of the random orientation of the motions were really satisfied.

For this purpose consider a great number of stars very near each other on the sphere, say all the stars of such a small constellation as the Southern Cross. For convenience' sake we will even assume them to be all apparently situate in the same point S (fig. 1, *P*) of the sphere, though not in space, because their distances would be different.

The peculiar proper motions of these stars will be distributed somewhat in the manner indicated in fig. 1, *P*.

In addition to this motion, which represents the real motion of the stars as seen projected on the sphere, they will have an *apparent* motion, the *parallactic* motion, which is due to the observer's own motion, or say the motion of the solar system, through space.

These parallactic motions, we all know, are directed away from the apex, which is the point where the sun's motion prolonged meets the sphere. For *all* the stars at S the parallactic motion will be directed along Sx.

The motions as really observed are the resultant of the peculiar and the parallactic motion.

Thus for the star whose peculiar motion is SB, let $S\beta$ be the parallac-

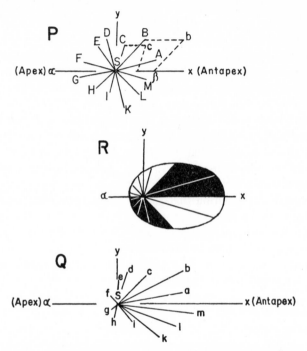

Fig. 1. The distribution of the peculiar proper motions.

tic motion, then the *observed* motion of that star will be Sb. Likewise the observed proper motion of the star having the peculiar motion SC will be Sc, and so on.

Making the composition for all the stars of fig. 1, P, we get the really observable motions distributed as in fig. 1, Q.

From this it must be evident that whereas, according to the hypothesis, the distribution of the *peculiar* motions would be radially symmetrical, this symmetry will be destroyed for the *observed* proper motions.

There will be a strong preference for motions directed towards the antapex. (See Q). One thing, however, must be clear, and we want no more for what follows; it is that there will remain a *bilateral symmetry*, the line of symmetry being evidently the line *aSx* through the apex, the star, and the antapex.

Near to this line, on the antapex side, the proper motions will be most numerous, and they will be greater in amount.

This evident condition of bilateral symmetry would furnish probably the best means of determining the position of the apex. For if from all our data about proper motion we determine these lines of symmetry for several points of the sky and prolong them, they must all intersect in two points, which are no other than the apex and the antapex.

In trying to realize this plan we must meet with the difficulty that on account of errors of observation and the restricted number of stars included in the investigation we must be prepared to find in reality no such perfect symmetry as theory demands. For the lines of symmetry we shall thus have to substitute *lines giving the nearest approach to symmetry*. Their position will depend, at least to a certain extent, on what we choose to consider as "the nearest approach to symmetry."

If we call the required line of symmetry the axis of the *x,* the line at right angles thereto the axis of the *y,* then we may, for instance, define that position of the *x*-axis as the line of greatest symmetry, which makes *zero* the sum of the *y*'s.

The lines of symmetry furnished by this definition, prolonged, will not pass through a single point; they will all cross a certain more or less extended area, the center of gravity of which might be taken as the most probable position of the apex.

Drawing great circles through this apex, we must necessarily find them diverging somewhat from the lines of best symmetry in different parts of the sky.

If, however, our hypothesis of random orientation is approximately true, the divergences will be small. The sum of the proper motions at right angles to these circles will be nearly zero in every part of the sky.

Not only that, but we will have further to expect that any other condition of symmetry will be approximately fulfilled, too, for every point of the sky. Such another condition will be, for instance, that on both sides of the great circles through the apex the total quantity of proper motion will be the same, or, again, that Σx shall be the same on both sides of these circles.

How the first of these conditions is satisfied is shown in fig. 2. This figure summarizes the more important points in regard to the question in hand. They show in compact form the results of a complete treatment of the proper motions of all the stars observed in both co-ordinates by Bradley (over 2,400 stars). These stars are distributed over two thirds of the whole of the sky. This surface has been divided up into twenty-eight areas.

From the stars contained on each area I have derived the distribution of the proper motions corresponding to the center of the area. How this was done need not here be explained. The whole of the materials were

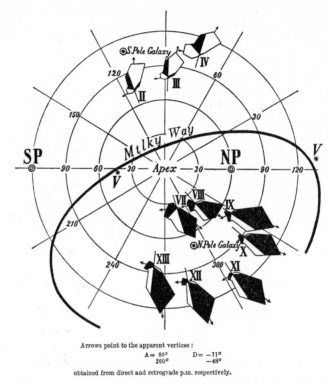

Arrows point to the apparent vertices :
A = 85°   D= −11°
260°   −48°
obtained from direct and retrograde p.m. respectively.

Fig. 2. Distribution of the proper motions for a particular region of the sky.

thus embodied in twenty-eight figures, like those of fig. 2, each of which shows at a glance the distribution of the proper motions for one particular region of the sky.

Not to overburden the plate, I have only included *ten* of the figures, for which the phenomenon to which I wish to draw your attention is most marked.

It is very suggestive that these lie all near to the poles of the Milky Way.

The figures have been constructed as follows: A line has been drawn making the angle of 15° with the great circle through the apex, the length of which represents the sum of all the proper motions making angles of between 0° and 30° with that circle.

In the same way the radius vector of 45° represents the sum of all the motions between 30° and 60°, and so on.

For the sake of uniformity all the results have been reduced to what they would have been had the total number of stars been the same for all the twenty-eight areas.

In order still better to show the important points in the distribution of the p.m. that part of the figures between the radii vectores making angles of *zero* and −60°, +60° and +180° have been blackened.

The position adopted for the apex is practically that found by a variety of methods all more or less akin to that described a moment ago.

If the random-distribution hypothesis were true, and if in consequence thereof the symmetry of our figures were complete, the blackened parts of the figures would have been equal to the corresponding lighter-tinted parts in the way of fig. 1, *R*.

The real state of things is something quite different, and, what is all-important, we see at once that the divergences are strikingly systematic. The figures at each pole of the Milky Way show them in nearly every particular of the same character. Near the North Pole the blackened parts are invariably much greater; at the South Pole the case is reversed.

At a first glance the difference of the more extensive parts on the side of the antapex is the most striking. As a matter of fact, however, the difference between the smaller parts is by no means less important. . . .

When we see that the motions of a certain group of stars converge to a same point on the sphere we conclude either that the real motions of the stars are in reality parallel, or that the motion is only apparent and due to a motion of the observer in the opposite direction. As long as we have no fixed point of reference we cannot decide between the two.

When we see *two* groups of stars converging towards two *different* points the latter explanation fails, at least for one of the groups, because the observer can have but one motion.

From the facts set forth in what precedes we must, therefore, at once conclude that one of our sets must have a real systematic motion in respect to the other.

We can even take one further step. As early as 1843, Bravais has shown, in a paper to which sufficient attention has not been paid, that, no matter how systematic the motions of any group of stars may be, we can determine the motion of the solar system with respect to the center of gravity of that group. Therefore, if for the present we take the center of gravity of all the Bradley stars as a fixed point of reference, we can determine the direction of the sun's motion. As far as I know no extensive determination of the solar apex has as yet been made, rigorously on the basis of Bravais' theory. But for reasons, into which we cannot enter here, the result can hardly differ from the best of our modern determinations, made by other methods. This position coincides with neither of the two points found just now.

We thus get a clear indication that we have to do with two star-streams, parallel to the lines joining our solar system to the two points mentioned.

That the method is not rigorous, that, therefore, the directions here found cannot lay claim to any great accuracy, may be left out of consideration for the present. But what is important to note is: (1) that the directions are only apparent directions; that is, directions of the motion relative to the solar system. (2) That if it be true that two directions of motion predominate in the stellar world, then, if we refer all our motions to the center of gravity of the system, these two main directions of motion must be in reality diametrically opposite. Some reflection must convince you that it must be so, and I will not, therefore, stop to demonstrate it.

For the sake of brevity I will call the points of the sphere towards which the star-streams seem to be directed the *vertices* of the stellar motion.

The *apparent* vertices were thus provisionally found to lie south of a *Orionis* and η *Sagittarii*. Knowing with some approximation the velocity of the sun's motion as compared with the mean velocity of the stars, it is easy to derive from the *apparent* positions of the vertices their *true* positions, which must lie at diametrically opposite points of the sphere.

Having once got what I considered to be the clue to the systematic divergences in the proper motions, and having at the same time obtained an approximation for the position of the vertices, I have made a more rigorous solution of the problem.

I will not here enter into the details of that solution. In order to prevent misconception, however, it will be well to state expressly that the existence of two main stream-lines does not imply that the real motions of the stars are all exclusively directed to either of the two vertices; there is only a decided *preference* for these directions. In my solution I have assumed that the frequency of other directions becomes regularly smaller as the angle with the main stream becomes greater, according to the most simple law of which I could think, which makes the change dependent on a single constant.

I have as yet only finished a first approximation to the solution. The result is that one of the two vertices lies very near to ζ *Orionis;* the other, diametrically opposite, is not near any bright star. They have been represented by the letter V in fig. 2. They lie almost exactly in the central line of the Milky Way. Adopting Gould's co-ordinates of the pole of this belt, I find the galactic latitude to be *two* degrees.

I will pass over the other quantities involved, but will only mention that the way in which I conducted the solution points to the conclusion that *all* the stars, without exception, belong to one of the two streams. To my regret I must pass over also the detailed comparison of theory and observation, because the detailed determination of the distribution of the proper motions from the data of our solution is such a laborious question that I have not yet made it, and would rather defer it till the real existence of the streams shall have been tested by other observations presently to be considered. I will only state that by this provisional solution the *total* amount of dissymmetry for our twenty-eight regions is reduced for the x components as well as for the y components to about *a third* of their amount in the hypothesis of random distribution of the directions. Moreover, they have lost their systematic character. . . .

Taking the evidence for what it is worth, we may say that it confirms the theory. The proof is not convincing, however, and I will conclude by giving expression to my hopes that those who are in a position to test the whole theory by more extensive and more reliable materials will not neglect to do so.

A few hundreds of stars, not pertaining to the *Orion* stars, and fainter than magnitude 3.5, must probably be sufficient for the purpose.

20. HISTORICAL NOTES ON ASTRO–PHOTOGRAPHY OF PRECISION

By Frank Schlesinger

Many of us can remember the time when photography was struggling for a foothold in astronomy as a method for making accurate measurement. In the early days of the subject and before proof to the contrary was forthcoming, it was only natural to believe that the direct measurement of objects in the sky, by means of the filar micrometer for example, should give more reliable results than those obtained by so indirect a process as exposing a plate to the sky, immersing it in succession in three kinds of chemical baths, and finally, after it had dried, measuring it in the laboratory. Although more than sixty years have now gone by since photography was first applied to such problems, it is only recently that the photographic plate has come into its own and has achieved general recognition as a precise method. Even to-day the accuracy and the great economy of the results that can be attained in this way are still not as well known as they ought to be. Certainly at the beginning of this century there was still a feeling akin to suspicion regarding the validity of photographic measurement, in spite of the early work of Rutherfurd, and of that brilliant group of men who came together to organize the work of the Astrographic Catalogue. One could give many instances of this feeling at that time, but I shall cite only one.

When the Yerkes Observatory began work thirty years ago, Professor Barnard was one of the original members of the staff, and for his principal work with the great refractor, which was at his disposal for two nights in each week, he decided to measure the relative places of a large number of stars in each of four globular clusters, the object being to set down accurate positions from which the relative motions in these objects could be determined by astronomers of the future. By 1900 Ritchey had obtained excellent photographs of these clusters with the same telescope; but still Barnard, who was as receptive of new ideas as anyone I have known, preferred to continue this work of triangulation by means of the filar micrometer, although he was well aware of the enormous amount of work this decision entailed. In 1903 it was my good fortune to come to the Yerkes Observatory, through the good offices of George Ellery Hale, to try astrometric work by means of photographs secured with the great refractor. I for one had no definite idea what accuracy to expect from such photographs, since up to that time all experience had been confined to plates taken with much smaller instruments, none of these, so far as I am aware, having a focal length as great as one-fourth of that of the Yerkes telescope. To answer this important question as soon as possible, I measured on photographs the same stars in the globular clusters that Barnard was measuring with the filar micrometer on the same telescope. He entered heartily into this experiment and, with his usual kindness, put at my disposal all the data necessary for a comparison.

It turned out that the measurement of one image on a plate had about one-third the probable error, and therefore about nine times the weight, of a complete micrometer measurement. I do not know which of us was the more surprised; perhaps it was I, for Barnard had a very modest notion of all his attainments, and I realized better than he did how high a standard his micrometer work had set. To reach a high plane of accuracy with the micrometer is a difficult matter requiring in any case years of patient practice, and even then it is not attained by all. On the other hand it is a characteristic of the photographic method (I do not know whether to call it an advantage) that an observer acquires in a few months, or even a few weeks, a standard of accuracy which years of experience do not enable him to surpass; and further, this standard seems to be about the same for one careful and conscientious observer as for another.

The economy of the photographic method is equally surprising. In the comparison that I have just described, the time spent at the telescope to secure the same result is at least one hundredfold greater with the micrometer, and the measurement on photographs of one hundred stars in each of four globular clusters is a task that can be carried out in a few weeks at most. It is worth mentioning that shortly after we made this experiment Barnard himself secured and measured photographs of his four clusters and continued his survey of them by their aid. . . .

The history of the measurement of stellar parallaxes has presented, more than any other department of astronomy, a continual struggle between the necessities of the problem and the methods for attacking it, very similar to the conflict that has gone on between heavier and heavier artillery and stronger and stronger armour-plate. A source of error having once been revealed, it is seldom that much time has elapsed until methods for eliminating it or avoiding it have been devised. After such improvements have been applied, new but smaller sources of error come to light to challenge our patience and ingenuity.

As a result of efforts embracing a score of years and half a score of telescopes, photography now enables us to determine from a series of fifteen to twenty plates, parallaxes whose true probable errors are $0''.008$ or lower, and that are free, under favourable circumstances, from systematic error to the extent of $0''.001$ or at most $0''.002$. . . .

We pass on to a consideration of the rôle that photography has played and can play in the determination of proper motions. This is a very different problem from that presented by stellar parallaxes, and a far less difficult one. Given two observations of a star at different epochs, a good determination of the proper motion can be made no matter how inaccurate the observations are, if only the interval between them is great enough. It has often been said that astronomy is a branch of physics. I do not think that this is true except in a somewhat narrow sense, although I would not go as far as does a colleague of mine in maintaining that physics is a branch of astronomy. Certainly the work that the two sciences demand may be very different in character. Physicists (and chemists too) are accustomed to begin and complete their own experiments. The astronomer, on the other hand, is continually referring to the work of his predecessors and repeating it in order to see what changes time has brought about; and he is continually making observations that can be of no use in his own life-time, but he expects that his successors will be glad to use them in their day. I do not know of any case in which a physicist has begun or completed an experiment of another generation. If he is interested in earlier observations at all it is only in the historical sense. I would remind you that Halley determined the first proper motions by comparing observations made in the seventeenth century with some made by Hipparchus and others at least eighteen centuries earlier; and that the wonderfully accurate work of Bradley, carried out nearly two centuries ago, still plays an important part in this subject. Equally striking examples might be drawn from other departments of our science, from any in fact in which the phenomena concerned unfold themselves very slowly. Though the inability to complete his own experiments is often a sore trial to the astronomer, he has the compensation of knowing that painstaking and honest observations are likely to survive as useful data for long periods, even though they may soon be excelled in accuracy.

Nearly half a century ago our predecessors set down in the *Astronomische Gesellschaft Catalogues* the places of about 130,000 stars in the northern hemisphere, with as great an accuracy as was feasible with so extensive an observing list. This work was, of course, all done by means of the meridian circle. A few years ago we began to realize that the time was approaching for us to repeat this work and so derive the proper motions of this vast number of stars. It was very natural that the practicability of dealing by photography with this repetition of the Gesellschaft catalogues should occur to us. At Allegheny Observatory and later at Yale we tried wide-angle fields ($5° \times 5°$) for this purpose and found that they gave unexpectedly good results. From two observations, one image on each of two plates, we were able to obtain positions with a probable error of $0''.16$. This is less than half that of the average of the original visual work in which, moreover, three observations were secured for each star. Even more gratifying is the economy of the photographic method. The number of hours spent at the telescope is reduced by a factor of ten or more, and the remainder of the process is so expeditious as to bring the re-observation of all our 130,000 stars well within the powers of a single institution instead of requiring the co-operation of more than a dozen. . . .

It is easily within our grasp to increase the number of fairly accurate proper motions. A simple comparison of a good place at this date with the one in the Gesellschaft zones will give a result whose probable error will average about $0''.008$. From this point of view two repetitions are unnecessary: a comparison of the mean of *two* modern places with the Gesellschaft place will reduce this probable error by only 4 per cent., the reason being that the modern place is so much more accurate than the old that almost all the uncertainty in the proper motion arises from the uncertainty in the earlier place.

The case is very different for very accurate proper motions. A full century must be allowed to elapse after the date of the original Gesellschaft observations in order to obtain from them proper motions whose average probable errors will be $0''.004$, and two centuries if we wish to attain $0''.002$ in this way. There can be no doubt that information as accurate as this will soon be indispensable to progress in the determination of the arrangement and the motions of stars. From this point of view we can hardly err on the side of doing too much in the effort to set down the positions of many faint stars at this epoch with the highest accuracy, especially in view of the ease with which photography now enables us to do such work.

21. ON THE ROTATION OF THE STARS

By G. Shajn and O. Struve

1. Captain W. de W. Abney was, we believe, the first to express (in 1877) the idea that the axial rotation of the stars could be determined from measurements of the widths of spectral lines. This opinion met with severe criticism on the part of H. C. Vogel, who pointed out that the great width of the hydrogen lines in certain stars could not possibly be due to rotation, since other lines in the same spectra usually appear narrow. It is now known, however, that in many stars all lines are broad, much broader in fact than the limiting width imposed by the resolving power of the spectrograph.

Professor Frank Schlesinger was the first to actually observe the rotation of stars. This he did by measuring the limb-effect in the eclipsing variables δ Librae and λ Tauri. The same problem was later discussed by G. Forbes.

In 1922 J. Hellerich published an important paper on the rotational effects as observed during the time of eclipse in a number of Algol-type variables.

A complete investigation of the rotational effect in β Lyrae and in Algol was carried out by R. A. Rossiter and Dean B. McLaughlin. More recently J. S. Plaskett has shown that the effect is also present in 21 Cassiopeiae, and McLaughlin has rediscussed the eclipsing variable λ Tauri on the basis of Ann Arbor spectrograms.

All the investigations on eclipsing binaries mentioned above prove that the absorption lines are actually widened by rotation. Adams and Joy have successfully predicted in a number of cases that spectroscopic binaries with very wide and diffuse lines have very short periods. Since it is probable that the rotational periods in spectroscopic binaries are equal to their orbital periods, it is clear that very short periods should be associated with wide and diffuse lines. A similar effect was also noticed by Antonia C. Maury in the spectroscopic binaries μ^1 Scorpii and V Puppis.

The contour of spectral lines in a rotating or pulsating star was investigated by H. Shapley and S. B. Nicholson in connection with Cepheid variables.

The important theoretical discussion by J. A. Carroll on the form of an absorption line in the spectrum of a rotating or expanding star unfortunately arrived only after most of our own work had been completed and prepared for publication. Both papers show that the intensities of spectral lines form a sensitive criterion for the detection of rotation. Additional sides of the problem in question are considered in this paper; we have particularly emphasized certain practical applications.

J. H. Jeans has recently shown that the flow of radiation, carrying with it momentum from one part of the star to another, seriously affects the internal motions of the stars, and in the first place their axial rotations. This factor of radiative viscosity, so called because of its analogy with the ordinary viscosity of gases, leads to the interesting result that the stars do not rotate as solid bodies, but that each spherical shell has its own angular velocity roughly proportional to r^{-2}. Only the innermost portions of the stars deviate markedly from this law. The period of rotation of the outer layer is approximately nine times longer than that of the inner layers. One would, therefore, expect that the effect of axial rotation upon the spectrum would be slight. This, however, may not be true for close double stars. Immediately after fission (if double stars really originate through fission) the periods of axial rotation and of orbital revolution must be identical.

There is also observational evidence in support of this equality. Thus in the β Lyrae stars the light-variation is continuous, indicating that the components are nearly in contact. A difference between the two periods would result in marked changes in the light curve from one revolution to the next, this being caused by the elliptical shape of the components. No such changes have been observed.

One of the stars in which Adams and Joy suspect rotational widening of the lines is the well-known variable W Ursae Majoris. Being a dwarf of spectral type F8p, the radii of its components cannot be very large. Adams and Joy found that the greatest semi-axis of each component is 540,000 km. or 0.78 in terms of the Sun's radius. The total mass of this system is very nearly that of the Sun, and the period is one of the shortest yet discovered, $0^d.334$. The lines are described as "so widened and weakened that measures for velocity and estimates of line-intensity can be made only with the greatest difficulty." It is suggested that "the unusual character of the spectral lines is due mainly to the rotational effect in each star, which may cause a difference of velocity in the line of sight of as much as 240 km. between the two limbs of the star." In this connection an interesting attempt was made by J. Schilt to consider W Ursae Majoris as a rotating star just preceding fission.

It can be shown that if the rotational velocity of W Ursae Majoris produces appreciable widening of the lines and decrease in their depth (or central intensity), a similar effect should exist in binaries of early spectral type and of somewhat longer period. . . .

[Here is omitted a mathematical statement of the problems associated with spectrum line contours and with stellar rotation.]

2. *Influence of Rotation upon Line-contour.*—There are a number of factors that influence the width of a spectral line and the distribu-

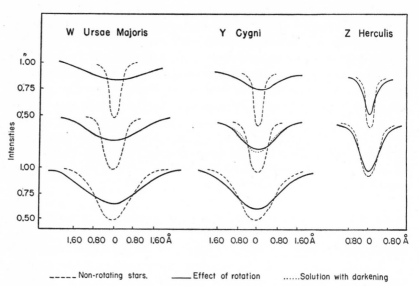

Fig. 1. The expected rotational effect for spectroscopic binaries.

tion of intensities in it: (1) Disturbing effects of neighboring atoms which interfere with the perfect periodicity of the absorbing atom so that its quantum states are not entirely sharp; (2) Doppler effect due to the temperature motion of the particles; (3) Doppler effect due to ascending and descending currents of matter; (4) Doppler effect due to rotation; (5) Compton scattering by free electrons having different velocities; (6) Rayleigh scattering; etc. . . .

The influence of the darkening at the limb, as will be seen below, is small. In one of the numerical examples we have used a darkening coefficient equal to 0.75 (see fig. 1; the intensity curve for Y Cygni, initial line-width 2 Å., drawn between the two curves for the rotating and non-rotating star, represents the solution with darkening at the limb).

The resulting curves will also be influenced by the character of the lines, from the point of view of their energy level. The light coming from the edges of the disc traverses thinner layers of the star's atmosphere and the line originates in higher and cooler regions. Accordingly certain lines must appear strengthened near the edges, while other lines are weakened and perhaps even reversed. It is also clear that the wings of certain lines may disappear near the edges. Obviously the low-temperature lines will be the most sensitive ones with respect to rotation.

To illustrate the order of magnitude to be expected for rotational effect we have made computations for a number of spectroscopic binaries under the assumption that the periods of rotation and of revolution are equal. The effect will, of course, depend upon the initial width and upon the central intensity of the line. We have carried through the computations for lines of 1, 2, and 4 Å. width. The greater values are encountered more frequently in the earlier spectral types, and it is in these that fast rotations are especially probable.

TABLE I. Central Intensity of Lines for Rotating Stars.
(The initial intensity was assumed to be $0^m.75$.)

	Width		
Star	1 Å	2 Å	4 Å
W Ursae Majoris	$0^m.12$	$0^m.24$	$0^m.43$
Y Cygni	.17	.35	.52
Z Herculis	.52	.69	
u Herculis		.23	

The results are shown in fig. 1 and in Table I. We assume that the initial intensity of the line is $0^m.75$, this representing the difference between the continuous spectrum and the center of the line.

The above results apply to the larger components of the binaries under consideration. Photographically the phenomenon will be more complicated, but the essential features will remain the same. There is not only a very pronounced widening but also a large decrease in the depth of the line. This latter should constitute a very sensitive criterion of stellar rotation. With our present-day instruments the effect should be noticeable for rotational velocities exceeding 30 km/sec. It is of interest to note that wide and narrow lines react in varying degrees to the effect of rotation.

It is readily seen that the effect of rotation depends also upon the

initial intensity of the line. Deeper lines will be more affected than
shallow ones. We have computed the contours of two lines in the spectrum of Y Cygni of the same initial width, 2 Å., but of different initial
central intensities, 0.50 and 0.20, in units of the intensity of the neighboring portion of the continuous spectrum. The results are shown in
fig. 2. The intensity-ratio of the two lines in question is not the same
in a non-rotating and in a rotating star:

	Intensity-ratio
For a non-rotating star	2.50
For a rotating star	1.29

Owing to rotation the unequal initial depths become more nearly uniform. This criterion is perhaps less applicable to spectroscopic binaries

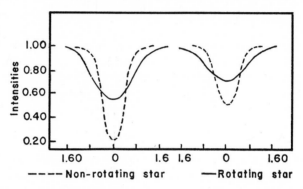

Fig. 2. Computed contours of two lines in the spectrum of Y Cygni.

since the superposition of the continuous spectrum of the secondary star
leads to a similar effect. However, in single stars, or in binaries where
there is no interference on the part of the fainter component, this difficulty does not exist.

In actual practice the effect of rotation can be studied in two different
ways. Consider a single spectral line. Its width and central intensity
are functions of many factors, rotation being one of them. In a given
number of stars there will be various intensities and line-widths, and it
is not *a priori* possible to distinguish the influence of rotation from that
of the other factors. We may, however, subdivide our stars into a number of groups, such that the expected rotations differ systematically
from one group to the next, while the influences of the other factors remain, on the average, constant. In that case rotation will stand out and
can be investigated separately. This method is purely statistical, and
we have followed it in section 3.

Since stellar spectra contain usually many lines, wide and narrow, intense and weak, it should, theoretically, be possible to investigate the effect of rotation for each star individually. Suppose we know, from the solar spectrum, the initial intensities and widths of certain lines. If the spectrum of a star of the solar type shows the narrow lines appreciably too weak, while the wide lines remain about the same, we may conclude that the weakening is due to rotation (provided, of course, that the particular line is not sensitive to changes in absolute magnitude).

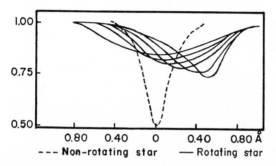

Fig. 3. Computed contours of a line of the primary component of Y Cygni for various phases.

An entirely different method can be used in eclipsing variables. We have mentioned before that the effect of rotation has actually been measured from the small variations in radial velocity at the epochs just preceding and just following the middle of eclipse. The line-contours observed during the same epochs will also be subject to rotational changes. The intensity distribution will be asymmetric and the degree of asymmetry will change with the phase. It should also be noted that during eclipse the central depths of the lines are less in a rapidly rotating star than in one having slow rotation. We have computed the contours of a line of the primary component of Y Cygni, for various phases (see fig. 3). We have assumed that (1) the star is spherical, (2) there is no darkening at the limb, (3) the spectrum of the principal star is not affected by that of the fainter. Our last assumption is invalid in the case of Y Cygni, so that, strictly speaking, our computations do not refer to that star. There are, however, many other eclipsing variables that sufficiently closely resemble this hypothetical object. The solution was obtained by a mechanical quadrature. . . . The computations refer to a line of initial intensity $0^m.75$. Table II shows the change in the central depth of the line as a function of phase.

TABLE II. Central Intensity of a Line of Initial Width 1 Å and
Intensity $0^m.75$ during Eclipse for Y Cygni.

Phase	Intensity
$0^d.129$	$0^m.33$
.110	.29
.089	.26
.070	.24
.052	.22
.018	.18

(The phases are counted from the middle of the eclipse.)

Investigations, made during an eclipse, of the line-contours in eclipsing variables would doubtless give useful results for the further study of their rotation. . . .

4. *The Rotational Effect in the Spectrum of Jupiter.*—To find what might actually be expected in the case of a rotating star, spectrograms of Jupiter were obtained in 1927 by means of a single-prism spectrograph attached to the 40-inch Simeis reflector. The rotation of Jupiter causes in the reflected sunlight a Doppler shift of about 25 km./sec. at the equatorial limb. The experiment consisted in comparing two spectrograms of Jupiter. The one was taken with the slit perpendicular to the equator and with the image of the planet accurately bisected by the slit. In this case there was, of course, no effect of rotational widening nor any decrease in line-depth. The next plate was taken with approximately the same exposure time, the image of the planet being oscillated uniformly across the slit by means of an electric motor for very slow motion, thus giving an integrated spectrum for the whole of the planet. This is probably a fair approximation to the spectrum of a rotating star, in spite of the notoriously strong darkening of Jupiter at the limb. The exposures, usually made near the meridian, were about 25 minutes long. The linear dispersion in the region measured was about 12.5 Å. per mm. The total number of plates was ten. The spectrograms were run through a Koch microphotometer with gear ratio 48.5 to 1.

The widths of a number of lines were measured on the microphotometer tracings. The results are affected by blending with other lines so that the precision is not great. However, since we are here concerned only with differential effects, these uncertainties, as well as instrumental or other errors, are of no consequence.

There is a slight widening of the spectral lines in the integrated spectrum, but its amount is of the same order of magnitude as the mean error, and is therefore not certain. Evidently an equatorial velocity of rotation of 25 km./sec. produces an amount of widening that is at the very limit of what the measurements can give. The rotation should probably be appreciably faster, perhaps of the order of 50 to 100 km./sec., to produce effects upon the line-widths that are definitely measurable.

The measures of line-intensities give very much better results. Since the exposure times of the spectrograms were such that the photographic densities of the continuous spectrum are the same, we have measured the depths of the lines in mm.'s and summarized the differences for each line.

The results show that the lines of the integrated spectrum are appreciably shallower than those which are not affected by rotation, but this difference is not very 'arge and is noticeable only after a careful study of the material.

We may thus conclude that the effect produced by a rotational velocity of the order of 25 km./sec. is about at the limit of what the measurements can reveal. Since in actual practice we shall never be able to compare with each other two spectra that are as much alike as two plates of Jupiter, the lower limit should be taken rather above 25 km./sec. However, there is every reason to believe that rotational velocities exceeding 50 km./sec. should be measurable without much difficulty. That such velocities are not infrequent, is shown by our statistical discussion of spectroscopic binaries. Whether or not single stars also frequently rotate with such high velocities remains to be seen.

22. A GREAT CATALOGUE OF STELLAR POSITIONS AND PROPER MOTIONS

By Ralph E. Wilson

The recent completion, through the joint efforts of the Carnegie Institution of Washington and the Dudley Observatory of Albany, of
the General Catalogue of the positions and proper motions of 33,342
stars marks an epoch in the history of astronomy of position. This work
employed an average of twenty persons continuously for over thirty
years and its total cost exceeded $1,000,000. The results appeared last
year in five volumes totaling 1,700 pages. Less than 1,000 copies have
been distributed and probably less than 100 of these will be extensively
used. The catalogue could hardly be considered a "best seller," but to
the investigators who will interpret the results in determinations of the
motions of the earth, sun and stars and in studies of the structure of the
stellar system, it presents a veritable mine of fundamental data. The
history of this undertaking is interesting.

During the century and a half following 1755, when Bradley made
the first accurate determinations of star positions, this class of observations constituted the main activity of most of the observatories of the
world. It consumed the energies of considerably more than half the
working force of astronomers during the nineteenth century. Govern-
ments rendered fairly adequate aid, because the practical uses of this
work, notably in geography and navigation, were evident. As a result,
by 1900 the libraries of all observatories were well stocked with catalogues containing hundreds of thousands of star positions, each independent of the others, some presenting accurate positions while others
were of little value even at the time they were published. It had become discouragingly apparent that further observation would be practically useless and the outlook for positional astronomy sorry indeed,
unless the raw material could be made a contributing factor in the
development of astronomical research.

The observer in this department of astronomy corresponds well with
the skilled collector in various branches of natural history. His aim is
to get as accurate positions as possible at the time of observation.

Owing to the lack of uniformity in the rotation of the earth, he observes star positions relative to a moving frame of reference. Each observer refers his positions to a standard equinox and equator near the date of his observations and, without knowledge of the motion of the reference frame and of the star itself, this tells us nothing definite about the position at any other time.

For any well-observed star we find a series of positions increasing or decreasing with the time. If the observations were all perfect, the plotted positions should lie on a straight line. Actually they do not, because each catalogue is subject to errors of observation, both accidental and systematic, the latter depending mainly upon the magnitude of the star or its position in the sky. A detailed study of each catalogue must be made to determine and eliminate the systematic errors and to estimate its value from the remaining accidental errors. When this has been done, the position of the star at the mean epoch and the "annual variation" in position may be derived from a plot directly, if approximate values only are desired, but for definitive results least-squares solutions based upon time, position and weight are necessary. Once the position at the epoch and its annual variation are known, the position at any other time is easily obtained. The annual variation is a result of observation and may be changed only by further observation. It is, however, a composite of two interdependent motions; one, the motion of the reference frame, precession; the other, the motion of the star across the sky, proper motion. Obviously, if we change the value of the precession, the proper motions are changed as a consequence.

It was quite possible in 1900 for an investigator interested in stellar positions to search all the catalogues available to him and, neglecting all consideration of systematic errors and weights, to derive approximate annual variations and consequent positions. However, not only was the process conducive to inaccuracy, and time consuming, but it lost all the value of the resulting proper motions, which to most students are now much more important than the positions. In 1900 not one-tenth of the available observations had been brought to bear upon the problems of proper motions. To quote Professor Russell, "the astronomical assets of 1900 could fairly be described as frozen, for a reason too familiar in these days—it cost too much to realize on them."

This state of affairs was changed mainly through the foresight of one man. In 1872 Lewis Boss was assigned the job of producing a set of uniform stellar declinations for use in the survey of the boundary between the United States and Canada from Minnesota to the Pacific. Becoming director of the Dudley Observatory in 1876, he continued with meager financial support the work he had earlier started on systematic star positions. He became so impressed with the importance of constructing a general catalogue of positions and proper motions which should make fruitful the preceding 150 years of effort in positional astronomy that he convinced the trustees of the Carnegie Institution of Washington of the scientific value of his plans and through them secured the necessary financial backing. The Department of Meridian Astronomy was founded in 1906 to carry out the plans, under the direction of Professor Boss until his death in 1912 and under his son, Benjamin Boss, from that time to the completion of the catalogue in 1937.

The program was fourfold. 1. The collection and reduction to a uniform system of all the extant observations of all stars brighter than visual magnitude 7.0 and of many fainter stars which appeared to have proper motions above the average. The original plans called for 20,000 stars, but over 13,000 others were added later. This part of the program involved a detailed study of nearly 250 catalogues to determine their systematic corrections and weights, the results of which in tabular form take up 170 pages of the first volume of the catalogue. The collection and systematic reduction of the observations was an enormous task, but once done the other material is for all time put in form for use with modern data and any revision of the fundamental system or the astronomical constants used may be incorporated in the final stages of new determinations of mean positions and proper motion without rediscussion of the older observations.

2. Reobservation of all the stars with the same instrument and as nearly as possible the same observers. Obviously modern observations would greatly strengthen the final results and the use of the same instrument in observations from north to south pole with a large overlapping zone near the equator would permit accurate determinations of the errors of observation. Since about one-fourth of the stars could not be observed from Albany, the plan involved the establishment of a temporary observing station in the southern hemisphere. The observations with the Olcott meridian circle were begun in Albany in 1907. In 1909 the instrument was dismantled and shipped to San Luis, Argentina, where a staff of ten men, under the leadership of Professor R. H. Tucker, of the Lick Observatory, accumulated 87,000 observations of position in two years, probably the greatest feat of intensive observing in the annals of astronomy. Observations were resumed in Albany in 1911 and continued in somewhat more leisurely fashion until the program was completed in 1918, 110,000 observations in all being made at that place.

3. Reduction of the Albany and San Luis observations. Anyone at all familiar with meridian work realizes that *minutes* at the telescope mean *hours* to be spent in the reductions leading to the final positions. Though the methods may be routine, the mere mechanical handling of the operations for 197,000 observations, reduced twice independently, is a job no institution would venture to undertake without a very definite end in view. The San Luis positions fill a 307-page volume, published in 1928; the Albany one of 430 pages, published in 1931. Naturally these had to be compared, evaluated and reduced to the same standard system as the older data.

4. Formation of the General Catalogue. A factor presenting serious difficulty in the completion of the catalogue arose from the activities of other observatories. Stimulated by the realization that positions would no longer be frozen assets, meridian observers produced many series of observations of later date than those made at Albany and San Luis which provided excellent material for better derivations of both mean position and proper motion. Many of these were studied and used in the catalogue, but as an apparently unending succession of them continued to appear, a closure date had to be set and no lists appearing after 1932 were considered. It was unfortunate that several excellent catalogues appearing soon after the closure date could not have been used, but such a situation would probably have to be faced at any epoch.

The great mass of data accumulated from 1755 to 1932 having been reduced to a system and properly weighted, definitive positions and annual variations were derived by least-squares solutions, a minimum of four being necessary for each star. Newcomb's values of the precessions were used. These were computed and subtracted from the annual variations to give the final values of the proper motions. All the significant data were collected on cards, from which the manuscript was made. In addition to the positions for the year 1950, the annual variations and proper motions, with their probable errors, the catalogue gives the magnitudes and spectral classes of essentially all the stars, numerous data regarding double and variable stars, and all the information necessary for incorporating added positional data. More than 6,700 southern stars were specially observed with a photometer at San Luis to make the list of magnitudes more accurate. Spectra not already published were obtained through the cooperation of Harvard College Observatory. Quoting Professor Russell again, "The completed volumes represent a transformation of the great mass of frozen assets into liquid and available form."

As to how useful the proper motions may be in studies of the structure of our stellar system, we can judge by some of the interesting results which followed the publication of a Preliminary General Catalogue in 1910. Produced in response to the insistent demand of astronomers for proper motions, this volume gave the required data for 6,188 stars the motions of which could be determined from the positions then available. It immediately became the standard authority on proper motions. It has been the basis for the better determinations from proper motions of the motion of the sun and of stars streaming, for the discovery of moving clusters with the resulting distances of their members, for determinations of mean distances of stars of all sorts. It has fixed the scale by which the distances of the extra-galactic systems have been measured, improved the value of the precessional constant and determined one component of the rotation of the galactic system. The General Catalogue presents better proper motions of the same stars and four times as many others. It may be expected to improve the quantitative results before obtained in all these fields, even if no new phenomena are uncovered.

An important use for the positions is already apparent. During the last decade it has become quite evident that for studies of the dynamics of our stellar system we need the motions of the faint as well as the brighter stars. As observations of stars fainter than the ninth magnitude become extremely difficult with a meridian circle, they must be made photographically. Strictly differential motions may be secured by comparison of plates separated by a sufficient time interval, but we must have the motions of the faint stars on the same system as those of the brighter ones. This requires satisfactory positions of the comparison stars necessary for reducing the plate positions to right ascension and declination. The stars of the General Catalogue offer the most extensive system of comparison stars now available. They provide an average of 0.8 stars per square degree, distributed fairly uniformly over the sky. The addition of possibly 1200 stars would fill the gaps, and the groundwork of the catalogue makes this addition a relatively simple task.

But the investigator is never completely satisfied with his data. The accumulation of observations goes on and, valuable as the positions and motions already presented are, they can even now be materially improved. The average proper motions of several sorts of stars, and of the faint stars in general, are of the same size as or less than the average probable errors of the motions of the General Catalogue. One good modern set of observations would considerably increase the accuracy of both positions and motions. Certainly all the weaker stars should be reobserved within the next ten years. Programs now under way, some of them already completed, will be of material aid. Once the observations have been secured, the revisions of the positions and motions will require comparatively little time, because we now have all the data necessary for incorporating the new data with the old.

Thus an estimate of the value of this General Catalogue must be based not only upon the rich supply of data it now furnishes for investigations of the structure of the stellar system, valuable as that is, but also on the fact that the assets of the past have been liquidated in such form that a future wealth of observations may be utilized without a complete reorganization of the capital structure. For this the research astronomer owes a debt of gratitude to the institutions and individuals who have cooperated in the completion of this great work.

23. EXPANDING STELLAR ASSOCIATIONS

By V. A. Ambartsumian

Up to recent years the astronomers have been engaged in the study of two types of "small" stellar systems, which belong, as collective members, to the population of the Galaxy. These are open and globular clusters. Double and multiple stars are related to open clusters. The circumstance that the energy of gravitational interaction in open clusters is of the same order as that of many ordinary visual binaries points to the kinship of these formations.

Recently, however, the author ascertained [1] that along with open and globular clusters there is in the Galaxy yet another type of stellar system, *stellar associations*,[2] which are of outstanding interest from the point of view of the problem of stellar evolution.

In the present paper we consider separate examples of stellar associations, different types of associations, and their properties.

The following may be given as examples of stellar associations:

(1) The group of variable stars of the T Tauri type and other stars connected with them in Taurus and Auriga. It is known that T Tauri type stars are met with only in some definite regions of the sky. In particular, 8 of them form an isolated group in the constellation of Taurus and Auriga, occupying in the sky an area 12° by 12°. With a distance to this group of the order of 100 parsecs, this means that the linear diameter of the group is of the order of 25 parsecs. Later in the same region of the sky Joy found a number of dwarf stars with bright lines in their spectra, this indicating that they are probably related to T Tauri type stars.

Here two important facts should be noted: (*a*) such a mutual close

---

[1] *Stellar Evolution and Astrophysics,* Erevan, 1947, pp. 12–16, 26–27.

[2] [The term "Stellar Associations" was already in use as one class of galactic cluster, *i.e.,* the most open variety (Shapley, *Star Clusters* [Harvard Observatory Monographs No. 2; McGraw-Hill, New York, 1930], p. 8 *et passim*). The concept of stars dispersing is not there introduced, however, and the term "Expanding Association" will serve to distinguish Ambartsumian's systems from Shapley's (which may in fact be expanding also, and certainly are slowly dispersing as a result of galactic rotation). See also the "Shoulder Effect" for Messier 67, in Shapley, *Star Clusters,* p. 99, and *Mount Wilson Contribution,* No. 117 (1916).]

proximity of stars of the considered group cannot be explained by chance; it is evident that here we have to do with a single system; (*b*) at the same time the density of the considered system of stars is so small that it could never have been detected as a cluster by direct observation, even if it were several times closer to us. Only the circumstance that the members of this association belong to a particular class of variable stars permitted its detection.

An exceptionally important characteristic of the system under consideration is its low space density. Even if we add to the 7 or 8 T Tauri type stars the above-mentioned 40 dwarfs with bright lines, assuming that they belong to the system, we find nevertheless that the space density in the system is much lower than the density of the galactic star field in which the considered association is embedded. It can even be assumed that, besides the dwarfs with bright lines, there are other dwarfs (without bright lines), several times greater in number, which belong to this system. However, even in this case the upper limit for the density is much lower than the density of the general star field.

According to known dynamical criteria this means that the considered system is unstable and should be disrupted by the action of the tidal forces of the general gravitational field of the Galaxy. Therefore, it should be assumed that the system is expanding in space.

The connection between stars of this system and diffuse matter, bright as well as dark, also deserves attention.

(2) Kukarkin's and Parenago's *General Catalogue of Variable Stars* (1948) lists in a small area around $\alpha = 18^h\ 40^m$, $\delta = +9°\ 0'$, 8 T Tauri type stars (or according to the terminology of the catalogue, RW Aur type stars). Three of these stars have question marks after the type of variability. We give the whole list.

Name	α		δ		Stellar magnitude	Type
V 637 Ophiuchi	$18^h31^m59^s$	$+\ 9°$	$51'.1$		$13^m.6–16^m.0$	T Tauri
V 681 Ophiuchi	32 34	$+\ 9$	06 .1		14 .2–15 .4	T Tauri?
V 643 Ophiuchi	32 34	$+\ 6$	16 .7		13 .6–15 .6	T Tauri?
V 645 Ophiuchi	33 02	$+11$	47 .1		14 .6–15 .6	T Tauri?
V 476 Aquilae	43 57	$+\ 7$	02 .3		13 .3–14 .2	T Tauri
V 480 Aquilae	45 43	$+\ 7$	00 .7		14 .0–16 .0	T Tauri
V 489 Aquilae	50 55	$+11$	56 .3		13 .1–14 .8	T Tauri
V 490 Aquilae	53 49	$+12$	51 .1		14 .1–15 .5	T Tauri

We see that all these stars are in a small area of the sky, 6° by 7°, not far from the galactic equator. Even if we leave out three stars, the type of which has yet to be confirmed, nevertheless the concentration of 5 T Tauri type stars in this region cannot be accidental and we have to do with members of some stellar system. At maximum brightness the variables in this system are 3 to 4 magnitudes fainter than the variables belonging to the association in Taurus. This apparently is evidence that the association in Aquila-Ophiuchus is remote.

(3) The group of O and B type stars and also red supergiants surrounding the Double Open Cluster in Perseus. This system was studied by Bidelman.[3] The observations prove without a doubt the existence of a group of supergiants of early and late types, which surround the χ and h Persei clusters. The Double Cluster is the nucleus of this association.

The whole system has a diameter of the order of 170 parsecs, while each of the clusters χ and h Persei has a diameter of the order of 10 parsecs (7 parsecs according to Oosterhoff). A characteristic peculiarity of the system is the presence of a large number of B type stars with bright lines. In particular there are at least five P Cygni stars (HD 12953, 13841, 14134, 14143, and 14818) in the system.

Even if we assume that the association as a whole contains tens of thousands of stars, nevertheless its mean density will be smaller than the density of the galactic star field. Therefore, without a doubt, this association also consists of stars moving apart in space. At the same time it should be noted that the two nuclei, χ and h Persei, which are ordinary open clusters, must be stable formations and their disruption can take place only in the manner characteristic for open clusters.

(4) The open cluster NGC 6231 is surrounded by a group of O and B type supergiants. The study of the radial velocities by Struve [4] shows that all these supergiants, together with the cluster, form one stellar association. Its distance from us is about 1000 parsecs. The diameter of the association is almost five times greater than that of the cluster and reaches approximately 30 parsecs. The circumstance that there are 2 Wolf-Rayet and 2 P Cyg stars in this association calls attention to itself.

It is needless to say that there can be no question of an accidental concentration of these stars around the cluster. In the given case we must again accept that the mean density of the association is small in comparison to the density of the galactic field. The association is unstable, although the open cluster NGC 6231, which is its nucleus, is possibly stable.

[3] *Astroph. Jour., 99,* 61 (1943).
[4] *Astroph. Jour., 100,* 189 (1944).

(5) Of special interest is the NGC 1910 system in the Large Magellanic Cloud. It consists of a large group of early-type supergiants, among which are P Cyg type stars including the famous star S Doradus. The diameter of this system is of the order of 70 parsecs, which many times exceeds the dimensions of ordinary galactic clusters.

(6) The stellar association in the Kapteyn SA 8 (center, $a = 1^h\ 00^m$, $\delta = +60°\ 10'$). The association is a group of faint O and B type stars and occupies an area in the sky with a diameter of 25 degrees. There are 2 B type stars with bright lines and a Wolf-Rayet star in the association. Apparently not less than 23 members of this association are OB type stars. It is necessary to note that the association is in a region that is poor in B type stars brighter than 8^m. Judging from the apparent magnitudes of the early-type stars, this association is at a distance of not less than 2000 parsecs. This gives a diameter of the order of 100 parsecs. The existence of this exceedingly interesting and distant association was established by Markaryan at the Burakan Observatory in 1948 on the basis of data in the Bergedorf catalogue. According to Markaryan the nucleus of this association is the open cluster NGC 381, which has a diameter of $7'$, corresponding to a linear diameter of not less than 4 parsecs.

THE MAIN CHARACTERISTICS OF STELLAR ASSOCIATIONS

From the above data the following general conclusions on stellar associations can be made:

(1) The associations are systems whose mean density is small in comparison to the density of the galactic stellar field. However, if we consider the partial concentration of stars of separate spectral types, the associations stand out sharply, owing to the relative abundance of stars of comparatively rare types. In some cases these are O and B type stars; in others, T Tauri type stars. As a result of their small density the associations cannot be in states termed, in stellar dynamics, as stationary. In distinction from globular and open clusters, the associations are nonstationary systems. Obviously the members of the association diverge in space, mixing in the course of time with stars of the field.

(2) Associations always have stars from which there is a continuous outflow of matter. In three of the six examples given above we meet with P Cyg type stars. In the fourth and sixth examples we find Wolf-Rayet stars among members of the associations. In the first two examples we meet with T Tauri type variables, the bright lines in the spectra of which have absorption components on the violet side, i.e.,

the same peculiarity as the bright lines in the spectra of P Cyg stars. Therefore it should be presumed that there is a continuous outflow of matter from these stars.

(3) In some cases the associations have nuclei in the form of open stellar clusters.

STELLAR ASSOCIATIONS IN THE LARGE MAGELLANIC CLOUD

It is known that the Large Magellanic Cloud is very rich in open clusters. At the same time the fact appears that the clusters of this Large Cloud have in some cases very large linear diameters (some tens of parsecs).[5] The above-mentioned example is the most striking. The distribution of diameters of open clusters in the Large Magellanic Cloud has however a minimum, which divides all the open clusters into two groups: (*a*) clusters with diameters in excess of 20 parsecs and (*b*) clusters with diameters smaller than 20 parsecs. Already this circumstance leads us to suspect that here we have to do with objects of two different types and scales. The presence, in at least some of the clusters of the first group, of P Cyg type stars makes us think that systems with diameters in excess of 20 parsecs are objects of the type of stellar associations met with in the Galaxy, while objects of the second group are ordinary open clusters.

The following considerations make this supposition an almost authentic fact. If we observe our Galaxy from some outer system, say from the Large Magellanic Cloud, the association surrounding χ and *h* Persei will distinctly stand out on the surrounding background, because of the presence in this association of a large number of supergiant stars. If we observe this system from inside the Galaxy, we meet with the fact that the [galactic] stars of low luminosity are projected on the field of the association. These stars are at a much smaller distance than the association and have, because of this, the same apparent stellar magnitudes as the supergiants belonging to the association. Therefore, the stars of the association are lost in the general background. An observer in the Large Magellanic Cloud would, without investigating the spectra, by direct observation detect this association as a cluster of supergiants with a diameter of 170 parsecs. The χ and *h* Persei clusters would appear to him as only compact groups in this immense system.

On the other hand if the system NGC 1910 were transferred from

5 Shapley, H., *Galaxies* (Blakiston, Philadelphia, 1945), p. 83.

the Large Magellanic Cloud to the Galaxy, to the place of χ and h Persei, it would be observed by us only as an association, i.e., it would not stand out as a noticeable condensation of stars if a separate study were not made of the distribution of early-type stars in this region of the sky.

Therefore, evidently all the giant systems in the Large Magellanic Cloud (about 15) are in reality stellar associations, the characteristic features of which were described in the previous paragraphs.

The Kinematics of Stellar Associations

As the forces of interaction of the stars in associations are small in comparison to the tidal forces of the general galactic field of force, we neglect the forces of mutual interaction for at least the peripheral members of the association.

When considering the motion of stars of the association in the galactic field of force, we should note that the differential effect of galactic rotation should lead to a mutual recession of members of the association. The velocity of mutual recession of two stars under the action of the galactic rotation effect is expressed through Oort's coefficient A by:

$$v_r = A r \ \sin \ 2(l - l_0).$$

In particular, a star at the periphery of an association with a radius R will move from the center with a velocity:

$$V_r = A R \sin \ 2(l - l_0).$$

This velocity will also be the velocity of increase of the radius of the system in a given galactic longitude l, i.e.,

$$\frac{dR}{dt} = A R \sin \ 2(l - l_0).$$

From this the ratio of the radii R_1 and R_2 for two moments of time t_1 and t_2 is

$$\ln \frac{R_2}{R_1} = A (t_2 - t_1) \sin \ 2(l - l_0).$$

Taking into account the value of A, we see that for $l - l_0 = 45°$, the radius will be doubled in a time of the order of 40 million years.

For this deduction we in fact assumed that all the stars of the association move in circular orbits around the galactic center. Actually it cannot be said beforehand what kind of galactic orbits the different stars have. However, if large initial relative velocities in the association are not assumed, the time necessary for doubling the radius will always be of the same order.

The obtained result, irrespective of the existence of other possible additional causes for expansion, leads to the conclusion that every association originated comparatively recently and consists of stars expanding in space from an initial volume in which members of the association were formed.

However if the expansion of associations was caused only by the differential effect of galactic rotation, the dimensions of the association would increase only in the galactic plane.

With regards to the possible expansion of associations in the direction perpendicular to the galactic plane under the influence of the difference in the periods of oscillating motion along the z-coordinate, it is necessary to state that this effect will act much more slowly. The reason for this is that the periods of oscillation for small amplitudes do not depend on the magnitude of the amplitude, i.e., on the initial conditions. In fact, when a star is at a height z above the galactic plane, the value w_z of the acceleration-component is determined by the integral:

$$w_z = -2\pi G \int_{-z}^{z} \rho(z)dz$$

where $\rho(z)$ is the density.

If z is small, then the variations of $\rho(z)$ within the limits $-z$ to $+z$ are relatively small and therefore

$$w_z = -4\pi G\rho(0)z$$

where $\rho(0)$ is the density in the galactic plane. We see that the acceleration is proportional to z. In other words, for small amplitudes we have to do with harmonic oscillations, the period of which does not depend on the amplitude.

As the observed associations are located in low galactic latitudes, the stars which belong to them should have approximately equal periods of oscillation along the z-coordinate. Therefore the considered effect should be very small in comparison with the differential rotation effect.

However, observations do not show especially strong flattening for the systems which were considered above as examples. This fact leads us to think that there is another cause for the expansion, which plays a much greater role than the differential effect of galactic rotation. Namely it remains to assume that the stars of the association were ejected in different directions with certain velocities from the initial volume in which they were formed.

These initial velocities should not be less than one km/sec, as otherwise, already for associations with dimensions of a few tens of parsecs, it would not be possible to disregard the differential rotation effect. On the other hand they should not exceed 10 km/sec, as otherwise when determining the radial velocities of the stars, for example in the association surrounding NGC 6231, this initial velocity would immediately be noticed.

If the initial velocity of expansion from the center is of the order of 5 km/sec, the differential effect of galactic rotation cannot predominate until the linear dimensions of the associations reach several hundred parsecs. However, such dimensions mean that the stars of the association will be completely dissolved among stars of the field, and that will be the end of the association. Consequently the flattening of the associations for such velocities will be slight, and therefore expansion velocities of the order of 5 km/sec are most plausible.

This leads to the conclusion that the stars, which compose the association of objects of the T Tauri type in Taurus-Auriga, were ejected from the above-mentioned initial volume several million years ago, and the stars composing the association surrounding Chi Persei 10 to 20 million years ago, and so on.

The moment when the association began to expand must be very close to the moment of star formation in the association itself, as the assumption that the system was in a stationary state for a long time and only later began to expand, would contradict stellar dynamics. From here we conclude that the age of the stars belonging to associations is measured by only millions, or in the extreme cases, by tens of millions of years.

This is in agreement with the fact that P Cyg type, Wolf-Rayet, or T Tauri type stars are found in associations. A star cannot be a P Cyg type star longer than 1 to 2 million years, as intense ejection of matter would lead to its complete disappearance. On the other hand P Cyg type stars, having the highest luminosities among all the known stars, have also apparently the largest masses. Even if there are other states, corresponding to larger or equal masses, their life-time must be very short, as such masses are extremely rare. P Cyg stars could not have evolved from stars of smaller mass. Therefore they must necessarily be counted as belonging to the youngest stars.

The Number of Stellar Associations in the Galaxy

At present it is difficult to answer with certainty the question on the number of associations in our Galaxy. If only those associations which have early-type supergiants are considered, then these can be detected at great distances (up to 2000–3000 parsecs). Therefore a considerable fraction of them should be accessible. It is very probable that the number of accessible associations of this type is measured by tens. This means that the total number of such associations in the Galaxy is of the order of 100.

With regards to associations composed of T Tauri type stars and other dwarfs with bright lines in their spectra, at present only two of them are known to us. However it is very important to note that so far they are detected by us at only very close distances. In a spherical volume with a radius of the order of 100 parsecs there is one such association. This means that their total number in the Galaxy is measured by thousands.

If we adopt this number as equal to, let us say, 10,000 and take into account that associations of this type can be observed as such during a time of the order of several million years, then in order to keep up the present number of these associations in the Galaxy, not less than one association composed of T Tauri type stars must be formed on the average every thousand years.

Problems of Star Formation

Some astronomers have suggested that all the stars in our Galaxy were formed simultaneously or almost simultaneously, several billions of years ago, i.e., at the epoch of formation of our Galaxy. In the light of the above-stated facts this supposition is completely refuted. The formation of stellar associations and the formation of stars in them from some other form of existence of matter takes place continuously, "almost before our eyes." The number of associations composed of T Tauri type stars, which have been formed during the life-time of the Galaxy, should be of the order of 10 million. We do not as yet know the average number of stars originating in one association, as we discern only the brightest members. However it should be assumed that this number is measured at least by hundreds. This means that at least a billion of stars in our Galaxy originated as a result of the formation of stellar associations from some other objects unknown to us.

Other Possible Types of Associations

It is very probable that the system of O and B type stars in Orion together with the Trapezium form one giant association with a diameter of more than 100 parsecs. The stars of the Trapezium and the open stellar cluster connected with it evidently form the nucleus of this association. The presence of a large diffuse nebula makes this system especially interesting. It deserves a thorough study.

The "moving cluster" in Ursa Major is a system with 32 members and a diameter of more than 200 parsecs. A group of 11 stars forms the nucleus of this system with a diameter of only 9 parsecs. However there are no direct indications showing that the stars belonging to it are young. The small number of members of this cluster is also striking. It is possible that the system is a remnant of a once rich association.

The sun is located inside this system, but as is known, it is not one of its members.

Conclusions

In the present study the presence in the Galaxy of a very large number of stellar associations is ascertained. These associations are stellar systems with a small density, unstable, and disintegrating in galactic space. Already the large role played by stellar associations in problems of stellar evolution is clear. They therefore deserve a most thorough study.

24. THE GALAXIES AS ANCHORS FOR STELLAR PROPER MOTIONS

By William Hammond Wright

Among the more important problems that invite the attention of the astronomer are those within the field of stellar dynamics; they relate to the structure of the Galaxy as a whole, and to the positions, masses, and velocities of its component stars. When one deals with positions and velocities, it is necessary first to establish a reference system—a system of coordinates if you will—with respect to which these positions and velocities can be measured. In our everyday experience we are, by force of circumstances, obliged to employ a variety of systems of reference for positions and motions: the surface of the earth, the walls of a room, the interior of a vehicle under acceleration and so on.

The astronomer is provided with no ready-made system that enables him to square himself precisely with the millions of stars, the planets, and the rapidly moving comets with which he is confronted, but has managed to use planes and lines that are determined by the stars on the assumption that these apparently slow moving objects, at least, move at random, that is to say as much in one direction as another, and this means that, taken as a whole, the system which in their aggregate they compose may be regarded as without sensible rotation. Coordinate systems based on this assumption have served very well in measuring the comparatively rapid motions of the planets and comets in their travels across the heavens, but how far may they be trusted in measuring the extraordinarily small movements of the stars themselves?

That this is a valid question will be realized when we recall that the revolutions of the planets about the sun are expressible in quantities of the order of degrees per day; the angular motions of the stars, on the other hand, are as a rule less than one second or arc per century. The ratio between these two quantities is, in round numbers, 130 million to one. It is therefore comprehensible that a framework of reference which does very well for planetary motions might not be altogether adequate for the study of the extraordinarily minute motions of the stars themselves.

Consider, for example, one of the more important problems in the field of stellar dynamics: the measurement of the rotation of the galactic system of stars. The most obvious approach to the solution of this problem is through the proper motions of the stars. But the coordinate system with respect to which these proper motions are measured has been determined on the hypothesis that the stars are moving at random, which is but a way of saying that the stellar system is without rotation. Recourse must then be had to special assumptions of one kind or another. Much ingenuity and labor have been directed toward the calculation of the galactic rotation, and there is no desire here to disparage in any way the results that have been achieved in that direction. Nevertheless it would seem desirable to approach the problem by a line of reasoning whose fundamental premise did not deny that the problem exists.

The whole question of stellar proper motions would be much simpler were it not for the precession of the equinoxes which, you will recall, imposes a conical motion of the earth's axis among the stars. This excursion of the axis has been likened to the motion of the axis of a dying top. The period of the motion is about 26,000 years. Now the unfortunate part of this is that the measurement of stellar positions is necessarily made with reference to the earth's axis and to the plane at right angles to it; that is to say the measurement is made with respect to a system of coordinates that is continually rotating with respect to the general system of the stars. As a consequence when observations of position are made at two different times, or epochs as they are customarily called, in order to determine the proper motion of a star, it is necessary to allow for the motion of the earth's axis during the interval between the two observations, especially since the amount of this motion will generally be a thousand or more times as great as the proper motion of the star itself. The amount of the precessional motion is about 50″ per annum. The cause of it is well understood, and its amount would be calculable from the figure and content of the earth, the mass of the Moon, and other astronomical data were they known with sufficient accuracy for the purpose. Unfortunately we do not know the shape and structure of the earth, nor the Moon's mass, well enough to support a calculation of the precessional motion. It has therefore been necessary to determine the precession from the stars themselves. To a logician this would appear to be begging the question, and it is. The saving grace—if in logic there be one—is that there are a great many stars in the sky. If we choose a sufficiently large value of the precession, all of the stars may, on paper, be made to revolve in one direction; too small a value will cause them to go in the opposite direction. For an intermediate value some stars will revolve one way and some the other. What is usually done is to choose a value of the precessional motion which will make as many stars seem to move one way as the other.

The precession was discovered in the year 130 B.C. by the Greek astronomer Hipparchus, who provided the first estimate of its amount; it has been re-evaluated many times since. The value now generally used is that derived by Simon Newcomb in 1898, and I venture to quote the opening paragraph of his discussion, as it closes with a note of prophecy which was soon to be fulfilled:

"In his determination of the elements of the four inner planets and the fundamental constants of astronomy the author was constrained to content himself with a provisional determination of the precessional constant, which he was afterward led to fear might prove too small. One reason for yielding to the pressure of circumstances in this connection was that the constant of precession is of such a nature that a small error in its determination will not seriously affect our general conclusions as to the positions and motions of the stars and planets, such an error being eliminated through the proper motions of the stars and the mean motions of the planets. Indeed, the fact of this elimination is one of the reasons why a satisfactory determination of the constant is difficult. There is, however, a class of researches which must come into prominence in the not distant future, to which the accurate determination of the precessional motion is a necessity. I refer to researches having for their object the determination of absolute and relative motions in the universe at large."

Some six years after the publication of Newcomb's paper, Kapteyn, from a study of proper motions in various parts of the sky, showed that the stellar motions are not, in fact, at random. He found preferential directional trends in various parts of the heavens, which he assumed to be due to the occurrence of two star streams. These preferential trends disappear of course when summed over the whole sky because the precessional constant has been so chosen as to make them do so. There is no occasion here to go into the theory of star streaming as developed by Kapteyn and his followers; the point to be emphasized is that Kapteyn's observations showed that the stars are *not* moving at random. If they are arbitrarily constrained to the hypothesis of random motion for the whole sky they break out locally. Is it logical to admit preferential motion in large areas of the sky, and at the same time to accept random motion as the fundamental hypothesis of dynamical astronomy? In any event does it not seem desirable to inquire whether some system of reference might be established which would circumvent the necessity of making any assumption whatever regarding the proper motions of the stars themselves?

In these circumstances one naturally turns to the extragalactic nebulae [galaxies] which lie far beyond the limits of our stellar system, and can probably be counted upon to be uninfluenced by the motions of the

Fig. 1. Carnegie telescope of the Lick Observatory.

stars that comprise that system. The matter has been the subject of some discussion in the literature, especially in recent years, on the part of astronomers in various parts of the world. This paper constitutes in effect an inquiry into the practicability of using the extragalactic nebulae as the material anchorages of a system of astronomical coordinates. A brief historical sketch of a current undertaking at the Lick Observatory directed toward that end may not be out of order.

I first entertained the idea of using the nebulae for the purpose just indicated in 1916, when it was coming to be realized that the spiral nebulae lie outside the Galaxy, but there was at that time no telescope at the Lick Observatory—nor in fact anywhere else—with which a research in that direction could be undertaken with hope of success. The

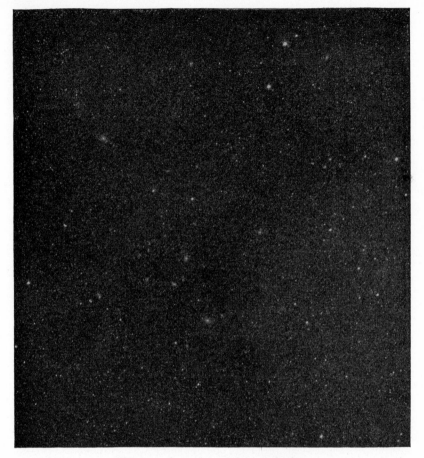

Fig. 2. A clustering of faint galaxies.

observations would require a photographic telescope, or camera, of relatively wide field, so as to include a sufficient number of catalogue
stars, as well as a lens of large linear and large angular aperture which
would record small faint nebulae in the necessary abundance. Indeed
it seemed questionable whether a lens of the requisite quality could actually be made. However, about the year 1920, Dr. Frank E. Ross designed a lens of comparatively large angular aperture which gives a
very fine field, and thereafter the problem was much in mind. In 1934
these considerations were brought to the attention of the Carnegie Corporation of New York by Dr. R. G. Aitken, then Director of the Lick
Observatory, and that organization very generously made provision for

the construction of a 20-inch (50 cm) photographic telescope and a suitable observatory in which to house it. The lens was calculated by Dr. Ross. The construction of the telescope and the development of the project have been greatly hampered by the war, and the disturbed conditions that are following it, but one tube of the telescope has been completed, and satisfactory photographs are now being taken with it. At the present time the full equipment of auxiliary instruments for the measurement of the photographs has not been completed, but enough apparatus has become available to test the various observational procedures at the critical points, and more particularly to check the effects of departures from routine observational practice that have been imposed by the unusual conditions of the problem. These critical checks are vital to the success of the undertaking, and will be referred to later in this paper. For the moment I trust that I may be permitted to pass them with the comments that they seem quite satisfactory, and to proceed with a general account of the observations.

Stated in simplest terms the undertaking includes the following procedures :

A set of photographs is taken which covers, piece by piece, all of the sky that can advantageously be observed from Mount Hamilton. After a suitable lapse of time, which will be many years, another set will be taken for comparison with the first. The future observer will take up a pair of photographs of the same region of the sky, one from the first and one from the second series. These will each show several thousand stars, of which perhaps 15 or 20 will be catalogued stars in whose motion we are especially interested. There will be, let us assume, 50 measurable nebulae on the photographs. In the interval between the taking of the plates the relative positions of the stars with respect to one another and to the nebulae will have changed. It is assumed that the nebulae, because of their enormous distances from our stellar system, will have remained substantially stationary on the celestial sphere. A correction is therefore made to all the measured positions—nebulae and stars alike—such as will cause the displacements of the nebulae to vanish. The resulting discrepancy in the position of each star, corrected where necessary for parallax, will then be regarded as its proper motion during the interval between the taking of the photographs. The question of precession of the equinoxes does not enter. . . .

In the foregoing comments I have tried to indicate the need that has arisen for a system of astronomical coordinates that is free from the assumption of random motion of the stars, and have outlined in a very informal way the theoretical basis for such a system. It is now in order to inquire whether the observations and technics suggested for realizing the proposed system of coordinates are competent; whether the plan will work. That the difficulties are serious is indicated by the fact that the proposed system, or something like it, has not actually become established. I have already taken the liberty of referring to my early interest in the problem some thirty years ago, and to the lack of suitable equipment for undertaking it. So obvious a notion as that of using the extragalactic nebulae to check the proper motions of the stars must have occurred to many persons. . . .

[From the Summary] The instruments and observational procedures employed in the work at the Lick Observatory are described as fully as seems necessary for the purposes of this paper. It appears that there are many thousands of nebulae distributed over approximately three quarters of the sky, whose positions can be measured photographically to the same order of precision as the places of stars are measured with the meridian circle. These nebulae are deemed sufficient to provide reference points for as many stellar proper motions as may be required. . . . It is anticipated that the proper motion stars will be limited to magnitudes 6.2 to 8.2, though material will be available for the calculation of the proper motions of stars between magnitudes 10.2 and 12.2.

The conclusion is drawn that the project is feasible, and will, with the collaboration of meridian astronomers, result in the establishment of a practicable system of coordinates positioned by the extragalactic nebulae.

V

THE SPECTRA OF STARS AND
NEBULAE

With the possible exception of the telescope itself, no astronomical instrument has excelled the simple spectroscope in importance for our knowledge of the material universe. It competes with the photometer, clocks, radiometers, polarimeters, and the photographic plate. In essence, astrophysics is based on the facts and suggestions from spectroscopy. Knowledge of the ages, temperatures, candle powers, motions, duplicity, and evolution of stars are derived from the spectra. The turbulence and chemical composition of nebulae are known through spectroscopy. The expanding of the universe is revealed by red shifts in the spectra of galaxies.

It was nearly a century ago that Norman Lockyer began serious exploitation of the meanings hidden in absorption lines in stellar spectra. His contemporaries in America and Europe helped in the unfolding of stellar analysis through studies of the distribution of intensity in the spectra of stars and nebulae and through their systematic observations of both absorption lines and the less common emission lines.

Various rough classifications of stellar spectra were advanced, especially at the Harvard Observatory where Professor E. C. Pickering and his assistants tried out various systems. Miss Antonia de C. Maury, with uncommon insight into the physical interpretation of stellar spectra, proposed and used a system that was later brilliantly analyzed by Hertzsprung (selection 42, below); but Miss Annie J. Cannon's work at Harvard on the Henry Draper foundation provided the practical spectral sequence, based chiefly on surface temperatures, that has become standard (selection 25). Much later W. W. Morgan and his colleagues at the Yerkes Observatory (selection 29) have usefully extended the Harvard plan, which is called the Henry Draper classification. Morgan provided what is essentially a second dimension that takes account of total luminosity (absolute magnitude) as well as temperature.

The pioneer work of W. S. Adams and Arnold Kohlschutter on the luminosity differences for stars of the same Henry Draper type led eventually to the method of getting the distances of stars from their apparent magnitudes and their spectral singularities. Their discoveries are sketchily presented in selections 26 and 27; and in the following selection is described an equally novel research wherein the mystery of the radiation from the Orion nebula and similar gaseous patches is solved. The hypothetical element Nebulium, which was held to account for the bright line radiation from diffuse nebulae, was shown to be chiefly the common elements oxygen and nitrogen behaving in uncommon ways.

In the nebulae the densities are extremely low, hence the uncommon behavior; in the collapsed white dwarfs the density on the other hand is extremely high, and again we find uncommon characteristics. As W. J. Luyten shows in selection 30, the degenerate white dwarfs, made of collapsed atoms and largely devoid of hydrogen fuel, are fairly common in our part of the universe, although few of them get into our records because their intrinsic faintness makes them hard to find and identify.

The interpretation of the stellar spectrum sequence, after the pioneer work of M. N. Saha,[1] was advanced by R. H. Fowler and E. A. Milne; unfortunately their papers are too technical for inclusion here.

[1] See introduction to part I, above.

25. PIONEERING IN THE CLASSIFICATION OF STELLAR SPECTRA

By Annie J. Cannon

The Henry Draper Memorial was established in 1886 at the Harvard College Observatory by Anna Palmer Draper, the widow of the distinguished investigator and astronomer, Dr. Henry Draper, who was Professor of Physiology and Chemistry at the University of the City of New York. In his undergraduate days, Dr. Draper became interested in photography, which was then in its infancy. A visit to Birr Castle, Parsonstown, in 1857, to see the famous six-foot reflector of the Earl of Rosse, is said to have increased his interest in astronomy, so that upon his return to America he constructed two reflecting telescopes for his private observatory at Hastings-on-Hudson. After the marriage of Dr. Draper to Anna Palmer in 1867, they resided in the summer at Dobbs Ferry, two miles distant from the Hastings Observatory. It was their custom to drive to the observatory in the evening, and so great was Mrs. Draper's interest that he never went without her. She assisted him by recording, calling out the time, and coating the glass plates. Dr. Draper made numerous photographs of the moon, of the Nebula of Orion, and was the first to photograph the absorption lines in the spectrum of a star. The spectrum of Vega, taken by him in August, 1872, showed four dark hydrogen lines. He also obtained the spectra of other bright stars, and of the Great Nebula of Orion. . . .

It will be remembered that the first photograph ever taken of a star was made at the Harvard Observatory under the direction of Professor W. C. Bond on July 17, 1850. In 1882, Professor E. C. Pickering's attention was called to the possibilities of celestial photography by Professor W. H. Pickering, and a small grant from the Rumford Fund of the American Academy enabled him to make some preliminary experiments. Two years later, by means of a larger grant from the Bache Fund of the National Academy, he procured a Voigtländer lens, 8 inches in aperture and with a focal length of about 45 inches. This telescope, as shown in Fig. 1, was mounted in such a way that polar stars are easily obtained, and no reversal is required when a star crosses

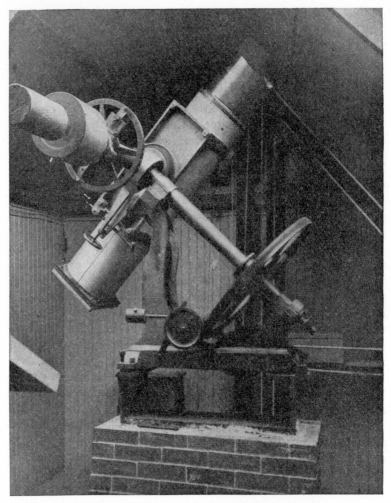

Fig. 1. The 8-inch Draper photographic camera.

the meridian. Professor Pickering commenced his photographic investigations of stellar spectra by means of a 13° prism placed before the object glass of this telescope. He had already made visual observations with a spectroscope attached to the 15-inch Equatorial and had discovered eighteen gaseous nebulae and three stars of Class *O,* often called the Wolf-Rayet type. The objective prism had been used previously by Secchi in his visual observations of the spectra of about 400 stars, and was at this time first applied to the photography of stellar spectra. It has the advantages that there is only a very small loss of light, and that spectra over the entire field are obtained. Any desired width can be given to the spectra by varying the rate of the clock. The original plan to obtain the spectra of all the brighter stars and of the fainter ones in certain regions proved too large for the Bache appropriation, but just at this critical time Mrs. Draper became interested and provided means for continuing the investigation as a memorial to her husband. With his characteristic breadth of vision, Professor Pickering laid out this great work, announcing in 1887, in the First Annual Report of the Henry Draper Memorial, that Mrs. Draper had already enlarged the scope of the work, "so that the final results shall form a complete discussion of the constitution and condition of the stars," including a catalogue of the spectra of all stars north of −24° of the sixth magnitude and brighter, with a detailed study of the brighter stars.

As soon as the photographs were obtained, the work of classifying the spectra was undertaken, and was soon placed under the charge of Mrs. Fleming. At once the greatest difficulty was encountered. How were the various kinds of stellar spectra shown on these photographs to be designated? The divisions into five types made by Secchi proved to be altogether inadequate to represent the numerous differences seen on the photographs. A new system had to be adopted which would permit the reader to understand the various aspects of the spectra as shown by the photographs. Therefore, the letters of the alphabet from A to Q were assigned to stellar spectra. This classification is purely empirical, being based wholly on the external appearances, without any idea of expressing differences of temperature or stages of evolution. Any system of designation which could assemble all similar spectra under one name was deemed to be sufficient, for, it was stated, that whenever a theory could be found to account for all the observed facts, any other name could be substituted and any other order assigned to the classes. The futility of attempting more at this early epoch is shown by the passing of Vogel's classification, in which the aim was made to explain the phase of development of each star. In the Draper classification, the letter A was assigned to spectra of [Secchi's] first type, showing the broad hydrogen lines, as in Sirius, the line K of calcium also generally being present. When other lines were seen, such as those at wave-lengths 4026 and 4471, the spectra were called B. The letters C and D were used to represent spectra of the first type, having certain peculiarities, such as double lines or bright bands, which were even then suspected to be instrumental rather than real. The letters E to L were assigned to spectra assumed to be of the second type, with the remark that Class F might be considered to be intermediate between the first and second types. The letters *M, N, O,* and *P* were given, respectively, to spectra of the third type, the fourth type, the fifth type, consisting mainly of bright lines, and the gaseous nebulae. *Q* was left for spectra so peculiar as not to be included under any of the former letters. The first classification of a large number of photographic stellar spectra was made according to this system by Mrs. Fleming and was published in Volume XXVII of the Harvard *Annals.* It was called the Draper Catalogue and contained 10,351 stars.

Realizing the importance of extending the investigations over the entire sky, in 1889 Professor Pickering sent an expedition to South America, at first in charge of Professor S. I. Bailey, and later of Professor W. H. Pickering. The 8-inch telescope was taken and, after trials at several places, was finally set up in Arequipa, Peru, where a permanent station was established. . . . The Observatory and a comfortable residence for the staff were erected at an elevation of 8,000 feet, at the bottom of an extinct volcano, El Misti, which is 19,200 feet high. Besides the 8-inch Bache Telescope, the 13-inch Boyden and 24-inch Bruce Telescopes were mounted later at Arequipa. To fill the place of the Bache Telescope in Cambridge, Mrs. Draper furnished a second, and nearly similar, instrument. . . .

While the general survey of the whole sky was made with the 8-inch telescopes, excellent spectra of the brighter stars were obtained with Dr. Draper's 11-inch photographic lens, having a focal length of 153 inches, which had been remounted in Cambridge. The brighter stars could thus be photographed with a dispersion of 8.00 cm., from $H\beta$ to $H\epsilon$, by means of four prisms nearly a foot square, placed before the object glass. Among the earliest results, it was stated in 1887 that the H line in α Cygni was found to consist of two components, and the lines $H\gamma$ and $H\delta$ were bright in o Ceti. Also that U Orionis which was at first supposed to be a Nova, had a spectrum similar to that of o Ceti, thus furnishing additional evidence that it is a variable of the same class. In 1888, photographs were taken of the spectrum of β Persei at the star's maximum and minimum light to show possible changes, but gave only negative results. These early data are interesting as marking the beginning of the extensive work of the Memorial on the spectra of the variable stars.

The detailed study of the spectra of the bright northern stars was assigned to Miss A. C. Maury, a niece of Dr. Draper. The photographs used by Miss Maury are excellent ones, taken with the 11-inch telescope, and of large dispersion, which enabled her to detect small peculiarities and make detailed studies of wave-lengths and intensities of lines. She formed 22 groups of spectra, using Roman numerals instead of letters to designate them, and calling attention to differences in the width of the lines by assigning the letters *a, b, c,* respectively, to spectra with medium, wide and narrow lines. Miss Maury's results were published in volume XXVIII, Part I, of the *Annals* [of the Harvard College Observatory] and formed a catalogue of 681 stars. One of Miss Maury's important discoveries was that the spectrum of β Lyrae changes in a remarkable manner, which, in her opinion, can be partially explained by the presence of a third body.

The revelations of these early photographs of stellar spectra were truly remarkable. It was almost as if the distant stars had really acquired speech, and were able to tell of their constitution and physical condition. Spectra of such stars as Arcturus were obtained on a scale so as to show 500 solar lines between the sodium line D and the calcium line K. These lines were compared with lines in the sun by means of Rowland's map of the solar lines. No one could do this patiently, line for line, noting in many cases the perfect agreement, and not be convinced that the distant star is a glowing body on the same order as our own luminary.

In 1888, Professor Pickering made the unique discovery that the lines in the spectrum of ζ Ursae Majoris, the familiar "Mizar," were double on one photograph, while certainly single on others. At first, it seemed possible that this doubling might be due to a photographic defect, about which he was so strenuously warned in those days. But there were the double $H\beta$, $H\gamma$, $H\delta$, $H\epsilon$, and even better seen, the fine line K of calcium and 4481, due to magnesium. Additional photographs soon confirmed the duplicity, and the first very close binary, since called spectroscopic binary, was discovered. A short time later, Miss Maury found the same peculiarity in the spectrum of β Aurigae, the second of these systems to be known. . . . Later, V Puppis was found to be a close binary by Professor Pickering, μ^1 Scorpii by Professor Bailey, ζ Centauri by Mrs. Fleming, and π Scorpii by the writer. All of these binaries consist of two stars nearly equal in brightness, revolving around a common center of mass, with rapid velocity, in periods from one to twenty days. There are now [1915] more than 300 spectroscopic binaries known, but in most of them one component is so much fainter that its lines are not visible on the photographs and its presence is revealed only by the variable velocity of the system in the line of sight.

One of the most interesting discoveries from the photographs of this memorial was made by Professor Pickering in 1897. He found that the spectrum of a second magnitude star, lettered ζ in the constellation of Puppis, contained, besides the well-known series of hydrogen lines, a second rhythmical series of absorption lines, at first supposed to be due to some substance unknown on the earth, but later assumed to be due to hydrogen under conditions unfamiliar to us. These lines were also found to be present in stars of Class O, but as emission lines. They were never produced in the laboratory until 1913, when Professor Fowler succeeded in obtaining them from a mixture of hydrogen and helium in a tube, and they are now generally believed to be due to helium.

The detailed classification of the spectra of the bright southern stars was undertaken by the writer in 1897. The results are published in Volume XXVIII, Part II, of the *Annals,* forming a catalogue of 1122 stars. The photographs for this investigation were taken at Arequipa with the 13-inch Boyden Telescope and have a dispersion somewhat greater than those taken in Cambridge with the 11-inch Draper Telescope. A modification of the system of letters used for the Draper Catalogue was adopted at this time. In various ways, some of the perplexing problems of the early days had already been solved. The stellar sequence was found to be in some respects less complex than was at first supposed.

The appearances for which some of the letters, such as C, D and E, had been assigned, were not confirmed by later and better photographs. Therefore, these letters were dropped from the sequence. In 1891, Professor Pickering wrote, "The principal question now outstanding is to determine what substance or substances cause the characteristic lines in the spectra of stars of the Orion type." This question was settled by Sir William Ramsay's discovery of helium in 1895, and the subsequent identification by Vogel of the lines characteristic of spectra of the Orion type with the new terrestrial element. Hence the so-called Orion stars, which were first known to prevail in that constellation, became helium stars. As it had been clearly proved by the Harvard Classification that these spectra precede Sirian spectra, it was necessary to place the letter B, which had been assigned to the Orion stars, before the letter A, or to change all the stars previously lettered A and B. Since several thousand had already been published, the change of the order of the letters was the only practicable course. This inversion of letters is variously regarded by astronomers as an advantage or a drawback to the system. The original letters that persisted were B, A, F, G, K, M, to represent the sequence as far as it was then established. This sequence is shown in Fig. 2. But, as was found in classifying the bright southern stars, the letter B could not stand for all the helium stars with their

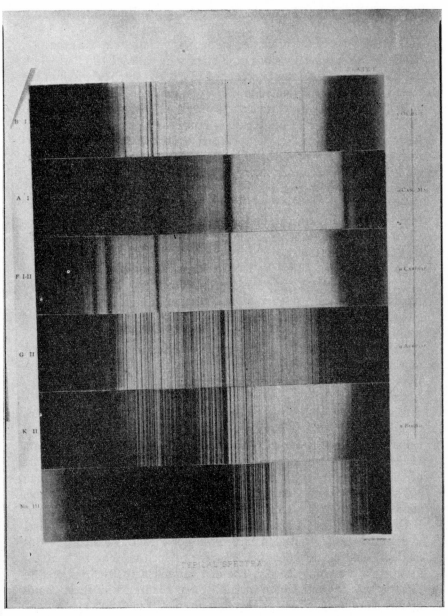

Fig. 2. Standard spectrum classes.

various intensities of lines and differences in number of lines present. Therefore, the writer adopted the plan of dividing into tenths the intervals between spectra represented by successive letters in the sequence. Thus the various subdivisions of *B* stars were called *B1A, B2A, B3A, B5A, B8A,* and *B9A,* later abbreviated to *B*1, *B*2, etc. The width of the lines was also carefully noted in this investigation, as had been done by Miss Maury for the northern stars, and remarks were used to designate those stars whose lines are certainly broad, as *a* Eridani, or certainly narrow, of division *"c,"* as *b* Puppis and ζ^1 Scorpii. Spectra having narrow lines have proved to be of unusual interest, due to the fact that stars with such spectra have great intrinsic brilliancy and generally very small proper motions. . . .

One of the most interesting results of this study of the excellent photographs of the southern stars was the subdivision of the spectra of Class *O* into a progressive series and the discovery of such spectra as that of 29 Canis Majoris which supplied the needed link between spectra of Class *O* and Class *B*. This spectrum resembles Class *O* in the presence of the "Pickering" series, first found in ζ Puppis, called, in the Harvard *Annals,* the "additional" hydrogen lines, and also resembles Class *B*0 in the helium and other absorption lines. Thus it happened again that the natural order of the alphabet must be broken, for *O* was then placed before *B* in the stellar sequence. Photographs with the objective prism have not established the position of stars of the fourth type, Class *N,* but Professor Hale's work indicates that they follow Class *M*.

Mrs. Fleming, in her regular survey of the large number of spectra photographed with the various instruments, discovered numerous peculiar and interesting objects. Among them may be mentioned 91 stars of the fifth type, whose spectra consist mainly of bright lines, 69 stars of the helium type with bright lines, 59 gaseous nebulae, more than 300 variable stars and ten new stars. The early photographs of variable stars which are bright at maximum, such as *o* Ceti, *R* Hydrae, *R* Leonis, *U* Orionis, and *R* Cassiopeiae, showed that their spectra are identical in essential points. Hence, in 1890, Professor Pickering stated that this method could be used for the discovery of such objects, without the necessity of watching the sky. The first variable actually discovered in this way was *R* Caeli, whose variability was confirmed by chart photographs, showing it to be of the seventh magnitude in October, 1889, and only the tenth magnitude in February, 1890. . . . A photograph, taken November 3, 1887, of a region in Perseus held a secret of its own for three years, for not until 1890, while making a careful examination of this plate, was it discovered by Mrs. Fleming that the spectrum of one star in this field showed bright lines. No trace of the object could be found on photographs before November, 1887, or after December, 1887. This object is called Nova Persei No. 1. This discovery was so long after the outburst that very little is known of its history, although the reality of the object was confirmed by several images on chart plates in November and December, 1887. The Draper Memorial photographs continued to yield new stars, as Nova Normae, discovered in 1893, Nova Carinae and Nova Centauri in 1895. The latter Nova had a spectrum resembling Class N, unlike any other new star so far photographed. This Nova appeared very near the nebula, N.G.C. 5253, and in this respect resembled S Andromedae, which appeared in 1885 in the Great Nebula of Andromeda. Nova Sagittarii, No. 1, appeared in 1897, and Nova Aquilae, No. 1, in 1898. Six years intervened before Nova Aquilae, No. 2, was discovered in 1905, and four years more before Nova Arae and Nova Sagittarii, No. 2, appeared in 1910. . . .

Within the last five years, an increased interest has arisen in the classification of stellar spectra. This is mainly due to the discoveries of Professor Campbell and Professor Kapteyn concerning the relation between the radial velocity of the stars and their spectral type, and of Professor Lewis Boss concerning a similar relation between proper motion and spectrum. Professor Campbell announced in 1910 that a comparison of the average motion of 17 km per second for stars of Classes G, K, and M with the motion of only 7 km per second for the B stars showed unmistakably that the radial velocities of stars are functions of their spectral types. Other interesting relations have also been found, such as a progressive increase in distance between components and in periods of binary stars, according to whether the type of spectrum is early or late.

Since the Draper Memorial collection contains photographs of stellar spectra over the entire sky, observations were commenced by the writer in October, 1911, for a New Draper Catalogue. A regular progress has been made since January, 1912, 5000 or more stars being classified each month. The classification is now completed for the northern stars and nearly as far as 18^h for the southern. The photographs were taken with the 8-inch telescopes at Cambridge and Arequipa and cover a region about 8° square. . . .

At present 199,196 spectra have been classified. About 150,000 of these are identified, the facts being copied on Library Bureau cards, which are placed in cases in the order of right ascension, where they may be consulted readily for any purpose. The spectra of more than 8,000 stars have already been sent to astronomers in England, Denmark, France, Germany, Holland and Italy, as well as in America, where they are being used for various purposes, such as color index, parallax, radial velocity, proper motion, and double star investigations. It is hoped that the observations for the New Draper Catalogue will be finished in six months, and that printing may be started soon after that time. When the classification is completed, a careful study will be made of the distribution of the various classes of stellar spectra, as a portion of the contribution of the Henry Draper Memorial to the greatest of all investigations, the constitution of the sidereal universe. The problem is so vast that we might despair of any completed result of our tasks were it not for the wonderful correlation, revealed during the last five years, among apparently disjointed investigations. That every fact is a valuable factor in the mighty whole should encourage each worker, remembering Argelander's words in 1844, when making an earnest plea for more zealous observations of the variable stars, " Each step brings us nearer the goal, and if we can not reach it, we can at least work so that posterity shall not reproach us for being idle, or say that we have not made an effort to prepare the way for them."

26. SPECTRAL PECULIARITIES RELATED
TO STELLAR LUMINOSITY

By Walter S. Adams and Arnold Kohlschütter

In the course of a study of the spectral classification of stars whose spectra have been photographed for radial velocity determinations some interesting peculiarities have been observed. The stars investigated are of two kinds: first, those of large proper motion with measured parallaxes; second, those of very small proper motion, and hence, in general, of great distance. The apparent magnitudes of the large proper motion, or nearer stars, are somewhat less on the average than those of the small proper motion stars, so that the difference in absolute magnitude must be very great between the two groups. The spectral types range from A to M.

The principal differences in the spectra of the two groups of stars are:

1. The continuous spectrum of the small proper motion stars is relatively fainter in the violet as compared with the red than is the spectrum of the large proper motion stars. The magnitude of this effect appears to depend on the spectral type, and increases with advancing type between F0 and K0.

2. The hydrogen lines are abnormally strong in a considerable number of the small proper motion stars. Thus six stars which show the well developed titanium oxide bands characteristic of Type M have hydrogen lines which would place them in types G4 to G6, and many others which show the bands strongly would be classified under type K from their hydrogen lines. That the spectra of these stars are not composite is shown by their radial velocities. The hydrogen lines in the spectra of the large proper motion stars which show the titanium oxide bands are without exception very weak.

3. Certain other spectrum lines are weak in the large proper motion stars, and strong in the small proper motion stars, and conversely. It is with the possibility of applying this fact to the determination of absolute magnitudes that the results given in this communication mainly have to deal.

I. Intensity of the Continuous Spectrum

A comparison of the intensity of the continuous spectrum of several pairs of stars of small and of large proper motion photographed upon the same plate was made recently by one of us, and showed a marked weakening relatively in the violet region for a majority of the small proper motion stars. With a view to supplementing these observations with the larger amount of material available in the radial velocity photographs we have calculated the densities at several points in the spectrum for a considerable number of these stars, and compared the resulting values for the stars of small with those of large proper motion.

The method employed, though by no means so accurate as a photometric measure of density, appeared to be as good as the character of the material would warrant. It is evident, of course, that in the case of an individual star the conditions of transparency under which the photograph was taken, the zenith distance, and some other factors might influence the result seriously. For the mean of a considerable number of photographs and stars, however, it would seem that these effects should counteract each other largely, or at least be similar for the two groups of stars under comparison. . . .

II. The Hydrogen Lines

The abnormal strength of the hydrogen lines in the spectra of certain of the small proper motion stars is of peculiar interest because of the possibility of selective absorption by hydrogen gas in interstellar space. The radial velocity affords a means of determining the origin of the additional absorption since it is highly improbable that the hydrogen in space would have the motion of the stars observed. Accordingly we have given especial attention to the determination of the radial velocities of these stars from the hydrogen lines as compared with other selected lines in the spectrum. The results obtained indicate that within the limits of error of measurement the hydrogen lines give essentially the same values as the other lines, and no differences have been found of an order to correspond to the abnormal intensity of the lines. . . .

III. The Relation of Line Intensity to Absolute Magnitude

Systematic differences of intensity for certain lines between stars of large and stars of small proper motion soon became evident in the course of the study of the spectral classification of these stars. In order to secure an accurate system of classification as well as to investigate these differences the following method was adopted. Pairs of lines were selected not far from one another in the spectrum and of as nearly possible the same intensity and character, and estimations were made of their relative intensities. For classification purposes a line decreasing in intensity with advancing type, such as a hydrogen line, was combined with a line increasing in intensity with advancing type, such as an ordinary metallic line. In addition to these pairs used for classification purposes several pairs were selected which included all lines suspected of systematic deviations in certain stars. . . .

Summary

We have found as a product of our investigations of the spectra of large and of small proper motion stars three phenomena which appear to have a distinct bearing upon the problem of the determination of the absolute magnitudes of stars.

1. The continuous spectrum of the small proper motion stars is decidedly less intense in the violet region relative to the red than the spectrum of the nearer and smaller stars. This effect appears to be a function of the spectral type, and so must be ascribed in part, at least, to conditions in the stellar atmospheres.

2. A considerable number of the small proper motion stars show hydrogen lines of abnormally great intensity. Measures of the radial velocity show the source of the additional absorption to be mainly, if not wholly, in the stars themselves.

3. Certain lines are strong in the spectra of the small proper motion stars, and others in the spectra of the large proper motion stars. The use of the relative intensities of these lines gives results for absolute magnitudes in satisfactory agreement with those derived from parallaxes and proper motions.

It seems very probable from physical considerations that the spectra of stars of quite different mass and size would differ considerably in certain respects even when the main spectral characteristics were the same. If the depth of the atmosphere for stars of similar spectral type is at all in proportion to the linear dimensions of the stars, we should expect the deeper reversing layers of the larger stars to produce certain modifications of the spectrum lines. Owing to the small scale of the stellar spectrum photographs, only the most marked changes could be distinguished, and among these the effect of the deep atmosphere upon the violet end of the spectrum should be especially prominent.

27. SPECTROSCOPIC PARALLAXES

By Walter S. Adams

Among the notable advances in astronomy during recent years there are few more striking than that made in our knowledge of the distances of the stars. From a total of about fifty parallaxes known with reasonable accuracy at the end of the nineteenth century the number has risen to about 2500 [in 1922], and additions are being made with extraordinary rapidity. In view of the fundamental importance of this constant for nearly all investigations dealing with stellar motions and the structure of our universe this progress may be regarded as one of the important contributions of the astronomers of the present day to the advancement of their science.

Three independent methods are now in regular use for the determination of stellar parallaxes, and of these two have been a development of the past few years:

1. The trigonometric method which has gained immensely in precision since the application of the photographic plate.

2. The so-called "hypothetical parallax" method which is applicable only to double stars, but which has furnished a large number of most valuable results for this class of objects.

3. The spectroscopic method which is based upon the effect of absolute magnitude on the intensities of certain spectral lines and on the distribution of light in the continuous spectrum. . . .[1]

The use of the conception of absolute magnitude which lies at the basis of the spectroscopic method of determining parallaxes has proved most fruitful in its results during recent years. If on physical grounds connected with spectral type, color, variability, or any other characteristic, conclusions may be drawn as to the intrinsic brightness of a star its distance may be determined however great it may be. In this way Shapley has derived the distances of star clusters some of which exceed 100,000 light-years, quantities which are far beyond the possibility of direct measurement.

---

[1] [Candle power and intrinsic luminosity are terms equivalent to the astronomer's "absolute magnitude."]

The principal effects of the intrinsic brightness of a star upon its spectrum appear to be two in number, and both were discovered by the Mount Wilson observers at about the same time in 1914. If a comparison is made between the spectra of two stars of nearly the same type, one of which is very bright intrinsically and the other faint, it is found that the brighter star is relatively deficient in violet light, that is, is slightly redder in color, and also that certain lines in its spectrum are either more or less intense than in the spectrum of the fainter star. Of these two effects the second is the more important as it lends itself to more definite measurement. The effect of absolute luminosity on color, however, may be recognized in the case of stars too faint to be photographed with a spectrograph of sufficient dispersion to show the lines satisfactorily. By this means Shapley determined the existence of giant stars of high luminosity among the redder stars in certain star clusters, and it seems probable that the method is capable of rather wide applications.

The method of determining the absolute magnitude of stars from the intensities of their spectral lines is purely an empirical one and is based upon stars of known parallax and absolute magnitude. A group of such stars of a given spectral type is selected and the intensities of the lines which vary with absolute magnitude are determined for each star. A curve is then constructed with absolute magnitude as one coordinate and line-intensity as the other. With the aid of this curve we may then read off the absolute magnitude of any star for which we know the intensities of the selected lines. As the method is used at Mount Wilson separate curves are constructed for the different spectral types based upon about 700 stars with parallaxes measured trigonometrically. The lines employed have been the two enhanced lines of strontium at $\lambda4077$ and $\lambda4216$ and the line at $\lambda4290$, all of which are more intense in the stars of high luminosity, and the calcium line at $\lambda4255$ which is stronger in the fainter stars. It may be stated as a general conclusion that the effect of great intrinsic luminosity is to strengthen the enhanced lines and those of hydrogen, while in stars of low luminosity the low temperature lines such as $\lambda4227$ of calcium are more prominent.

The principal results of the use of the spectroscopic method at Mount Wilson are contained in a list of the parallaxes of 1646 stars published recently. Included among these are nearly 1000 for which trigonometric parallaxes are available and a considerable portion of those for which hypothetical values have been computed. Since the methods used are so radically different it is of considerable interest to compare the results for the stars common to the three systems. This comparison is

shown in the following summary, the differences given being spectroscopic parallaxes minus trigonometric or hypothetical.

	Observatories	Number of stars	Difference
Spectroscopic parallaxes minus Trigonometric	Allegheny	296	$+0''.002$
	McCormick	209	$-0\ .001$
	Mount Wilson	79	$-0\ .003$
	Yerkes	98	$+0\ .002$
Spectroscopic parallaxes minus Hypothetical	Jackson & Furner	81	$+0\ .001$

The accordance of these results is such as to lead to the conviction
that the systematic errors present must be small in amount. A discussion by Strömberg has shown that the trigonometric parallaxes as determined at these various observatories require systematic corrections
varying from $+0''.003$ to $-0''.003$ and that the spectroscopic absolute
magnitudes should be multiplied by a factor ranging from about 0.9 to
1.1. The effect of this factor would rarely exceed $0''.002$ for parallaxes
of the average size. An important characteristic of the results is the
fact that the probable error while constant for the trigonometric values
is proportional to the amount of the parallaxes in the case of those derived by the spectroscopic method. Accordingly we have the unusual
but favorable condition present that the values which are most accurate on one system are least so upon the other, and a suitable combination of the two should yield good results over a wide range of values
of the parallax.

The applications of these results to stellar problems are far-reaching
in character. It is evident that a knowledge of the intrinsic brightness
of a sufficiently large number of representative stars will enable us to
draw important conclusions as to the distribution of stars according to
absolute magnitude and the probable order of their development. Moreover the knowledge of distance taken in conjunction with radial velocity and proper motion provides the means for deriving actual motions
in space: and these in turn furnish the material for discussions of
stream and group motion and numerous problems connected with the
dynamics of our stellar system.

28. SOLUTION OF THE MYSTERY OF NEBULIUM

By Ira S. Bowen

In the spectrum of the gaseous nebulae several very strong lines are observed that have not been reproduced in the laboratory. At first these lines were ascribed to some unknown element "nebulium." From the character of the other elements, H, He, C, N, and O, known to be present in the nebulae it was quite certain that these lines must be emitted by some element of low atomic weight. The development of present ideas concerning atomic structure, however, leaves no place for an unknown element of low atomic weight. Further, Wright's studies of the relative intensities of these lines in a large number of nebulae show that the various lines behave in such widely different ways that they can hardly all come from the same element.

All these considerations lead to the conclusion, expressed by H. N. Russell, that "it is now practically certain that they must be due not to atoms of unknown kinds but to atoms of known kinds shining under unfamiliar conditions." This unfamiliar condition Russell suggests to be low density.[1]

One type of line, which would be possible only under conditions of very low density, is that produced by an electron jump from a metastable state. A metastable state may be considered to be one from which jumps are very improbable, i.e., one whose mean life before spontaneous emission is very long. Under laboratory conditions the mean time between impacts of a given atom with other atoms or with the walls, is, even in the most extreme cases, only 1/1000 second. Consequently, an atom in a metastable state will, in general, be dropped down to a lower state by a collision of the second kind long before it will be able to return directly with the emission of radiation. In the nebulae, however, where the mean time between impacts is variously estimated at from 10^4 to 10^7 seconds, such atoms will return spontaneously and radiate a line with the frequency corresponding to the difference in energy between the metastable state and the final state. Since the probability of emission of these lines is very small, the probability of their absorption

---

[1] Russell further states the reason why new lines might be emitted in a gas of low density as follows: "This would happen, for example, if it took a relatively long time (as atomic events go) for an atom to get into the right state to emit them, and if a collision with another atom in this interval prevented the completion of the process."

must also be small. Thus these lines should not be observed in absorption.

As stated above, H, He, C, N, and O are the only elements known to exist in the nebulae. C_I, N_I, N_{II}, O_I, O_{II}, and O_{III} are the only ions of these elements that have metastable states so placed that jumps from them would give rise to lines in the region of wave-lengths that is observable in nebulae. Since the low stages of ionization of these elements are not observed in the nebulae, C_I, N_I, and O_I can at once be eliminated. In a four-valence-electron system such as N_{II} and O_{III} the normal configuration of two 2s- and two 2p-electrons is characterized by ^3P-, ^1D-, ^1S-terms. Of these, $^3P_0$ is the stable ground state while the rest are metastable since any jump from them involves zero change in the azimuthal quantum number. In a five-electron system such as O_{II} the normal configuration of two 2s- and three 2p-electrons has ^4S-, ^2D-, ^2P-terms of which ^4S is the stable state and ^2D and ^2P are metastable.

The only cases where the differences between these terms are accurately known [2] are ^1D—^1S of O_{III} and ^2D—^2P of O_{II}. These are 22,916 and 13,646 frequency units which correspond to wave-lengths of 4362.54 and 7326.2 A, respectively. Two of the strongest nebular lines are found at 4363.21 and 7325 A. The deviation between calculated and observed values corresponds to about 3 frequency units. Since the foregoing wave-lengths were calculated from the difference in frequencies of lines in the region between 500 and 800 A, this corresponds to an error of only about .01 A in these lines. The group at 7325 A should have three components with an extreme separation of about 10 A. As the only observations of this line were made with an instrument having a dispersion of 600 A per millimeter, it is not surprising that the line was not resolved.

The ^4S—^2D-group in O_{II} can be predicted roughly from the difference between the term values. As no intercombinations between quartets and doublets have been observed, this difference depends solely on the independent adjustment of the terms by series formulae and therefore is only approximate. This difference predicts a pair at 27,157 and 27,175 frequency units, or 3681.25 and 3678.81 A. The two strongest ultra-violet nebular lines are 3728.91 and 3726.16 A. In view of the uncertainties mentioned above, the agreement in position and separation are both satisfactory.

Two other nebular pairs are at 5006.84, 4958.91 and 6583.6, 6548.1 A. The frequency separation of these are 193 and 82.3 cm^{-1}, respectively, while the separation of $^3P_1$—$^3P_2$ in O_{III} is 192 cm^{-1} and in N_{II} 82.7 cm^{-1}. This quite certainly identifies these pairs as $^3P_2$—$^1D_2$ and $^3P_1$—$^1D_2$ of O_{III} and N_{II}. In the case of N_{II} a check on this identification is possible

[2] Bowen, *Physical Review*, 29, 231, 1927.

as A. Fowler and L. J. Freeman [3] have found intercombinations between singlets and triplets which enable them to fix certain of the singlet levels relative to the triplet levels. The foregoing identification of the pair of nebular lines as ^3P—^1D enables one to fix the position of ^1D relative to these same triplet levels and then to calculate accurately the frequency of the combinations between the ^1D-level and certain of the singlet levels found by Fowler and Freeman. These calculations predict that the combination with the ^1P-level of the $s^2p\cdot s$-configuration should give rise to a line at 746.98 A and with the ^1D of $s^2p\cdot d$ to a line at 582.15 A. Two strong nitrogen lines that have not previously been classified occur on plates obtained in this laboratory at 746.97 and 582.16 A.

Of the other possible jumps from the metastable state of these ions ^4S—^2P of O_{II}, $^3P_1$—^1S of O_{III}, and probably $^3P_1$—^1S of N_{II} fall below 3300 A where they cannot be observed in the nebulae. ^1D—^1S of N_{II} should occur with a wave-length somewhat greater than 5500 A where photographic observations are difficult except in the case of the strongest line. Thus all of the lines expected on this hypothesis are found, and in so doing all but two or three of the strong nebular lines are explained.

These nebular lines constitute the first direct experimental evidence for the idea, expressed above, that metastable states are not absolutely metastable but are states whose mean life is long before spontaneous radiation begins, the electron being able to return from them after the proper lapse of time even under conditions where the atom cannot be affected by surrounding atoms.

[The following table from the *Publications of the Astronomical Society of the Pacific 39*, 297 (1927)] "includes all of the lines of an easily observed wave-length which would be excited by this mechanism in N II, O II, and O III. Only two or three of the strong nebular lines are left unidentified."

TABLE I. Identification of Nebular Lines

I. A.	Source	Series Designation
7325.	OII	^2D—^2P
6583.6	NII	$^3P_2$—^1D
6548.1	NII	$^3P_1$—^1D
5006.84	OIII	$^3P_2$—^1D
4958.91	OIII	$^3P_1$—^1D
4363.21	OIII	^1D—^1S
3728.91	OII	^4S—$^2D_3$
3726.16	OII	^4S—$^2D_2$

[3] *Proceedings of the Royal Society, 114*, 662, 1927.

29. THE TWO–DIMENSIONAL CLASSIFICATION
OF STELLAR SPECTRA

By William W. Morgan, Philip C. Keenan and Edith Kellman

The *Atlas of Stellar Spectra* and the accompanying outline have been prepared from the viewpoint of the practical astronomer. Problems connected with the astrophysical interpretation of the spectral sequence are not touched on; as a consequence, emphasis is placed on "ordinary" stars. These are the stars most important statistically and the only ones suitable for large-scale investigations of galactic structure. The plan of the *Atlas* can be stated as follows:

a) To set up a classification system as precise as possible which can be extended to stars of the eighth to twelfth magnitude with good systematic accuracy. The system should be as closely correlated with color temperature (or color equivalent) as is possible. The criteria used for classification should be those which change most smoothly with color equivalent.

b) Such a system as described under (a) requires a classification according to stellar luminosity, that is, the system should be two-dimensional. We thus introduce a vertical spectral type, or luminosity class; then, for a normal star, the spectrum is uniquely located when a spectral type and a luminosity class are determined. The actual process of classification is carried out in the following manner: (1) an approximate spectral type is determined; (2) the luminosity class is determined; (3) by comparison with stars of similar luminosity an accurate spectral type is found.

As it may not be immediately apparent why an increase in accuracy in spectral classification is desirable, a short digression on some problems of stellar astronomy will be made.

The problem of stellar distribution in the most general sense does not require any spectroscopic data. Stars of all types and temperatures may be considered together, and some general features of the distribution of stars in the neighborhood of the sun can be found. For this purpose a certain frequency distribution of stellar luminosities must be assumed. This luminosity function has a large dispersion and must be varied with galactic latitude. In addition, there are certain regional fluctuations in the frequency of stars of higher luminosity in classes B, A, and M.

As a result of these considerations (and because of difficulties with interstellar absorption) the general method has very definite limitations; the large dispersion of the luminosity function means we must have a large sample, and this in itself precludes detailed analyses of limited regions. In addition, there is evidence of clustering tendencies for stars of certain spectral types—a cluster or star cloud might be well marked for stars of type A, for example, and be not at all apparent from a general analysis of star counts.

There is, then, for certain kinds of problems a great advantage in the use of spectral types of the accuracy of the *Henry Draper Catalogue*. Consider, for example, the stars of classes B8–A0 as a group. The dispersion in luminosity is far less than in the case of the general luminosity function, and the space distribution of stars of this group can be determined with a correspondingly higher accuracy. In addition, we are able to correct for systematic errors due to interstellar absorption from observations of the color excesses of these stars. We have thus gained in two particulars: we have limited at one time the dispersion in luminosity and in normal color.

The further refinement of a two-dimensional classification makes possible an even greater reduction in the dispersion in absolute magnitude for a group of stars. The mean distance of a group of stars of the same spectral type and luminosity class can be determined with great precision, even when the group consists of a relatively small number of stars. Even for individual stars distances of good accuracy can be derived. A corresponding gain is made in problems concerned with intrinsic colors and interstellar absorption.

In the fifty-five prints which make up the accompanying atlas an attempt has been made to show most of the common kinds of stellar spectra observed in stars brighter than the eighth magnitude. The dispersion selected is intermediate between that used for very faint stars, where only a few spectral features are visible, and the larger-scale slit spectra which show a multitude of details. A sufficient number of lines and bands are visible to allow an accurate classification to be made, both by temperature and by luminosity equivalent, while the relatively low dispersion makes it possible to observe bright and faint stars in a uniform manner and avoids the possibility of appreciable systematic differences in their classification.

A small one-prism spectrograph attached to the 40-inch refractor was used to obtain the plates. The reduction of collimator to camera is about 7; this makes it possible to use a fairly wide slit and still have good definition in the resulting spectra. The spectrograph was designed by Dr. Van Biesbroeck and constructed in the observatory shop by Mr. Ridell. The camera lens was constructed by J. W. Fecker, according to the design of Dr. G. W. Moffitt. The usable spectral region on ordinary blue-sensitive plates is from the neighborhood of K to $H\beta(\lambda\lambda3920-4900)$.

The dispersion used (125 A per mm at $H\gamma$) is near the lower limit for the determination of spectral types and luminosities of high accuracy. The stars of types F5–M can be classified with fair accuracy on slit spectra of lower dispersion, but there is probably a definite decrease in precision if the dispersion is reduced much below 150 A per mm.

The lowest dispersion capable of giving high accuracy for objective-prism spectra is greater; the limit is probably near 100 A per mm. The minimum dispersion with which an entirely successful two-dimensional classification on objective prism plates can be made is probably near 140 A per mm. This value was arrived at from a study of several plates of exquisite quality taken by Dr. J. Gallo, director of the Astronomical Observatory at Tacubaya, Mexico; for plates of ordinary good quality the limit is probably considerably higher.

The *Atlas* and the system it defines are to be taken as a sort of adaptation of work published at many observatories over the last fifty years. No claim is made for originality; the system and the criteria are those which have evolved from a great number of investigations. Specific references to individual investigations are, as a rule, not given.

By far the most important are those of the investigators at Harvard and Mount Wilson. The idea of a temperature classification is based on the work of Miss Maury and Miss Cannon at Harvard and of Sir Norman Lockyer. We owe to Adams the first complete investigation of luminosity effects in stellar spectra. If we add to this the work of Lindblad on cyanogen and the wings of the Balmer lines in early-type stars and the investigations of the late E. G. Williams, we have the great majority of the results on which the new classification is based. References to individual papers are given in the *Handbuch der Astrophysik*.

The present system depends, then, to a considerable extent on the work of these investigators, combined with data which were not available until recently. These data are of two kinds: accurate color equivalents for many of the brighter stars and accurate absolute magnitudes for a number of the same stars. These results have been used to define the system of classification more precisely, both in the temperature equivalents and the luminosity class. The most important of the determinations of color equivalents for this purpose are the photoelectric colors of Bottlinger and of Stebbins and his collaborators and the spectrophotometric results of the Greenwich Observatory and those of Hall.

The absolute magnitudes used depend on a variety of investigations. There are the classical catalogue of trigonometric parallaxes of Schlesinger; the catalogue of dynamical paralaxes of Russell and Miss Moore; various cluster parallaxes, principally due to Trumpler; and, in the case of the stars of earlier class, parallaxes from interstellar line intensities and from the effects of galactic rotation.

Throughout the discussion emphasis will be laid on the "normal" stars. A number of peculiar objects are noted; but the main aim of the investigation has been to make the classification of the more frequent, normal stars as precise as possible for the use of the general stellar astronomer. This investigation is not concerned with the astrophysical aspects of stellar spectra or with the spectra of the dwarfs of low luminosity. Relatively few of the latter are met with among stars brighter than the eighth magnitude, and their classification can be considered as a separate problem.

There appears to be, in a sense, a sort of indefiniteness connected with the determination of spectral type and luminosity from a simple inspection of a spectrogram. Nothing is measured; no quantitative value is put on any spectral feature. This indefiniteness is, however, only apparent. The observer makes his classification from a variety of considerations—the relative intensity of certain pairs of lines, the extension of the wings of the hydrogen lines, the intensity of a band—even a characteristic irregularity of a number of blended features in a certain spectral region. To make a quantitative measure of these diverse criteria is a difficult and unnecessary undertaking. In essence the process of classification is in recognizing similarities in the spectrogram being classified to certain standard spectra.

It is not necessary to make cephalic measures to identify a human face with certainty or to establish the race to which it belongs; a careful inspection integrates all features in a manner difficult to analyze by measures. The observer himself is not always conscious of all the bases for his conclusion. The operation of spectral classification is similar. The observer must use good judgment as to the definiteness with which the identification can be made from the features available; but good judgment is necessary in any case, whether the decision is made from the general appearance or from more objective measures.

The problem of a classification according to luminosity is a difficult one. In the first place, lines or blends which may be useful at one spectral type may be quite insensitive at another. In fact, some lines which show a positive absolute-magnitude effect for some spectral classes may show a negative one for others. This is true for certain lines of H, Sr II, and Ba II.

Besides the variation with spectral type, there is also a very marked change in appearance with the dispersion of the spectrograms used. Some of the most useful indicators of absolute magnitude are lines and blends which can be used only with low dispersion. The hydrogen lines, for example, show marked variations with absolute magnitude in spectra as early as B2 and B3 on plates of low dispersion; with higher dispersion the wings which contribute to the absolute-magnitude effect are not apparent to the eye, and the lines look about the same in giants and dwarfs. In stars of classes G2–K2 the intensity of the CN bands in the neighborhood of $\lambda 4200$ is one of the most important indicators of absolute magnitude. The band absorption has a different appearance on spectrograms of high and low dispersion, and it is doubtful whether high-dispersion plates show the luminosity effects of CN as well as those of low dispersion.

On the other hand, a considerable number of sensitive line ratios are availiable on high-dispersion spectra which cannot be used with lower dispersion. One of the most sensitive lines to absolute-magnitude differences for the F8–M stars is Ba 4554; this line is too weak to be observed on low dispersion spectra. A number of the other ratios found by Adams to be sensitive indicators of absolute magnitude are also too weak to be used with low dispersion.

These considerations show that it is impossible to give definite numerical values for line ratios to define luminosity classes. It is not possible even to adopt certain criteria as standard, since different criteria may have to be used with different dispersion. In the *Atlas* some of the most useful features for luminosity classification have been indicated, but it should be emphasized that each dispersion has its own problems, and the investigator must find the features which suit his own dispersion best.

The luminosity classes are designated by Roman numerals; stars of class I are the supergiants, while those of class V are, in general, the main sequence. In the case of the B stars the main sequence is defined by stars of classes IV and V. For the stars of types F–K, class IV represents the subgiants and class III the normal giants. Stars of class II are intermediate in luminosity between the supergiants and ordinary giants.

30. WHITE DWARFS AND DEGENERACY

By W. J. Luyten

The discovery by Adams at Mount Wilson in 1915 that the faint star revolving around Sirius, the brightest star in the sky, was *white* not only caused a minor revolution in astronomical thinking but it forged another important link in the chain of events which eventually led to the atomic bomb.

In a sense this faint star had been discovered before it had been *seen* for its existence had been proved by Bessel from the gravitational pull it exerts upon the bright star in whose motion in the sky it produces small oscillations. By 1915 the paths in which the two stars revolved around each other were well known, also the distance of the pair from us—some 8½ light-years—and from this information we could, in turn, derive the masses of the two stars, and the luminosity, i.e., the total amount of light they send into space. Sirius itself proved to be a very luminous white star, much heavier than the sun, but quite in line with all other known white stars in the sky.

The little companion turned out to give only 1/400 part of the light of the sun, and to have a mass, in other words, to contain a total amount of matter, very nearly equal to that of the sun. Taken by themselves these figures appeared little out of ordinary, for we know literally thousands of stars, called dwarfs, with masses smaller than that of the sun, and with luminosities anywhere down to one ten-thousandth of that of the sun. But—these dwarfs are *red* and when Adams found the companion of Sirius to be *white* things began to hum. For if a star is whiter than the sun, it must also be hotter, and this, in turn, means that it gives more light per square inch than the sun does. Now if the companion to Sirius gives three times as much light per square inch as the sun does, and yet gives a total light 400 times less, then it must have a total surface 1200 times smaller than that of the sun for the same reason that if a patch of dazzlingly white snow gives much less light than a gray ledge of rock, it must be very much smaller in area. But if the surface is 1200 times smaller, the diameter must be 35 times less and the volume 40,000 times less. And, finally, if in this small volume there is packed nearly as much substance as there is in the sun—the mass is nearly the same—then this star must be unbelievably dense: a cubic inch must "weigh" 2500 times as much as gold, one *ton* per cubic inch.

Even astronomers, used though they are to large figures, could not accept such a conclusion, right away; the figures were too staggering to be believed without further evidence. Could there be an error somewhere? What had happened? But while we pondered these things, two more stars of the same kind were found; small, extremely faint, and white and all three were named "White Dwarfs." Then, a few years later came the answer to the riddle, and, as usual in science, the explanation was so simple that astronomers began to ask themselves why they hadn't thought of it before. It was the genius of Eddington that solved our problem.

To understand fully Eddington's solution we must make a little excursion into physics and consider the structure of matter. By delving into the secret of the atom, physicists had come to the conclusion that atoms, the smallest particles of matter that still possess a chemical identity, are not themselves really indivisible, but are, in turn, made up of very much smaller particles revolving around each other, one might almost say a miniature solar system. In the center is the nucleus, an extremely tiny particle, one millionth of one millionth of an inch in size, that carries a positive electric charge, and around this revolve a number of particles with negative charges, the electrons, about the same size as this nucleus. Hydrogen, the simplest of all substances, has just one electron, helium has two, and the more complex substances like the heavy metals may have up to 90 or more, arranged in definite layers. Under ordinary conditions these electrons revolve around the nucleus in paths that are about one hundred-millionth of an inch in size, but when it gets very hot, then the atoms begin to fly about so fast that eventually the electrons are all knocked off and the whole structure of the atom collapses. Where before the "size" of the atom was the size of the orbit of the outermost electron, now, when all the electrons have gone off into space, the "size" of the atom becomes merely the size of the particles themselves. Sir Oliver Lodge has so well put it: normally, "matter" can be compared to a lot of flies buzzing in a cathedral but when the temperature goes up—and inside a star it may reach hundreds of millions of degrees—then the cathedral walls collapse, and all we have left is the flies. And, although we can build only a few cathedrals in a large city-square, we could obviously pack an enormously large number of flies in the same space. Hence, the terrific density of stars where the atoms have been "stripped" of their electrons. We can even calculate that those densities might go up to not merely a thousand times, but nearly one billion times that of gold.

Meanwhile, as the theoretical astronomers were groping for the solution, the observers were busy trying to discover more white dwarfs— but without much luck. One more was found in 1922 but we weren't even sure of that one for nearly 15 years and about the only thing we were sure of was that most of these white dwarfs were so faint that we couldn't reach them with the then existing spectroscopic equipment. Another conclusion was that if they are such feebly luminous stars then we should search for them especially among those stars that *appear* faint, and for which we have strong suspicion that they are also *near*, since the combination of these two characteristics point to a low luminosity. Now to determine, by actual measurement, the distance of a star is an extremely tedious and difficult job, but fortunately we don't really need to do that. All that we need is to get some indication that a star is *probably* near and that is made evident by observing the motion of a star across the sky. Any star that appears to move fast must either have a terrific speed in miles per second or it must be near us in space, and since stars of the first kind are excessively rare, we need only to find *faint* stars that have a large *proper motion:* among those we shall have the best chance of finding white dwarfs.

Partly with this search in view the writer began, in 1927, his general survey of the whole sky, looking for stars of large proper motion. First at Harvard, then at the Harvard Southern Station, at Bloemfontein, South Africa, and later at Minnesota, plates taken with the Harvard 24-inch Bruce telescope were examined in the blink microscope and among some 60,000,000 stars examined about 100,000 stars of appreciable proper motion were found, including some 3,000 stars with large proper motion. Using the army language, this completed the "screening test": these 3,000 stars must almost all be nearby stars of low luminosity, thus constituting the largest potential source of white dwarfs. All that remained was to determine whether these stars were *red* or *white;* the great majority belong to the first group; they are ordinary red dwarfs, smaller, cooler, less luminous, and somewhat more dense than the sun, but on the whole rather comparable to it; while in the second group we would certainly find the white dwarfs—the end of the long trail.

Before arrangements could be completed to observe the colors or spectra of these 3,000 stars of large proper motion, more powerful spectroscopes became available on the 36-inch telescope at Lick, the 100-inch at Mount Wilson, and the 82-inch McDonald telescope. When this equipment was used on the proper motion stars already published, principally by Ross and Wolf, a number of new white dwarfs were identified among them, chiefly by Kuiper.

The systematic color program finally was inaugurated in 1939 with the 36-inch Steward reflector at Tucson, with the active aid, first of Dr. E. F. Carpenter, and later of Dr. P. D. Jose. Since the latitude of Tucson is 32° N, no star farther south than declination −48° could be observed from there, and the aid of some observatory in the southern hemisphere had to be enlisted Here we were fortunate to secure the enthusiastic cooperation of Dr. Martin Dartayet of the Cordoba Observatory, Argentina, and in the spring of 1945 we could say, for the first time, that the white-dwarf survey among stars of large proper motion had been completed insofar as the stars with motions larger than 0″.5 annually were concerned. In the meantime many stars with smaller proper motions had also been observed and at the time of writing about 80 white dwarfs, in all, have become known, 48 of them having been found at the University of Minnesota. It must be added here that very recently Zwicky, at the California Institute of Technology, has inaugurated a new and very ingenious method of discovering white dwarfs. Instead of laboriously finding the proper motions first, Zwicky takes plates in blue and yellow light in areas where there is either heavy obscuration, or where, because of high galactic latitude no very distant stars are expected to show. If, in either case, a faint white star is found on the plates the probability is great that it is near, and a white dwarf.

Among those 80 white dwarfs 14 are components of binaries; these will ultimately become of great importance since it is only in such cases that we can determine the actual masses, and hence the real densities. We shall probably have to wait a long time, perhaps decades, for this information since the orbital motion of these stars is slow and cannot be quickly determined with any accuracy.

Perhaps the most interesting among these is a binary where both components are white dwarfs, the only one of this type thus far known, a faint star situated in the constellation Antlia, just barely visible from our northern latitudes. It appears to be of the fourteenth magnitude and has a motion of about one-third of a second of arc per year (it would take just 5,000 years to move as much as the diameter of the full moon). Its components are separated by approximately three seconds of arc and appear to revolve around each other in 200–300 years. The story of the discovery of this binary shows again how cooperation is the life-blood of all scientific research: we first found the proper motion of the star on plates taken at Harvard; from plates taken at the Steward

Observatory at Tucson we suspected it to be white and double; from Mount Wilson these suspicions were confirmed; and finally its orbital motion and spectrum were determined from McDonald plates.

With 80 white dwarfs known we may begin to analyze their properties statistically and determine what these stars are like, how they behave, and, incidentally, use them as a crucial test of two divergent, fundamentally different conceptions of the structure of matter. Thus, Eddington, continuing his theoretical work, predicted that the smaller a white dwarf was, the more massive it would be, whereas Chandrasekhar holds that there should not be a great range in their masses. If Eddington is right, the very white dwarfs would be the smallest. They should also be the most massive, and hence we could expect them to have the slowest motions. While some extremely white or perhaps one should say "blue" white dwarfs have been announced by Kuiper, on the basis of their spectra, our own observations of color at Tucson do not substantiate this, since the color indices of these stars—which appear to be more reliable indications of the star's whiteness than the very small-dispersion spectra—are not particularly white; and so we cannot as yet draw any definite conclusion as to whether Eddington or Chandrasekhar is right, though the evidence seems to favor the latter's point of view.

By and large the white dwarfs appear to move in much the same way as ordinary stars of large proper motion—which was to be expected since they were found mostly among these very same large-proper-motion stars. Just how large a fraction of all stars in space they make up is not yet determined definitely; first indications were of the order of one per cent but it now appears probable that we shall have to revise this estimate considerably upwards, and increase it probably to at least three or four per cent. Another factor which may plague our interpretations is the fact that whereas we started out thinking that a dwarf star had to be either "red" or "white" we have now definitely established the existence of "intermediate" stars—whiter than ordinary stars of the same luminosity but yellower than the white dwarfs. Thus, while attempting to solve the mystery of the white dwarfs, another problem has been uncovered, that of the "intermediates." In science, this is the usual course of events, and thus it always should be, for in science we never come to the end of the trail. There are always more worlds to conquer, always more frontiers to cross.

[A later survey of White Dwarfs by Dr. Luyten appears in *Vistas in Astronomy 2*, 1048–1056 (1956).]

31. IRON IN THE STARS [1]

By Paul W. Merrill

As long ago as 2500 B.C. iron was probably known to the Egyptians; by the time of Christ it was in common use. Its introduction into warfare caused concern perhaps similar to that recently aroused by the use of the atomic bomb. Pliny the Elder, A.D. 23–79, wrote, "It is by the aid of iron that we construct houses, cleave rocks, and perform so many other useful offices of life. But it is with iron also that wars, murders, and robberies are effected, and this, not only hand to hand, but from a distance even, by the aid of weapons and winged weapons, now launched from engines, now hurled by the human arm, and now furnished with feathery wings. This last I regard as the most criminal artifice that has been devised by the human mind; for as if to bring death upon man with still greater rapidity, we have given wings to iron and taught it to fly." Curiously enough, Pliny was somewhat consoled by the knowledge that iron is subject to rapid corrosion. "Nature, in conformity with her usual benevolence," he wrote, "has limited the power of iron by inflicting upon it the punishment of rust; and has thus displayed her usual foresight in rendering nothing in existence more perishable than the substance which brings the greatest dangers upon perishable mortality."

During the past century man's ever increasing utilization of the properties of iron atoms has extended into investigations of the heavenly bodies. Iron is one of the three most important elements in the physical study of the stars, the other two being hydrogen and calcium. Helium perhaps would come fourth although carbon, oxygen, titanium and others also are of great consequence.

Although our knowledge of cosmic chemistry has been gained largely through the power of the spectroscope, some important items have been derived by other means. Among all the celestial bodies, meteorites are the only ones of which samples can be examined in the laboratory for chemical and isotopic composition. In these objects from interplanetary—if not from interstellar—space, iron is the most important constituent. Alloyed with nickel and with smaller amounts of cobalt and

[1] [Interstellar iron has been detected by Theodore Dunham and Walter S. Adams *Pub. Ast. Soc. Pac. 53,* 341 (1941).]

copper, it forms from 70 to 94 per cent of the so-called iron meteorites; it is also abundant in "stony" meteorites.

Iron is one of the numerous elements whose atoms do not all have the same atomic weight. Most iron atoms have 56 units of atomic mass, but about 6 per cent have 54 units; smaller percentages have 57 or 58 units. Although these isotopes have practically the same chemical and spectroscopic properties, they are distinguishable in the mass spectrometer. An important question bearing on the nature and origin of the chemical elements is this: Is the proportion of the isotopes of any element the same in various heavenly bodies as in the crust of the earth? Extensive investigations with the mass spectrometer show that the distribution of iron isotopes in meteorites is very closely the same as in terrestrial iron. The facts establish a reasonable presumption that the relative abundance of the stable isotopes of iron is uniform throughout the solar system; and it may well be uniform throughout the stellar system although of this we have no direct evidence.

It is only when cosmic matter is in gaseous form that the spectroscope can analyze it with ease and certainty. This is not a severe limitation, however, because a large part of the matter in the universe is gaseous. It is true that a very small fraction of the whole mass is subject to our inspection, but these bits—interstellar gas, nebulae, and the outer layers of stars—are probably pretty good samples of the whole. Thus our present extensive information concerning the distribution and behavior of iron throughout the universe comes from the photographed spectra of the sun, stars, and other objects. The interpretation of astronomical spectra is based upon detailed laboratory knowledge of the light emitted by iron and other elements when in gaseous form. This knowledge began to accumulate about 1861 when Kirchhoff recorded definite data concerning the intricate line spectra of iron and other metallic elements. Since then, through the efforts of a multitude of physicists and astronomers, the wave lengths of thousands of iron lines have become known to an accuracy of one part in several million; and thanks to the quantum theory the relationships of the lines to the various electronic transitions within the atom are now classified according to an elaborate but extremely useful scheme.

Each chemical element has a distinctive atomic number equal to the number of units of electric charge carried by the atomic nucleus. For hydrogen, the lightest element, the atomic charge is 1; for uranium, the heaviest element, 92; for iron, 26. All the atoms prefer to be electrically neutral; hence they normally attract and retain in their outer structure just enough negative electrons to balance the positive charge of the nucleus. The iron atom, for example, regularly has 26 negative electrons.

When an electron is detached, the permissible energy levels in the atom undergo rearrangement, and a quite different line spectrum results. Thus the iron atom has altogether 26 different sets of lines corresponding to atoms in which 0 to 25 electrons are missing. The bare nucleus (26 electrons missing) does not produce line spectra. The designations of the various sets of lines are as follows:

Electrons Present	Electrons Missing	Spectrum
26	0	Fe I
25	1	Fe II
24	2	Fe III
23	3	Fe IV
.	.	.
.	.	.
.	.	.

The complexity of the entire line spectrum of iron, once it is understood and reduced to an orderly system, offers remarkable opportunities for interpreting conditions prevailing in stellar atmospheres.

The fundamental physical classification of stars—a sequence of surface temperature—was originally based largely on the behavior in stellar spectra of the dark lines of hydrogen and calcium. Because of their great intensity, these lines are especially suitable for classifying spectra on small-dispersion objective-prism photographs. On spectrograms of higher dispersion, iron lines would serve well throughout the sequence except near the high-temperature end where helium lines supply the chief criteria.

In spectra of the coolest stars, lines arising from the normal undisturbed "ground" state of the neutral atom, *Fe* I, are stronger than other iron lines; as the temperature rises in passing to hotter stars, lines from higher energy levels exhibit gradually increasing relative intensities. At a certain stage, however, ionization sets in, and lines of the ionized atom, *Fe* II, begin to appear, while all neutral lines become progressively weaker. After a relatively brief interval the singly ionized lines in turn weaken and disappear as double ionization begins. Lines of doubly ionized iron, *Fe* III, never become conspicuous in typical stellar spectra, partly because of the increasing opacity of the highly ionized reversing layers. No dark lines of *Fe* IV, *Fe* V, or of higher states of ionization have been detected; this is because the strong lines retreat into the far ultraviolet where the opacity of the earth's atmosphere prevents our observing them.

The chasms between the stars are so scarcely populated by atoms that collisions are infrequent, and the undisturbed atoms find their encircling electrons all settling down in the lowest energy levels for the same general reason that drops of water run down hill and eventually come to rest in the ocean. The atomic process is much the more rapid, however, requiring only a split second—a second split into about 100,000,000 parts.

In previous paragraphs we have been considering *dark* lines formed by atoms which lie between the observer and a stellar surface (photosphere) which sends out a luminous background of continuous spectrum. The *bright* lines emitted by the atoms themselves are sometimes observable in the light from stars. Bright hydrogen lines are common, and bright iron lines are probably next in frequency.

Bright lines of neutral iron, notably λ4202 and λ4308, are characteristic post-maximum features of the spectra of red long-period variables of class Me such as o Ceti and R Leonis. Bright lines of ionized iron are conspicuous in the solar chromosphere and prominences and in many hot B-type stars with extensive incandescent atmospheres. Some very peculiar and interesting objects have been found among the emission-line B stars. One of these, XX Ophiuchi, was called "the iron star" because in 1921–1923 the most noteworthy features of its spectrum were the bright lines of ionized iron. As if in rebuke for this gratuitous appellation, the star in 1925 changed its spectrum largely into one of absorption, lines of ionized *titanium* becoming stronger and more numerous than those of iron. In later years, however, it has resumed the earlier characteristics.

The circumstances concerning a special class of lines called "forbidden," indicated by square brackets, e.g., [Fe II], are of special interest. Strongly forbidden lines are emitted in appreciable intensity only by atoms that are left undisturbed—free from collisions—for a relatively long time. This means that if the atoms have ordinary kinetic velocities they must be very far apart, or that in any given volume there must be very few atoms. Hence the intensity of light emitted per unit volume will be low, and an enormous volume may be required to emit lines strong enough to be observed. Here astrophysics has the advantage over laboratory spectroscopy; columns of gas thousands or millions of miles long are available for experiment. Forbidden lines became of great importance in astrophysics when Bowen, about 1927, identified the chief lines in the spectra of gaseous nebulae with forbidden lines of oxygen and nitrogen. Bright lines of [Fe II] and [Fe III] are now well known in many celestial objects; moreover, forbidden lines of several higher states of ionization are becoming of increasing astrophysical importance.

The remarkable southern variable star η Carinae—an old nova—has made a contribution to our knowledge of the spectrum of iron. Its spectrum, photographed by the Harvard observers near Arequipa, Peru, by Sir David Gill at the Cape of Good Hope, and in more detail in 1912 by Moore and Sanford at the Chile station of the Lick Observatory, is a most unusual one consisting chiefly of numerous emission lines. Many of these lines were readily ascribed to hydrogen, ionized iron, and other elements, but a number of strong lines remained unidentified. In 1928 it was found [by Dr. Merrill] that these were forbidden lines of ionized iron [*Fe* II]. They have not yet been produced in the laboratory, but their wave lengths measured in η Carinae and other Be stars fit so precisely into the pattern of the energy levels of the singly ionized iron atom that one cannot doubt the identification. The reasoning is illustrated by the following crude analogy. Suppose that you discover lying in the street a pile of small boards of odd shapes, no two alike. Taking them into a nearby house, you find that every one fits snugly into a hole in the floor, and that no holes are left unfilled. You would conclude that you knew where the boards came from.

Bright lines observed in the spectum of the variable star RY Scuti (the strongest line being $\lambda4658$) were identified some years ago by Edlén and Swings as forbidden lines, [*Fe* III], of the *doubly* ionized iron atom. The knowledge of the energy levels of the ion necessary for this identification was built up from laboratory measurements of lines in the extreme ultraviolet observable only with the vacuum spectrograph.

As the ionization of iron proceeds, more and more energy is required to detach each successive electron. In the "hurly-burly" of stellar interiors atoms are badly mangled. In observable stellar atmospheres, however, iron atoms with more than two electrons missing are rare, but they have been detected in a few stars. Lines of [*Fe* V], [*Fe* VI], [*Fe* VII] are all well marked in the spectra of so-called symbiotic stars such as AX Persei and CI Cygni. These objects present a strange and as yet mysterious combination of features of high and of low excitation.

Interest in the forbidden lines of even more highly ionized atoms was given a tremendous impetus not long ago by the results of Grotrian and Edlén on the probable identification of the lines of the solar corona [1]—a tantalizing problem which had baffled astronomers for many years.

---

[1] [Edlén's epochal research is described by him in Selection 8 above.]

These forbidden lines have not been found in laboratory spectra but their wave lengths can be computed—with various degrees of accuracy from analysis of the "permitted" lines that have been observed.

Lines of [*Fe* X], [XI], [XIII], [XIV], and [XV] are believed to correspond to lines in the solar corona. The most important are the famous green line λ5303 [*Fe* XIV], and a strong red line λ6374 [*Fe* X] both of which have been detected in two stars, RS Ophiuchi and T Pyxidis. Astronomers are much concerned to understand how the solar atmosphere finds sufficient energy to detach so many electrons from the atom. Several tentative suggestions have been made concerning the physical process at work, but no hypothesis has yet found general acceptance.

The foregoing account of iron in the stars should serve to emphasize the close connection between laboratory and astronomical science and thus to illustrate the underlying unity of all knowledge.

VI

VARIABLE STARS

J udged in terms of the number of workers and the number of publications, the subject of variable stars leads all other fields of astronomy. This is not only because amateur astronomers and observatories with small equipment can usefully contribute valuable observations of the light changes in the stars, but because stellar variation, whether of light, speed, position, or spectrum, is our gateway to stellar evolution, both in observation and in theory. Without straining we could treble the length of this part of the Source Book, but much of the observational labor that could be reported really consists in piling measures on measures, improving the accuracy of the light curves of periodic variables, finding the next decimal place in the length of the period, or elucidating the radial velocity variations. The work is valuable but essentially routine.

There is not much to say about the immense variable star catalogues of this half century except that they exist. The largest at present is maintained for the International Astronomical Union. It is located in Moscow, in the charge of B. V. Kukarkin and P. P. Parenago. Recording all the discoveries (now in the tens of thousands), and the improvements in positions, periods, and magnitudes when reported, is a major labor.

In the following chapters we do not report on the programs of variable star discovery in America, Russia, Germany, and Holland, or on the development of eclipsing star theories that make it possible to go from light curve to full information about such double stars. The work has not been described by the investigators in a way suitable for this volume. It must suffice here to point to theoretical and computational investigations by Alexander Roberts, H. N. Russell, Harlow Shapley, Sergei Gaposchkin, F. B. Wood, Z. Kopal, M. B. Shapley, and J. E. Merrill. Some of the future needs of the eclipsing binary theory, however, are sketched in selection 36 by Kopal; and the spectral and velocity variations of cepheid variables are well accounted for in a large research by A. H. Joy (selection 34).

In the further study of stellar variation as a guide to stellar evolution we have a serious photometric and spectroscopic responsibility, and also the promise of rich returns.

32. DISCOVERY OF THE PERIOD–MAGNITUDE RELATION [1]

By Henrietta S. Leavitt

A Catalogue of 1777 variable stars in the two Magellanic Clouds is given in *Harvard Annals 60*, No. 4. The measurement and discussion of these objects present problems of unusual difficulty, on account of the large area covered by the two regions, the extremely crowded distribution of the stars contained in them, the faintness of the variables, and the shortness of their periods. As many of them never become brighter than the fifteenth magnitude, while very few exceed the thirteenth magnitude at maximum, long exposures are necessary, and the number of available photographs is small. The determination of absolute magnitudes for widely separated sequences of comparison stars of this degree of faintness may not be satisfactorily completed for some time to come. With the adoption of an absolute scale of magnitudes for stars in the North Polar Sequence, however, the way is open for such a determination.

Fifty-nine of the variables in the Small Magellanic Cloud were measured in 1904, using a provisional scale of magnitudes, and the periods of seventeen of them were published in *Harvard Annals 60,* No. 4, Table VI. They resemble the variables found in globular clusters, diminishing slowly in brightness, remaining near minimum for the greater part of the time, and increasing very rapidly to a brief maximum. Table I gives all the periods which have been determined thus far, 25 in number, arranged in the order of their length. The columns contain the Harvard Number, the brightness at maximum and at minimum as read from the light curve . . . and the length of the period expressed in days. . . . A remarkable relation between the brightness of these variables and the length of their periods will be noticed. In *Harvard Annals 60,* No. 4, attention was called to the fact that the brighter variables have the longer periods, but at that time it was felt that the

<hr>

[1] For measurement of great distances, the period-luminosity relation was revolutionary. It was developed from Miss Leavitt's discovery. The period of a cepheid variable is an indicator of its absolute magnitude and a clue to its distance.

From TABLE I. Periods of Variable Stars in the Small Magellanic Clouds

H. V.	Max.	Min.	Period	H. V.	Max.	Min.	Period
1505	14^m.8	16^m.1	1^d.25336	1400	14^m.1	14^m.8	6^d.650
1436	14.8	16.4	1.6637	1355	14.0	14.8	7.483
1446	14.8	16.4	1.7620	1374	13.9	15.2	8.397
1506	15.1	16.3	1.87520	818	13.6	14.7	10.336
1413	14.7	15.6	2.17352	1610	13.4	14.6	11.645
1460	14.4	15.7	2.913	1365	13.8	14.8	12.417
1422	14.7	15.9	3.501	1351	13.4	14.4	13.08
842	14.6	16.1	4.2897	827	13.4	14.3	13.47
1425	14.3	15.3	4.547	822	13.0	14.6	16.75
1742	14.3	15.5	4.9866	823	12.2	14.1	31.94
1646	14.4	15.4	5.311	824	11.4	12.8	65.8
1649	14.3	15.2	5.323	821	11.2	12.1	127.0
1492	13.8	14.8	6.2926				

number was too small to warrant the drawing of general conclusions. The periods of 8 additional variables which have been determined since that time, however, conform to the same law.

The relation is shown graphically in Figure 1, in which the abscissas are equal to the periods, expressed in days, and the ordinates are equal

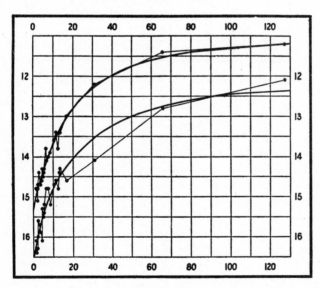

Fig. 1. The period-magnitude relationship for 25 cepheids in the Small Magellanic Cloud.

to the corresponding magnitudes at maxima and at minima. The two resulting curves, one for maxima and one for minima, are surprisingly smooth, and of remarkable form. In Figure 2, the abscissas are equal to the logarithms of the periods, and the ordinates to the corresponding magnitudes, as in Figure 1. A straight line can readily be drawn among each of the two series of points corresponding to maxima and minima, thus showing that there is a simple relation between the brightness of the variables and their periods. The logarithm of the period increases

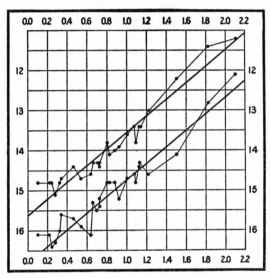

Fig. 2. Magnitudes and logarithms of the periods.

by about 0.48 for each increase of one magnitude in brightness. . . . It should be noticed that the average range, for bright and faint variables alike, is about 1.2 magnitudes. Since the variables are probably at nearly the same distance from the Earth, their periods are apparently associated with their actual emission of light, as determined by their mass, density, and surface brightness.

The faintness of the variables in the Magellanic Clouds seems to preclude the study of their spectra, with our present facilities. A number of brighter variables have similar light curves, as UY Cygni, and should repay careful study. The class of spectrum ought to be determined for as many such objects as possible. It is to be hoped, also, that the parallaxes of some variables of this type may be measured. Two fundamental questions upon which light may be thrown by such inquiries are whether there are definite limits to the mass of variable stars of the cluster type, and if the spectra of such variables having long periods differ from those of variables whose periods are short.

The facts known with regard to these 25 variables suggest many other questions with regard to distribution, relations to star clusters and nebulae, differences in the forms of the light curves, and the extreme range of the length of the periods. It is hoped that a systematic study of the light changes of all the variables, nearly two thousand in number, in the two Magellanic Clouds may soon be undertaken at this Observatory.

33. THE PULSATION HYPOTHESIS

By Harlow Shapley

INTRODUCTION

The purpose of the present discussion is an attempt to investigate the question of whether or not we should abandon the usually accepted double-star interpretation of Cepheid variation. In addition to the brief statement of some general considerations and correlations of the many well known characteristics of Cepheid and cluster variables, certain recently discovered properties of these stars are discussed in greater detail, because chiefly upon them are based the conclusions reached in this study.

It seems a misfortune, perhaps, for the progress of research on the causes of light-variation of the Cepheid type, that the oscillations of the spectral lines in nearly every case can be so readily attributed, by means of the Doppler principle, to elliptical motion in a binary system. The natural conclusion that all Cepheid variables are spectroscopic binaries has been the controlling and fundamental assumption in all the recently attempted interpretations of their light-variability, and the possibility of intrinsic light-fluctuations of a single star has received little attention.

From the very first there have been serious troubles with each new theory. Considered from the spectroscopic side alone, the Cepheids stand out as unexplainable anomalies. There are persistent peculiarities in the spectroscopic elements, such as the low value of the mass function, the universal absence of a secondary spectrum, and the minute apparent orbits. Practically the only thing they have in common with ordinary spectroscopic binaries is the definitely periodic oscillation of the spectral lines, which permits, with some well known conspicuous exceptions, of interpretation as periodic orbital motion. Adding, then, to the spectroscopic abnormalities the curious relations between light-variation and radial motion, the difficulties in the way of all the proposed simple solutions seem insurmountable. Geometric explanations of the light-variation fail completely, and little better can be said of the hypotheses that involve partly meteorological and partly orbital assumptions.

The writer can offer no complete explanation of Cepheid variability as a substitute for the existing theories that are shown to be more and more inadequate. At most, only the direction in which the real interpretation seems to lie can be pointed out, and an indication given of the strength of the observational data that would support the theory developed along the lines suggested. The principal results of a rather extensive investigation, further details of which it is hoped can be published in subsequent papers in the near future, are outlined in the following paragraphs. The main conclusion is that the Cepheid and cluster variables are not binary systems, and that the explanation of their light-changes can much more likely be found in a consideration of internal or surface pulsations of isolated stellar bodies.

THE ESSENTIAL IDENTITY OF CEPHEID AND CLUSTER VARIABLES

The subdivision of the short-period variables into the cluster type and the Cepheid is an artificial one. This proposition scarcely needs proof, although the assumption of the essential similarity of the two groups is important in the following discussion. Practically all writers on the subject are more or less inclined to accept this view. The definition of the cluster-type variable is, in fact, by some merely "short-period Cepheid." Others, including Hartwig and Kron, have considered only those with rapidly decreasing brightness and constant light at minimum as "antalgol" or cluster-type variables. Kron calls the shortest-period variable known [at that time, XX Cygni, period $3^h 14^m$] a Cepheid, and Hertzsprung designates as Cepheids only those variables whose periods are greater than a day. The writer proposes to adopt, merely as a convenience, the latter practice, arbitrarily calling the Cepheids of periods less than a day cluster-type variables; for there is at present no evidence of real difference between the two classes in the nature or probable causes of the light and velocity variations. Hertzsprung calls attention to the maxima in the frequency-curve of the periods at twelve hours and at seven days, and notes also that the longer-period Cepheids are in the galaxy, while the shorter-period Cepheids or cluster variables are apparently distributed more at random over the sky. Making the reasonable assumption that the data, though rather meager, are sufficient, nevertheless, to establish the reality of both phenomena, these conditions do not impeach the hypothesis that the light and velocity variations of the long- and short-period Cepheids are attributable to the same causes, and that the only modifications necessary in an explanation of one, to make it applicable to the other, are those depending on the length of the periods and other gradative characteristics, such as differences of spectral type and relative speed of light-change at corresponding phases. Among the several arguments that tend to prove the inherent similarity of the two groups of Cepheids, the following are the most important.

a) For RR Lyrae, period 13.6 hours, which is commonly classified as a cluster-type variable, the spectroscopic orbit by Kiess resembles in all details the peculiar orbits characteristic of the longer-period Cepheids. The light-curve is typical of cluster variables in all its properties.

b) From the photometric standpoint, Graff and Bottlinger have found no essential differences between light-curves of cluster and Cepheid types, and insist on the artificiality of the division into two classes. Very few, if any, of the cluster-type variables have rigorously constant light at minimum phase, as Plummer, among others, has shown. In fact, it was partly for this reason that Hartwig abandoned, in the *Vierteljahrsschrift* catalogue, the former term "antalgol" and the former distinction between cluster and Cepheid variables.

c) Russell's harmonic analyses of the mean light-curves of typical cluster variables and typical Cepheids indicate the necessity of analogous interpretations of the two.

d) An unpublished investigation by the writer of the relation between the periods and spectral types of all variables shows the existence of a continuous property from the longest-period Cepheids to the shortest-period cluster variables.

e) The shift of the maximum intensity in the spectra toward the violet with increasing light is a property common to both classes.

[Here follows a discussion of the irregularities in the period-length of Cepheid variables and irregularities in their light-curves, presented as arguments against the binary hypothesis.]

Changes in Color and Spectral Type

A third argument against the binary interpretation of Cepheids is the difficulty such theories would have in explaining the periodic change of the spectral type, though it must be admitted that to a certain extent Duncan's hypothesis, if otherwise acceptable, could account for spectral changes through the medium of atmospheric absorption. The evidences of the change of spectral type with changing light, though not well known nor generally recognized, are decisive and important. Schwarzschild, Wirtz, and more particularly Wilkens have demonstrated for Cepheids of longer period that the range of light-variation is greater in the photographic than in the visual part of the spectrum. The photographic work of Martin and Plummer suggests similar results for cluster-type variables, while recent simultaneous photographic and photovisual observations by Mr. Seares and the writer at Mount Wilson establish the fact definitely. The shift of the maximum intensity in the spectra of Cepheids, discovered by Albrecht, has been confirmed by Kiess and other Lick observers. These two factors—the greater photographic range and the shift of the maximum intensity— would suggest as an underlying and common cause a change in the spectral type. Albrecht and Duncan have observed that Wright's spectrograms of η Aquilae suggest a later type of spectrum at minimum than at maximum. The Harvard classification of TT Aquilae at maximum is G, at minimum, K.

For the cluster-type variables there is more direct evidence of distinct and continuous change. At the writer's request Miss Cannon has examined some of the Harvard spectrograms of certain cluster variables. For RR Lyrae no definite change was recorded on the plates examined, and similarly for XZ Cygni, but the spectrum, when faint, was extremely uncertain. For SW Andromedae the spectrum was of type A at maximum and clearly of a redder type at minimum. The most conclusive results, however, are obtained from the series of spectrograms taken by Mr. Pease in July of this year with the 60-inch reflector of the Mount Wilson Observatory. The variable RS Bootis, period $9^h.1$, shows a continuous change of spectral type from F0 at minimum to B8 at maximum. One consequence of this result is that hereafter the classification of all Cepheid and cluster-type spectra must be made with due specification of the corresponding phase of light-variation. Another difficulty in the spectral changes, that must not be overlooked in attempting a complete explanation of the Cepheid phenomena, is a peculiarity observed by Albrecht on his plates of Y Ophiuchi and T Vulpeculae. Various lines showed large irregular shifts, which are not progressive with the phase of the star in its light-period.

CONCERNING EXISTING HYPOTHESES

The fourth principal argument against the binary interpretation of Cepheids is the inadequacy of all the existing double-star hypotheses. To many this is not only the best argument but is sufficient in itself. A detailed criticism of these attempted explanations is unnecessary, for this has been generously provided by the proposers of the theories themselves, as well as by others, including Campbell, Plummer, Brunt, Kiess, and Ludendorff. There is one point, however, that has not been considered, which is of prime importance in the discussion of Cepheid phenomena. Russell and Hertzsprung have independently shown that the Cepheids are stars of small peculiar motions and small parallaxes, and hence of great absolute brightness. The former finds a mean absolute magnitude of −2.4 and the latter of −2.3, that is, the average Cepheid (the spectrum is of solar type) is nearly 700 times as bright as the sun. It is reasonable to assume that the Cepheids and the sun have a comparable surface brightness. The average Cepheid, then, has a volume between fifteen and twenty thousand times as great as that of the sun.

Interpreted as spectroscopic binaries these giant stars move in orbits whose apparent radii average less than one-tenth the radii of the stars themselves. In order that the radii of the real orbits may greatly exceed those of the apparent orbits, the inclinations must be very small, a condition which cannot be supposed to exist generally for Cepheid orbits. The difficulty in applying the hypotheses of Eddie, Loud, Duncan, and Roberts is therefore immediately apparent. Moreover, if the mass of the average Cepheid is admitted to be as much as five times the solar mass, the density is still astonishingly low—hardly three ten-thousandths that of the sun. Considering the low average value of the mass function derived from the orbits of the Cepheids, and taking a random distribution of the orbital inclinations, the non-luminous second body, to which Duncan's theory assigns the extensive atmosphere that must envelope the giant primary, has about one-tenth of the mass and therefore must move with an average apparent orbital velocity of about 200 km a second. Remembering the size of the primary star compared with its orbit, we know that the mass of the secondary must be still smaller and the velocity higher to separate the stars.

A SUGGESTED EXPLANATION OF CEPHEID VARIATION

In the face of all these difficulties, it seems appropriate to abandon completely the attempts to interpret Cepheids on the basis of a binary-star assumption. It has been shown by Russell that the light-variations cannot be explained satisfactorily by the uniform rotation of a single spotted star; the light-change must be intrinsic, and not just apparent. The explanation that appears to promise the simplest solution of most, if not all, of the Cepheid phenomena is founded on the rather vague conception of periodic pulsations in the masses of isolated stars. The vagueness of the hypothesis lies chiefly in our lack of knowledge of the internal structure of stellar bodies, and not in the difficulty of explaining the observed facts if once we assume the stars to be ideally gaseous figures of equilibrium. Moulton has considered the matter of explaining certain types of stellar variation from this point of view, but his conclusions are scarcely applicable to Cepheid variables in the light of our present knowledge of their peculiar properties. According to him, the light-change should be due to the heat generated by the oscillation of a spherical star from an oblate to a prolate form, there being a maximum of light-emission every time the star passes through its mean spherical figure. The period of velocity variation, then, should be double that of the light-change, and this, of course, does not conform with known conditions. It is to this phenomenon of pulsating stellar masses, however, that the writer would ascribe the light and velocity variation of Cepheid and cluster variables, and the theoretical work of Moulton, Jeans, Emden,[1] and others on the properties of gaseous spheres already justifies the conclusion that such oscillations are both possible and probable. They might arise, as Moulton suggests, from the collision with masses of only planetary dimensions, from the near approach of two stars, or in other ways.

Without any pretense of explaining clearly or fully on this hypothesis all the properties of Cepheid variation that have given the double-star theories such hopeless difficulty, a few points favoring the pulsation suggestion will be summarily stated. There will exist originally, as the result of the initial disturbance, a great number of oscillations of different periods. The character of these various vibrations will depend on the nature of the stellar structure. For the ideal homogeneous fluid mass investigated by Kelvin, and for the polytropic gaseous sphere defined and studied by Emden, the period of vibration of each type is independent of the volume and mass and depends only on the mean density and the order of the harmonic term defining the oscillation. For any given mean density the most important oscillation is that corresponding to the second-order harmonic. Its period is the longest, its amplitude the greatest, and it may persist with inappreciable change in period almost indefinitely, while the oscillations of higher order are more rapidly destroyed by friction.

If, then, we attribute the principal light-change in a Cepheid variable

[1] [In this regard R. Emden's *Gaskugeln* (Teubner, Leipzig, Berlin, 1907) is significant, especially the paragraphs under "Die Pulsierende Sonne."]

to this principal oscillation, and if we are willing to adopt Emden's polytropic gaseous sphere as a stellar model, we can compute at once the density of each individual variable. Obtained in this way the densities are probably of the right order of magnitude, whatever function of the radius, within reasonable limits, the density is assumed to be; and for an incompressible homogeneous fluid they would be only 2.5 times as large. The densities in terms of the sun for all the Cepheids of known periods and spectra have been derived in this manner, with the results given in Table I.

TABLE I

Type	No. Stars	Mean Period	Mean Density
M	3	33	0.000006
K	9	18	0.000020
G	31	11	0.000056
F	31	6	0.000200
A	9	0.4	0.04
B	1	0.19	0.2

The extremely low densities for the Cepheids of the redder spectral types until recently might have thrown serious doubt on the hypothesis that demands such abnormally low values. But now for two reasons we are ready to accept as possible these supposedly impossible densities. In the first place, as Hertzsprung has pointed out in his proof that the Cepheids of solar spectral type are giant stars, the average mean density must be of the order of 6×10^{-5}, if the masses are comparable with that of the sun. They may be larger, but from our knowledge of stellar masses in general we are inclined to believe that they are not more than ten times that of the sun, which is sufficient to prove the point. In the second place, the densities of several long-period eclipsing binaries of types G and K are now available for comparison. For instance: RX Cassiopeiae, type K0, mean density 5×10^{-4}; W. Crucis, type Gp, mean density 3×10^{-6}; SX Cassiopeiae, type G3, mean density 5×10^{-4}; RZ Ophiuchi, type F8, density of one component 10^{-3}. For the A- and B-type spectra, the Cepheid densities are, of course, entirely normal compared with eclipsing star densities.

As previously stated, the Cepheids without doubt are enormously large. Their small observed velocity variations, even if attributed altogether to motion in the line of sight and not at all to pressure-shifts, are not larger than might arise from a radial oscillation through but a small fraction of their mean diameters. In the central mass of the star the period of the supposed pulsation should, of course, be perfectly regular, but its effect need by no means be regular on the radiating surface.

We may suppose that, because of the internal vibration, the photosphere of the star is periodically scattered or broken through by the rush of hotter gases from the interior. Maximum light and maximum velocity of approach would obviously be approximately synchronous, and their coincidence would naturally be independent of the direction of the observer in space. Ludendorff's correlation of range of light and range of velocity is highly significant in this connection. The essentially harmonic nature of the oscillation at the surface of the star would easily lend itself to interpretation as elliptic motion, though non-elliptic motion need not be unexpected, nor the anomalous behavior of certain spectral lines. In stars in which the initial disturbance is of recent origin, the presence of secondary oscillations could be expected, which would affect the light as well as the velocity.

It should be noted as an important factor in the explanation of Cepheid variation, that a change in the spectrum of a given radiating surface from one type to the next will change the visual brightness of that surface by approximately one stellar magnitude, and the color range by four-tenths of a magnitude. These quantities correspond very closely to what is observed in cluster variables, and suggest that, if desired, it is unnecessary to go farther for the explanation of the light-variation than to suppose that a surface of approximately constant area progressively changes its spectral type as the result of a periodic flow and ebb of heat. That the light-change should be of a more explosive character for the cluster-variables than for the longer-period Cepheids would be expected because of their higher mean densities.

Various other details suggesting the possibility of the above interpretation could be cited,[2] but this sketch of the pulsation argument will suffice for the present, since the purpose of the paper is not so much to advance an alternative theory as to question the validity of the spectroscopic binary hyothesis.

[2] [For the first theoretical development of the pulsation hypothesis see papers by A. S. Eddington in the *Monthly Notices, Roy. Ast. Soc. 79*, 2–22 (1918); *79*, 177–188 (1919); see also *101*, 182–194 (1941).]

34. RADIAL VELOCITIES OF CEPHEID VARIABLE STARS

By Alfred H. Joy

Radial-velocity determinations for variable stars of the δ Cephei type are of especial value for extending our knowledge of the activities taking place in the atmospheres of unstable stars as well as for obtaining data concerning the motions of distant stars and the movement of the galaxy as a whole. In general, Cepheids are situated close to the galactic plane and are well distributed in longitude.

The characteristics of individual velocity-curves of several of the brighter Cepheid variables have been known for a number of years through the results obtained, mostly at the Lick Observatory, from three-prism spectrograms. Previous to 1920 about a dozen stars had been observed in detail and the relationship of the light- and velocity-changes carefully studied.

In planning a comprehensive program of spectroscopic observation of stars of this type, it seemed advisable to survey the whole group down to the faintest limit of magnitude possible with the instruments available rather than to continue the practice of intensive observation of certain selected stars. It was hoped that ten, or even fewer, spectrograms of each star, if properly distributed in phase, would be sufficient to determine a reliable normal velocity for the star and give a first approximation to the form of the velocity-curve.

Up to the present time spectrographic velocity-observations for 29 variable stars of the δ Cephei type have been published, exclusive of Polaris, which is peculiar in its behavior. They are as follows:

η Aql	DT Cyg	W Sgr
U Aql	β Dor	X Sgr
FF Aql	ζ Gem	Y Sgr
RT Aur	W Gem	RV Sco
l Car	T Mon	SZ Tau
SU Cas	S Mus	R TrA
TU Cas	S Nor	S TrA
δ Cep	Y Oph	T Vul
X Cyg	κ Pav	U Vul
SU Cyg	S Sge	

Twenty stars were observed at the Lick Observatory by 14 different observers, ·12 at Mount Wilson, mostly by Sanford, 2 at Michigan, and 1 each at the Pulkowa and the Cape observatories. In a few cases observations were made at more than one observatory. The present investigation adds to this number 126 stars and leaves only 35 stars, mostly fainter than fourteenth magnitude, north of declination −40° for which no observations are now available.

The observations and reductions have been kept as uniform as possible during the period of time involved. One-prism spectrographs were used, and the camera employed depended on the brightness of the star so that the exposure-times could be limited to three hours or less.

The spectra of Cepheids are known for their wide deep lines and for the presence of numerous strong lines of ionized atoms. These characteristics are especially favorable for the use of low dispersion. It was found that most of the lines used could be seen and readily measured on small-scale spectrograms, if attention was given to the proper exposure of the photographic plate. Although the observations of each star usually extended over several years, the resulting velocities, when sufficient in number, may be used to plot a mean velocity-curve, if the period is reasonably constant. It is necessary to depend upon the period of variation given by photometric observers. The photographic results, especially those of Robinson of the Harvard Observatory, have been used if available. In most cases preference has been given to elements based on data covering as long a period of time as possible. A number of the elements are apparently in need of further observations for confirmation or revision. . . .

[Here follows some sixty pages of observational tables and light- and velocity-curves. A sample of the curves is shown in Figure 1, where the velocity-curves for six variables are those with observations plotted as open circles.]

Discussion.—In any statistical use of the data presented it should be kept in mind that the observations are too few to determine unique curves and that, in many cases, their features have been purposely adapted to preconceived notions obtained from well-determined curves previously published and by experience in drawing other curves in this present series. It is rather surprising to find that for 90 stars of the list the simple form of the δ Cephei curve, which is practically made up of two straight lines, represents the observations as well as any other. . . .

Hertzsprung has studied the forms of light-curves of Cepheids in relation to period and has suggested that certain irregularities are char-

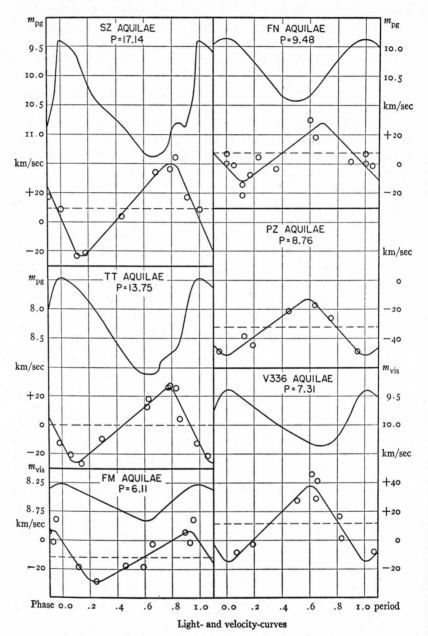

Fig. 1. For six cepheid variables a sample of the light and velocity curves.

acteristic of certain definite periods. Robinson, who had at hand a much larger collection of photometric material, states that for periods less than about seven days the relation between light-curve and period is slight, but that for periods between 8.5 and 30 days a relationship certainly exists which has but few exceptions. The velocity-curves . . . show little or no tendency to conform to set rules. In most cases the number of observations is not sufficient to show irregularities, but there is apparently a larger percentage of irregular curves among the stars with longer periods. Half the curves showing irregularities have

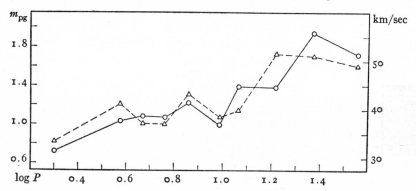

Fig. 2. Amplitudes in magnitude (full line) and velocity plotted against the logarithms of the periods.

periods greater than 13 days, while of 36 stars with periods between 7 and 11 days only two have irregular velocity-curves.

The period is the fundamental characteristic of Cepheid variation. Numerous investigators have shown the relation of various physical properties of these stars to the period of the light-variation. The most striking relationships are those correlating period with luminosity, spectrum, light-range, and velocity-range.

In view of the large increase in the amount of data now available, it will be worth while to reconsider the relation of period to range and eccentricity of light- and velocity-curves, and the lag of velocity extremes.

In Figure 2 the photographic magnitude-range and the velocity-range have been plotted as ordinates and the logarithms of the periods as abscissae. The points represent normal places for ten groups taken according to the logarithm of the period. The numbers of stars in each group from left to right are: 6, 18, 18, 24, 24, 18, 12, 22, 5, 5. With the exception of a drop at about 10 days, there is, in the mean, a fairly steady increase in range, with increasing period, from 0.7 to 1.9 mags. in light and from 33 to 50 km/sec in velocity. The decrease in range shown by the last group may or may not be real. It is based on only 5 stars and is considerably influenced by the small range of 1 Carinae both in light and velocity.

In statistical studies of stars of this kind a large dispersion is found when individual stars are considered. The scatter diagram (Fig. 3)

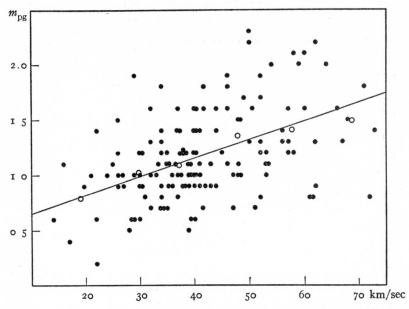

Fig. 3. Scatter diagram showing the correlation between magnitude-range and velocity-range.

showing the correlation between magnitude-range- and velocity-range will serve to illustrate the amount of spread which may be expected. The dots represent individual stars, and the circles are the normal points for groups chosen according to velocity-range.

The same groups which were used in Figure 2 have been employed in Figure 4 to show the relation between period and steepness or eccentricity of the light- and velocity-curves. The mean values of the time elapsing between minimum and maximum light $(M - m)$, and between maximum velocity of recession and approach (ϵ), expressed in percentage of the period, are plotted as ordinates. The well-known symmetry of the light-curves of stars of the ζ Geminorum type causes the notable drop in the curve of $M - m$ in the region of phase 10 days,

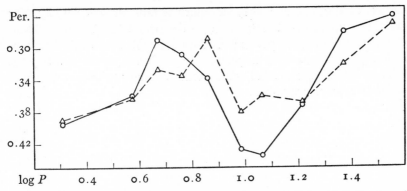

Fig. 4. Relation between period and asymmetry of light (full line) and velocity curves.

but this drop is not so prominent in the diagram showing the steepness of the velocity-curves. In general, there is an increase of eccentricity with period in both curves.

The curves of Figure 5 show the relation of period to the lag of the velocity extremes at maximum and minimum light. The same groups are used as for the preceding figures, and the lags are given in percentages of the period. The differences between the curves are, of course, related to the differences in eccentricity shown in Figure 4. There is apparently an increase of lag with lengthening period up to

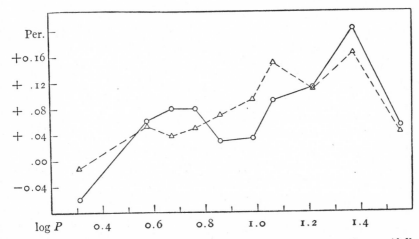

Fig. 5. Relation of period to lag of velocity extremes at maximum (full line) and minimum light.

periods of about 30 days, beyond which it falls off considerably. The lag varies up to about 20 percent of the period and averages 6.9 at maximum light and 7.4 percent at minimum light. These values may be affected by inaccuracy of the photometric elements used in drawing the velocity-curves and also to some extent by the arbitrary manner in which the curves are drawn.

35. SIX–COLOR LIGHT CURVES OF DELTA CEPHEI AND POLARIS

By Joel Stebbins

DELTA CEPHEI

The variable star Delta Cephei, the prototype of its class, has been observed so long and so often—visually, photographically, and photo-electrically—that little new can be expected from additional observations, no matter what their quality. Nevertheless, during the summer of 1941, when the new six-color photometer on the 60-inch reflector was working satisfactorily, a short series of measures was taken on half a dozen nights just to determine the amplitude of variation of this star in short and long wave lengths. Then in the two succeeding summers further observations were taken as opportunity offered—enough to fill out a satisfactory light-curve for each of six colors. . . .

POLARIS

Ever since the light-variation of Polaris was established by E. Hertz-sprung in 1911, this star has stood out at one end of the sequence of Cepheid variables as having the smallest amplitude of change. Compared with a photographic range of 0.17 mag. for Polaris, no other star in H. Schneller's catalogue for 1941 is clearly of the same type, with a variation of less than 0.30 mag. The continuous variables of small range, like Beta Cephei and 12 Lacertae, have early B-type spectra; and ordinarily they do not repeat their changes accurately even in successive cycles. Likewise, Delta Scuti, though of spectrum F4, is irregular in its variation. Polaris, however, seems to be a true variable of the Delta Cephei or Zeta Geminorum type, repeating its light and spectral changes in the period of 3.96 days continuously for years.

After the light-curves in six colors for Delta Cephei itself had been determined with a photocell on the 60-inch reflector at Mount Wilson, we planned to do the same for Polaris when opportunity offered. Accordingly, the star was measured on eight nights in the summer of 1944 to fix the amplitudes of the simple sine-curves which characterize

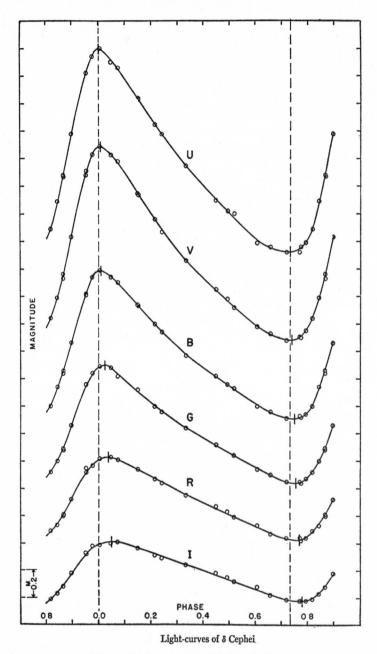

Light-curves of δ Cephei

Fig. 1. Light curves of Delta Cephei.

the variation. As expected, there was a progressive decrease in the amplitudes with increasing wavelength from 3530 A to 10,300 A. Because of the small total variation it would be very difficult to establish a slight retardation of phase like that of Delta Cephei for the longer wave lengths. . . .

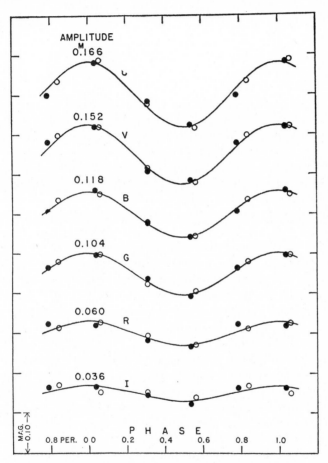

Light-curves of Polaris: open circles, 1944; solid circles, 1945

Fig. 2. Light curves of Polaris.

36. SOME UNSOLVED PROBLEMS IN THE THEORY OF ECLIPSING VARIABLES

By Zdenek Kopal

Out of the many unsolved problems in the theory of close binary systems, two are singled out for a brief discussion.[1] The first one concerns the light and velocity curves of binaries the components of which are nearly or actually in contact. As long as these components are sufficiently far apart for the disturbing action of one star upon the other to be regarded as that of a mass point, the light and radial velocity curves exhibited by such systems between eclipses as well as within minima are now known in full generality.

What happens, however, if the components are brought closer together until they are eventually in contact? What will their light and velocity curves then look like? The form of the components even under such conditions continues, fortunately, to be known; for free surfaces of centrally condensed stellar configurations are unlikely to depart to any appreciable degree from the Roche model. What will, however, be the distribution of brightness over the surfaces of such configurations? What will, in particular, be the law according to which this brightness will vary with the angle of foreshortening?

The answer will not be known until solutions of the equation of radiative transfer in spherical coordinates are available. Will the gravity-darkening, characteristic of moderately distorted systems, continue to hold good, or will this effect eventually disappear as a consequence of the equalization of the temperature gradient beneath the surface caused by internal convection currents? The quest for an answer in this case would take us deep in stellar hydrodynamics and the outcome is difficult to foresee. Not until all these questions are answered, however, shall we be in a position to determine with certainty whether or not

---

[1] [In the symposium on Eclipsing Binaries where Dr. Kopal's paper was given there were also papers by H. N. Russell, Otto Struve, R. M. Petrie, and Harlow Shapley. In my paper it was shown that the W Ursae Majoris variables, with components practically or actually in contact, were by far the most common variable star, and possibly of greatest service, as Dr. Kopal intimates, to the theories of the origin of double stars and of stellar evolution.]

the observed light curve of any known eclipsing system indicates that its components are in actual bodily contact.

Another outstanding problem which confronts persistently every investigator of eclipsing variables is that of the notorious asymmetries of light curves of many such systems—a phenomenon which the ordinary equilibrium theory of tides leaves wholly unexplained. Such cases are, moreover, too numerous to be regarded as exceptional. It is tempting to conjecture that the origin of observed asymmetries may in some way be related to the possibility of a resonance of orbital period with nonradial oscillations of the constituent components. What are the free periods of nonradial oscillations of stellar configurations? How are they affected by axial rotation of the stars? Is a resonance between orbital period and that of some mode of nonradial free oscillation really possible; and if so, what would be all its observable consequences? Answers to all these questions are still almost completely lacking; though their importance for a deeper understanding of various phenomena in close binary systems cannot be overrated.

VII

STELLAR STRUCTURE

It is amazing what scientific inquiry has been able to do with points of light at astronomical distances. In this part and the next some outstanding observations and speculations are presented to show how from the study of these apparently simple luminous dots we can speak with confidence of many phases of stellar structure. We now know much of the sizes, masses, and densities of stars, their temperatures inside and out, their candle powers, the turbulence of their gaseous atmospheres, their chemical compositions, and the transmutation of atoms deep in their interiors. This last is an atomic evolution, which as a by-product provides the long-continued solar radiation that has made life on this planet possible.

To produce so much important knowledge of distant objects, we have had the advantage, first, of being near an average star that is in no sense peculiar, and second, of benefiting from the invention of photographic emulsions, spectroscopes, and photocells. We have by no means exhausted the store of available information; the radio telescopes, photon counters, magnetometers, interferometers, high altitude instruments, and other developments of the future will undoubtedly continue to enrich greatly our knowledge of stellar structure. And not the least of these tools will be astrophysical theory and inspired interpretations such as those offered in this part and the next by Baade, Chandrasekhar, Eddington, Hertzsprung, Kuiper, and Russell.

In 1947, H. W. Babcock at Mount Wilson reported the results of his first work on stellar magnetism. The article (*Publications of the Astronomical Society of the Pacific 59*, 112 [1947]) is too specialized for the general reader, but the introduction to his report has a place in this part on stellar structure.

Observations carried out within the past year strongly support the theory that rapidly rotating stars of early type possess general magnetic fields much more intense than that attributed to the sun by G. E. Hale. In particular, the metallic-line A-type star 78 Virginis exhibits an integrated Zeeman effect characteristic of a general magnetic field having a strength of at least 1500 gauss at the pole. Similar fields of the same order of magnitude have been found for λ Equulei, β Coronae Borealis, HD 125248, and a few other stars. . . . It is significant that the observed magnetic polarities of the three stars just named are opposite in sign to the polarity of 78 Virginis. The intensity and extent of the stellar magnetic fields, comparable in strength to the fields of sunspots, leads us to infer that magnetic and electric phenomena occur on a far grander scale in and near some of the early-type stars than in the sun.

The general magnetic field of the sun is comparatively very weak, probably not in excess of one gauss.

37. THE DIAMETER OF BETELGEUX

By Albert A. Michelson and Francis G. Pease

[Although the project has not been followed through in recent years, the construction and use of interferometers for the measurement of small angles, such as those subtended by close double stars or by the diameters of giant stars, awakened much interest in the 1920's when Dr. Albert Michelson and associates at Mount Wilson made some ingenious interferometer experiments, one of which was the measurement of the apparent diameter of Alpha Orionis. The work was of limited value, since only a few stars are within practical range, and the measures, even if eventually simple to make, could tell nothing of the linear diameter of a star without knowledge of the parallax; and the linear diameter is the important parameter we seek. The parallaxes of the stars in our neighborhood have such relatively large probable errors that it is better to estimate the sizes from spectroscopic and photometric data than by way of angular measures. The linear diameter of many eclipsing binaries, for example, can be estimated with fair success, and with much the same accuracy whether near or far.

[The uncertainty in the value of the linear diameter of Betelgeux, given below by Michelson and Pease, is very large, not only because of errors in the angular measures and in the assumed parallax, but also because the asssumption of a uniformly luminous disk is, as the authors recognize, an approximation of convenience.

[A description of the instrument and the analysis of the preliminary results are not suited for reproduction in this place. An abstract, briefly reporting the method and first result, will suffice to salute this pioneer venture.]

TWENTY-FOOT INTERFEROMETER FOR MEASURING MINUTE ANGLES

Since pencils of rays at least 10 feet apart must be used to measure the diameters of even the largest stars, and because the interferometer results obtained with the 100-inch reflector were so encouraging, the construction of a 20-foot interferometer was undertaken. A very rigid

Fig. 1. The interferometer beam on the 100-inch reflector.

beam made of structural steel was mounted on the end of the Cassegrain cage, and four 6-inch mirrors were mounted on it so as to reduce the separation of the pencils to 45 inches and enable them to be brought to accurate coincidence by the telescope. The methods of making the fine adjustments necessary are described, including the use of two thin wedges of glass to vary continuously the equivalent air-path of one pencil. Sharp fringes were obtained with this instrument in August, 1920.

Diameter of α Orionis.—Although the interferometer was not yet provided with means for continuously altering the distance between the pencils used, some observations were made on this star, which was known to be very large. On December 13, 1920, with very good seeing, no fringes could be found when the separation of the pencils was 121 inches, although tests on other stars showed the instrument to be in perfect adjustment. This separation for minimum visibility gives the angular diameter as $0''.047$ within 10 percent, assuming the disk of the star uniformly luminous. Hence, taking the parallax as $0''.018$, the linear diameter comes out 250×10^6 miles.

38. THE INTERIOR OF A STAR

By Arthur Stanley Eddington

The sun belongs to a system containing some 3,000 million stars.[1] The stars are globes comparable in size with the sun, that is to say, of the order of a million miles in diameter. The space for their accommodation is on the most lavish scale. Imagine thirty cricket balls roaming the whole interior of the earth; the stars roaming the heavens are just as little crowded and run as little risk of collision as the cricket balls. We marvel at the grandeur of the stellar system. But this is probably not the limit. Evidence is growing that the spiral nebulae are 'island universes' outside our own stellar system. It may well be that our survey covers only one unit of a vaster organization.

A drop of water contains several thousand million million million atoms. Each atom is about one hundred-millionth of an inch in diameter. Here we marvel at the minute delicacy of the workmanship. But this is not the limit. Within the atom are the much smaller electrons pursuing orbits, like planets around the sun, in a space which relatively to their size is no less roomy than the solar system.

Nearly midway in scale between the atom and the star there is another structure no less marvelous—the human body. Man is slightly nearer to the atom than to the star. About 10^{27} atoms build his body; about 10^{28} human bodies constitute enough material to build a star.

From his central position man can survey the grandest works of Nature with the astronomer, or the minutest works with the physicist. To-night I ask you to look both ways. For the road to a knowledge of the stars leads through the atom; and important knowledge of the atom has been reached through the stars.

The star most familiar to us is the sun. Astronomically speaking, it is close at hand. We can measure its size, weigh it, take its temperature, and so on, more easily than the other stars. We can take photographs of its surface, whereas the other stars are so distant that the largest tel-

[1] [Subsequent researches on the population of the Milky Way indicate a much larger number, perhaps 10^{11}.]

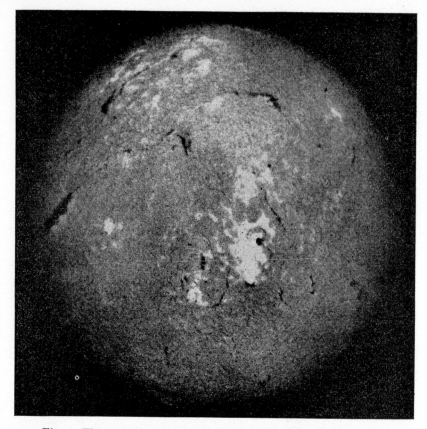

Fig. 1. The surface of the sun—a Sacramento Peak photograph.

escope in the world does not magnify them into anything more than points of light. Figs. 1 and 2 [2] show recent pictures of the sun's surface. No doubt the stars in general would show similar features if they were near enough to be examined.

I must first explain that these are not the ordinary photographs. Simple photographs show very well the dark blotches called sunspots, but otherwise they are rather flat and uninteresting. The pictures here shown were taken with a spectro-heliograph, an instrument which looks out for light of just one variety (wave-length) and ignores all the rest. The ultimate effect of this selection is that the instrument sorts out the different levels in the sun's atmosphere and shows what is go-

[2] Fig. 1 is from a photograph taken by Mr. Evershed at Kodaikanal Observatory, Madras. Fig. 2 is from the Mount Wilson Observatory, California.

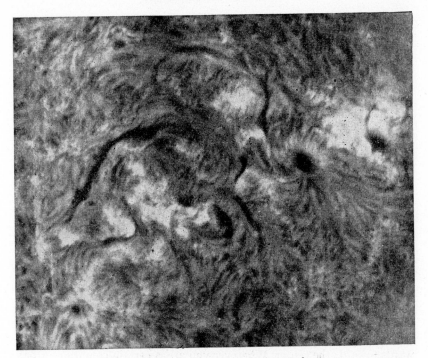

Fig. 2. Another view of the solar surface.

ing on at one level, instead of giving a single blurred impression of all levels superposed. Fig. 2, which refers to a high level, gives a wonderful picture of whirlwinds and commotion. I think that the solar meteorologists would be likely to describe these vortices in terms not unfamiliar to us—'A deep depression with secondaries is approaching, and a renewal of unsettled conditions is probable.' However that may be, there is always one safe weather forecast on the sun; cyclone or anticyclone, the temperature will be *very warm*—about 6,000° in fact.

But just now I do not wish to linger over the surface layers or atmosphere of the sun. A great many new and interesting discoveries have recently been made in this region, and much of the new knowledge is very germane to my subject of 'Stars and Atoms.' But personally I am more at home underneath the surface, and I am in a hurry to dive below. Therefore with this brief glance at the scenery that we pass we shall plunge into the deep interior—where the eye cannot penetrate, but where it is yet possible by scientific reasoning to learn a great deal about the conditions.

Temperature in the Interior

By mathematical methods it is possible to work out how fast the pressure increases as we go down into the sun, and how fast the temperature must increase to withstand the pressure. The architect can work out the stresses inside the piers of his building; he does not need to bore holes in them. Likewise the astronomer can work out the stress or pressure at points inside the sun without boring a hole. Perhaps it is more surprising that the temperature can be found by pure calculation. It is natural that you should feel rather sceptical about our claim that we know how hot it is in the very middle of a star—and you may be still more sceptical when I divulge the actual figures! Therefore I had better describe the method as far as I can. I shall not attempt to go into detail, but I hope to show you that there is a clue which might be followed up by appropriate mathematical methods.

I must premise that the heat of a gas is chiefly the energy of motion of its particles hastening in all directions and tending to scatter apart. It is this which gives a gas its elasticity or expansive force; the elasticity of a gas is well known to everyone through its practical application in a pneumatic tyre. Now imagine yourself at some point deep down in the star where you can look upwards towards the surface or downwards towards the center. Wherever you are, a certain condition of balance must be reached; on the one hand there is the weight of all the layers above you pressing downwards and trying to squeeze closer the gas beneath; on the other hand there is the elasticity of the gas below you trying to expand and force the superincumbent layers outwards. Since neither one thing nor the other happens and the star remains practically unchanged for hundreds of years, we must infer that the two tendencies just balance. At each point the elasticity of the gas must be just enough to balance the weight of the layers above; and since it is the heat which furnishes the elasticity, this requirement settles how much heat the gas must have. And so we find the degree of heat or temperature at each point.

The same thing can be expressed a little differently. As before, fix attention on a certain point in a star and consider how the matter above it is supported. If it were not supported it would fall to the center under the attractive force of gravitation. The support is given by a succession of minute blows delivered by the particles underneath; we have seen that their heat energy causes them to move in all directions, and they keep on striking the matter above. Each blow gives a slight boost upwards, and the whole succession of blows supports the upper material in shuttlecock fashion. (This process is not confined to the stars; for instance, it is in this way that a motor car is supported by its tyres.) An increase of temperature would mean an increase of activity of the particles, and therefore an increase in the rapidity and strength of the blows. Evidently we have to assign a temperature such that the sum total of the blows is neither too great nor too small to keep the upper material steadily supported. That in principle is our method of calculating the temperature.

One obvious difficulty arises. The whole supporting force will depend not only on the activity of the particles (temperature) but also on the number of them (density). Initially we do not know the density of the matter at an arbitrary point deep within the sun. It is in this connexion that the ingenuity of the mathematician is required. He has a definite amount of matter to play with, viz. the known mass of the sun; so the more he uses in one part of the globe the less he will have to spare for other parts. He might say to himself, 'I do not want to exaggerate the temperature, so I will see if I can manage without going beyond 10,000,000°.' That sets a limit to the activity to be ascribed to each particle; therefore when the mathematician reaches a great depth in the sun and accordingly has a heavy weight of upper material to sustain, his only resource is to use large numbers of particles to give the required total impulse. He will then find that he has used up all his particles too fast, and has nothing left to fill up the center. Of course his structure, supported on nothing, will come tumbling down into the hollow. In that way we can prove that it is impossible to build up a permanent star of the dimensions of the sun without introducing an activity or temperature exceeding 10,000,000°. The mathematician can go a step beyond this; instead of merely finding a lower limit, he can ascertain what must be nearly the true temperature distribution by taking into account the fact that the temperature must not be 'patchy.' Heat flows from one place to another, and any patchiness would soon be evened out in an actual star. I will leave the mathematician to deal more thoroughly with these considerations, which belong to the following up of the clue; I am content if I have shown you that there is an opening for an attack on the problem.

This kind of investigation was started more than fifty years ago. It has been gradually developed and corrected, until now we believe that the results must be nearly right—that we really know how hot it is inside a star.

I mentioned just now a temperature of 6,000°; that was the temperature near the surface—the region which we actually see. There is no serious difficulty in determining this surface temperature by observation; in fact the same method is often used commercially for finding the temperature of a furnace from the outside. It is for the deep regions out of sight that the highly theoretical method of calculation is required. This 6,000° is only the marginal heat of the great solar furnace giving no idea of the terrific intensity within. Going down into the interior the temperature rises rapidly to above a million degrees, and goes on increasing until at the sun's center [3] it is about 40,000,000°.

Do not imagine that 40,000,000° is a degree of heat so extreme that temperature has become meaningless. These stellar temperatures are to be taken quite literally. Heat is the energy of motion of the atoms or molecules of a substance, and temperature which indicates the degree of heat is a way of stating how fast these atoms or molecules are moving. For example, at the temperature of this room the molecules of air are rushing about with an average speed of 500 yards a second; if we heated it up to 40,000,000° the speed would be just over 100 miles a second. That is nothing to be alarmed about; the astronomer is quite accustomed to speeds like that. The velocities of the stars, or of the meteors entering the earth's atmosphere, are usually between 10 and 100 miles a second. The velocity of the earth travelling round the sun is 20 miles a second. So that for an astronomer this is the most ordinary degree of speed that could be suggested, and he naturally considers 40,000,000° a very comfortable sort of condition to deal with. And if the astronomer is not frightened by a speed of 100 miles a second, the experimental physicist is quite contemptuous of it; for he is used to handling atoms shot off from radium and similar substances with speeds of 10,000 miles a second. Accustomed as he is to watching these express atoms and testing what they are capable of doing, the physicist considers the jog-trot atoms of the stars very commonplace.

Besides the atoms rushing to and fro in all directions we have in the interior of a star great quantities of ether waves also rushing in all directions. Ether waves are called by different names according to their wave-length. The longest are the Hertzian waves used in broadcasting; then come the infra-red heat waves; next come waves of ordinary visible light; then ultra-violet photographic or chemical rays; then X-rays; then Gamma rays emitted by radio-active substances. Probably the shortest of all are the rays constituting the very penetrating radiation found in our atmosphere, which according to the investigations of Kohlhörster and Millikan are believed to reach us from interstellar

[3] [Subsequent research on the theory reduces this temperature to about one-half the value given.]

space. These are all fundamentally the same but correspond to different octaves. The eye is attuned to only one octave, so that most of them are invisible; but essentially they are of the same nature as visible light.

The ether waves inside a star belong to the division called X-rays. They are the same as the X-rays produced artificially in an X-ray tube. On the average they are 'softer' (i.e. longer) than the X-rays used in hospitals, but not softer than some of those used in laboratory experiments. Thus we have in the interior of a star something familiar and extensively studied in the laboratory.

Besides the atoms and ether waves there is a third population to join in the dance. There are multitudes of free electrons. The electron is the lightest thing known, weighing no more than 1/1,840 of the lightest atom. It is simply a charge of negative electricity wandering about alone. An atom consists of a heavy nucleus which is usually surrounded by a girdle of electrons. It is often compared to a miniature solar system, and the comparison gives a proper idea of the *emptiness* of an atom. The nucleus is compared to the sun, and the electrons to the planets. Each kind of atom—each chemical element—has a different quorum of planet electrons. Our own solar system with eight [4] planets might be compared especially with the atom of oxygen which has eight circulating electrons. In terrestrial physics we usually regard the girdle or crinoline of electrons as an essential part of the atom because we rarely meet with atoms incompletely dressed; when we do meet with an atom which has lost one or two electrons from its system, we call it an 'ion.' But in the interior of a star, owing to the great commotion going on, it would be absurd to exact such a meticulous standard of attire. All our atoms have lost a considerable proportion of their planet electrons and are therefore *ions* according to the strict nomenclature.

IONIZATION OF ATOMS

At the high temperature inside a star the battering of the particles by one another, and more especially the collision of the ether waves (X-rays) with atoms, cause electrons to be broken off and set free. These free electrons form the third population to which I have referred. For each individual the freedom is only temporary, because it will presently be captured by some other mutilated atom; but meanwhile another electron will have been broken off somewhere else to take its

[4] [Pluto was later discovered; the corresponding element would be fluorine.]

place in the free population. This breaking away of electrons from atoms is called *ionization,* and as it is extremely important in the study of the stars I will presently show you photographs of the process.

My subject is 'Stars and Atoms'; I have already shown you photographs of a star, so I ought to show you a photograph of an atom. Nowadays that is quite easy. Since there are some trillions of atoms present in the tiniest piece of material it would be very confusing if the photograph showed them all. Happily the photograph exercises discrimina-

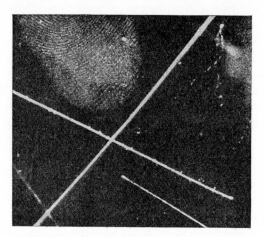

Fig. 3. Helium atoms flashed across the field of view.

tion and shows only 'express train' atoms which flash past like meteors, ignoring all the others. We can arrange a particle of radium to shoot only a few express atoms across the field of the camera, and so have a clear picture of each of them.

Fig. 3 [5] is a photograph of three or four atoms which have flashed across the field of view—giving the broad straight tracks. These are atoms of helium discharged at high speed from a radio-active substance.

I wonder if there is an under-current of suspicion in your minds that there must be something of a fake about this photograph. Are these really the single atoms that are showing themselves—those infinitesimal units which not many years ago seemed to be theoretical concepts far outside any practical apprehension? I will answer that question by asking you one. You see a dirty mark on the picture. Is that somebody's thumb? If you say 'Yes,' then I assure you unhesitatingly that these

[5] I am indebted to Professor C. T. R. Wilson for Figs. 3–6.

streaks are single atoms. But if you are hypercritical and say 'No. That is not anybody's thumb, but it is a mark that shows that somebody's thumb has been there,' then I must be equally cautious and say that the streak is a mark that shows where an atom has been. The photograph instead of being the impression of an atom is the impression of the impression of an atom, just as it is not the impression of a thumb but the impression of the impression of a thumb. I don't see that

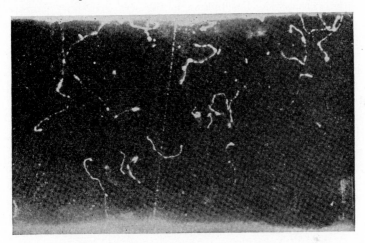

Fig. 4. The tracks of numerous electrons.

it really matters that the impression is second-hand instead of first-hand. I do not think we have been guilty of any more faking than the criminologist who scatters powder over a finger-print to make it visible, or a biologist who stains his preparations with the same object. The atom in its passage leaves what we might call a 'scent' along its trail; and we owe to Professor C. T. R. Wilson a most ingenious device for making the scent visible. Professor Wilson's 'pack of hounds' consists of water vapour which flocks to the trail and there condenses into tiny drops.

You will next want to see a photograph of an electron. That can also be managed. The broken wavy trail in Fig. 3 is an electron. Owing to its small mass the electron is more easily turned aside in its course than the heavy atom which rushes bull-headed through all obstacles. Fig. 4 shows numerous electrons, and it includes one of very high speed which on that account was able to make a straight track. Incidentally it gives away the device used for making the tracks visible, because you can see the tiny drops of water separately. . . .

A cynic might remark that the interior of a star is a very safe subject to talk about because no one can go there and prove that you are wrong. I would plead in reply that at least I am not abusing the unlimited opportunity for imagination; I am only asking you to allow in the interior of the star quite homely objects and processes which can be photographed. Perhaps now you will turn round on me and say, 'What right have you to suppose that Nature is as barren of imagination as you are? Perhaps she has hidden in the star something novel which will upset all your ideas.' But I think that science would never have achieved much progress if it had always imagined unknown obstacles hidden round every corner. At least we may peer gingerly round the corner, and perhaps we shall find there is nothing very formidable after all. Our object in diving into the interior is not merely to admire a fantastic world with conditions transcending ordinary experience; it is to get at the inner mechanism which makes the stars behave as they do. If we are to understand the surface manifestations, if we are to understand why 'one star differeth from another star in glory,' we must go below—to the *engine-room*—to trace the beginning of the stream of heat and energy which pours out through the surface. Finally, then, our theory will take us back to the surface and we shall be able to test by comparison with observation whether we have been badly misled. Meanwhile, although we naturally cannot prove a general negative, there is no reason to anticipate anything which our laboratory experience does not warn us of.

The X-rays in a star are the same as the X-rays experimented on in a laboratory, but they are enormously more abundant in the star. We can produce X-rays like the stellar X-rays, but we cannot produce them in anything like stellar abundance. . . . In the star you must imagine the intensity multiplied many million-fold, so that electrons are being wrenched away as fast as they settle and the atoms are kept stripped almost bare. The nearly complete mutilation of the atoms is important in the study of the stars for two main reasons.

The first is this. An architect before pronouncing an opinion on the plans of a building will want to know whether the material shown on the plans is to be wood or steel or tin or paper. Similarly it would seem essential before working out details about the interior of a star to know whether it is made of heavy stuff like lead or light stuff like carbon. By means of the spectroscope we can find out a great deal about the chemical composition of the sun's atmosphere; but it would not be fair to take this as a sample of the composition of the sun as a whole. It would be very risky to make a guess at the elements preponderating in the deep interior. Thus we seem to have reached a deadlock. But now it turns out that when the atoms are thoroughly smashed up, they all behave nearly alike—at any rate in those properties with which we are concerned in astronomy. The high temperature—which we were inclined to be afraid of at first—has simplified things for us, because it has to a large extent eliminated differences between different kinds of material. The structure of a star is an unusually simple physical problem; it is at low temperatures such as we experience on the earth that matter begins to have troublesome and complicated properties. Stellar atoms are nude savages innocent of the class distinctions of our fully arrayed terrestrial atoms. We are thus able to make progress without guessing at the chemical composition of the interior. It is necessary to make one reservation, viz. that there is not an excessive proportion of hydrogen.[6] Hydrogen has its own way of behaving; but it makes very little difference which of the other 91 elements predominate.

The other point is one about which I shall have more to say later. It is that we must realize that the atoms in the stars are mutilated fragments of the bulky atoms with extended electron systems familiar to us on earth; and therefore the behaviour of stellar and terrestrial gases is by no means the same in regard to properties which concern the size of the atoms.

To illustrate the effect of the chemical composition of a star, we revert to the problem of the support of the upper layers by the gas underneath. At a given temperature every independent particle contributes the same amount of support no matter what its mass or chemical nature; the lighter atoms make up for their lack of mass by moving more actively. This is a well-known law originally found in experimental chemistry, but now explained by the kinetic theory of Maxwell and Boltzmann. Suppose we had originally assumed the sun to be composed entirely of silver atoms and had made our calculations of temperature accordingly; afterwards we change our minds and substitute a lighter element, aluminium. A silver atom weighs just four times as much as an aluminium atom; hence we must substitute four aluminium atoms for every silver atom in order to keep the mass of the sun unchanged. But now the supporting force will everywhere be quadrupled, and all the mass will be heaved outwards by it if we make no further change. In order to keep the balance, the activity of each particle must be reduced in the ratio $\frac{1}{4}$; that means that we must assign throughout the aluminium sun temperatures $\frac{1}{4}$ of those assigned to the silver sun. Thus

[6] [Thirty years later the study of stellar interiors has led us to believe in hydrogen dominance in the constitution of stars.]

for unsmashed atoms a change in the assigned chemical composition makes a big change in our inference as to the internal temperature.

But if electrons are broken away from the atom these also become independent particles rendering support to the upper layers. A free electron gives just as much support as an atom does; it is of much smaller mass, but it moves about a hundred times as fast. The smashing of one silver atom provides 47 free electrons, making with the residual nucleus of the atom 48 particles in all. The aluminium atom gives 13 electrons or 14 particles in all; thus 4 aluminium atoms give 56 independent particles. The change from smashed silver to an equal mass of smashed aluminium only means a change from 48 to 56 particles, requiring a reduction of temperature by 14 per cent. We can tolerate that degree of uncertainty in our estimates of internal temperature; it is a great improvement on the corresponding calculation for unsmashed atoms which was uncertain by a factor 4.

Besides bringing closer together the results for different varieties of chemical constitution, ionization by increasing the number of supporting particles lowers the calculated temperatures considerably. It is sometimes thought that the exceedingly high temperature assigned to the interior of a star is a modern sensationalism. That is not so. The early investigators, who neglected both ionization and radiation pressure, assigned much higher temperatures than those now accepted.

RADIATION PRESSURE AND MASS

The stars differ from one another in mass, that is to say, in the quantity of material gathered together to form them; but the differences are not so large as we might have expected from the great variety in brightness. We cannot always find out the mass of a star, but there are a fair number of stars for which the mass has been determined by astronomical measurements. The mass of the sun is—

2000000000000000000000000000 tons.

I hope I have counted the 0's rightly, though I dare say you would not mind much if there were one or two too many or too few. But Nature *does* mind. When she made the stars she evidently attached great importance to getting the number of 0's right. She has an idea that a star should contain a particular amount of material. Of course she allows what the officials at the mint would call a 'remedy.' She may even pass a star with one 0 too many and give us an exceptionally large star, or with one 0 too few, giving a very small star. But these deviations are rare, and a mistake of two 0's is almost unheard of. Usually she adheres much more closely to her pattern.

How does Nature keep count of the 0's? It seems clear that there must be something inside the star itself which keeps check and, so to speak, makes a warning protest as soon as the right amount of material has been gathered together. We think we know how it is done. You remember the ether waves inside the star. These are trying to escape outwards and they exert a pressure on the matter which is caging them in. This outward force, if it is sufficiently powerful to be worth considering in comparison with other forces, must be taken into account in any study of the equilibrium or stability of the star. Now in all small globes this force is quite trivial; but its importance increases with the mass of the globe, and it is calculated that at just about the above mass it reaches equal status with the other forces governing the equilibrium of the star. If we had never seen the stars and were simply considering as a curious problem how big a globe of matter could possibly hold together, we could calculate that there would be no difficulty up to about two thousand quadrillion tons; but beyond that the conditions are entirely altered and this new force begins to take control of the situation. Here, I am afraid, strict calculation stops, and no one has yet been able to calculate what the new force will do with the star when it does take control. But it can scarcely be an accident that the stars are all so near to this critical mass; and so I venture to conjecture the rest of the story. The new force does not *prohibit* larger mass, but it makes it risky. It may help a moderate rotation about the axis to break up the star. Consequently larger masses will survive only rarely; for the most part stars will be kept down to the mass at which the new force first becomes a serious menace. The force of gravitation collects together nebulous and chaotic material; the force of radiation pressure chops it off into suitably sized lumps.

This force of radiation pressure is better known to many people under the name 'pressure of light.' The term 'radiation' comprises all kinds of ether-waves including light, so that the meaning is the same. It was first shown theoretically and afterwards verified experimentally that light exerts a minute pressure on any object on which it falls. Theoretically it would be possible to knock a man over by turning a searchlight on him—only the searchlight would have to be excessively intense, and the man would probably be vaporized first. Pressure of light probably plays a great part in many celestial phenomena. One of the earliest suggestions was that the minute particles forming the tail of a comet are driven outwards by the pressure of sunlight, thus accounting for the fact that a comet's tail points away from the sun. But that particular application must be considered doubtful. Inside the star the intense stream of light (or rather X-rays) is like a wind rushing outwards and distending the star.

THE INTERIOR OF A STAR

We can now form some kind of a picture of the inside of a star—a hurly-burly of atoms, electrons, and ether-waves. Dishevelled atoms tear along at 100 miles a second, their normal array of electrons being torn from them in the scrimmage. The lost electrons are speeding 100 times faster to find new resting places. Let us follow the progress of one of them. There is almost a collision as an electron approaches an atomic nucleus, but putting on speed it sweeps round in a sharp curve. Sometimes there is a side-slip at the curve, but the electron goes on with increased or reduced energy. After a thousand narrow shaves, all happening within a thousand millionth of a second, the hectic career is ended by a worse side-slip than usual. The electron is fairly caught, and attached to an atom. But scarcely has it taken up its place when an X-ray bursts into the atom. Sucking up the energy of the ray the electron darts off again on its next adventure.

I am afraid the knockabout comedy of modern atomic physics is not very tender towards our aesthetic ideals. The stately drama of stellar evolution turns out to be more like the hair-breadth escapades on the films. The music of the spheres has almost a suggestion of—jazz.

And what is the result of all this bustle? Very little. The atoms and electrons for all their hurry never get anywhere; they only change places. The ether-waves are the only part of the population which accomplish anything permanent. Although apparently darting in all directions indiscriminately, they do on the average make a slow progress outwards. There is no outward progress of the atoms and electrons; gravitation sees to that. But slowly the encaged ether-waves leak outwards as though through a sieve. An ether-wave hurries from one atom to another, forwards, backwards, now absorbed, now flung out again in a new direction, losing its identity, but living again in its successor. With any luck it will in no unduly long time (ten thousand to ten million years according to the mass of the star) find itself near the boundary. It changes at the lower temperature from X-rays to light-rays, being altered a little at each re-birth. At last it is so near the boundary that it can dart outside and travel forward in peace for a few hundred years. Perhaps it may in the end reach some distant world where an astronomer lies in wait to trap it in his telescope and extort from it the secrets of its birth-place.

It is the leakage that we particularly want to determine; and that is why we have to study patiently what is going on in the turbulent crowd. To put the problem in another form; the waves are urged to flow out by the temperature gradient in the star, but are hindered and turned back by their adventures with the atoms and electrons. It is the task of mathematics, aided by the laws and theories developed from a study of these same processes in the laboratory, to calculate the two factors— the factor urging and the factor hindering the outward flow—and hence to find the leakage. This calculated leakage should, of course, agree with astronomical measurements of the energy of heat and light pouring out of the star.

39. THE EMPIRICAL MASS–LUMINOSITY RELATION

By Gerard P. Kuiper

The discovery of the mass-luminosity relation has been of great importance to the progress of astronomy. The relation has been used in statistical astronomy and in double star astronomy and has been a central problem of theoretical astrophysics. Since for most stars no direct determination of mass can be made, the use of the mass-luminosity relation is the only method of estimating the total mass of the known stars per volume of space—an important dynamical quantity. In binary star statistics the observable Δm and a rough knowledge of the absolute magnitude can now be used in obtaining a statistically useful determination of the mass ratio, which is of great cosmogenetic interest. Investigations such as a recent one on Epsilon Aurigae show the need of having the relation well established. But particularly the theoretical importance, both in abstract form and in numerical form for particular stars, is clear from Eddington's work and from more recent developments [B. Strömgren, S. Chandrasekhar].

1. *Historical.*—The idea that the mass and the luminosity are correlated seems to have developed gradually as the knowledge of parallaxes increased. Even today the knowledge of masses is to a large extent limited by the knowledge of accurate parallaxes. The fact that one of the best known stars, Sirius B, does not follow the general mass-luminosity relation must have delayed the discovery of that relation.

Halm was probably the first to state explicitly the existence of a statistical relation between intrinsic brightness and mass. His conclusion was essentially based on the correlation of mass with spectral type and of spectral type with luminosity. The first relation was established mainly from double-line spectroscopic binaries. The result was therefore partly accidental, because double-line spectroscopic binaries are very rare among giants, so that Halm was essentially dealing with main-sequence stars only. In fact, the relation was not recognized in Russell's paper of 1913, although in 1914 Russell found evidence for a definite correlation, which was obtained by comparing absolute magnitudes derived from hypothetical parallaxes with absolute magnitudes derived by methods of stellar statistics.

Hertzsprung, in 1918, gave the relation

$$\log \ m = -0.06(M_v - 5),$$

in fair agreement with modern data. He also gave the corresponding formula for the mass ratio of a binary derived from Δm, and the first formula for dynamical (not hypothetical) parallaxes. Van Maanen shortly afterward emphasized that the mass-luminosity relation is independent of spectral type.

Many other investigations followed, most of which have been reviewed by Lundmark. Of all these, Eddington's results are undoubtedly the best known, particularly because Eddington used the relation in connection with his well-known theoretical investigations.

The most recent study of the subject is that by P. P. Parenago, which was received after most of the present investigation was completed. In some ways the two investigations run parallel: for instance, in the use of bolometric corrections derived from radiometric observations and the estimation of weights of the individual mass determinations. But the present study uses several improved orbits, mass ratios, and Δm values and does not use spectroscopic parallaxes. . . .

2. *The present investigation* is limited to those individual stars for which the three fundamental quantities—the mass m, the radius R, and the luminosity L—may be obtained from the observations with fair precision. The derivation of the radius, in addition to the other two quantities, does not require any extra information, since radius and luminosity are simply related by the effective temperature, which is needed in any case before the eclipsing binaries can be used. In view of the fact that all three parameters are required in theoretical work, we shall give them explicitly as the final result of the investigation.

Since all the emphasis is laid on accurate data for individual stars (as required in theoretical work), no attempt is made here to supplement the results by a statistical treatment of other data of smaller individual accuracy. Such a statistical treatment would be useful for determining the trend of the mass-luminosity relation for the giants and the O stars, where few data of high individual accuracy are available.

A close co-operation between theory and observation will certainly be the quickest and least wasteful method of solving the fundamental problems of stellar structure and evolution. In this connection it would seem that the visual binaries deserve preference in double-star observations and in determinations of parallax and mass ratio. At least four to six determinations for each object will be needed in order to make the mass determinations sufficiently accurate. Photometric and spectroscopic measures of double-line spectroscopic binaries that are also

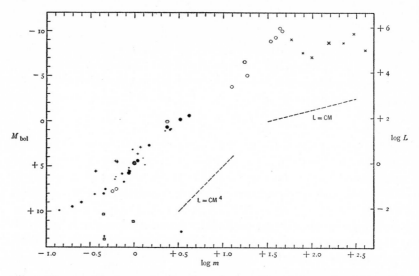

Fig. 1. The empirical mass-luminosity relation. Closed dots: visual binaries; open circles: spectroscopic binaries; vertical crosses: Hyades; diagonal crosses: Trumpler's stars; squares: white dwarfs.

eclipsing variables are obviously the second group of observations especially needed. Finally, investigations on the stellar temperature scale are required for the computation of luminosities from radii and vice versa.

Many important direct or indirect contributions to the knowledge of stellar masses have recently become available: Schlesinger's *Catalogue of Parallaxes*; Boss's *General Catalogue of 33342 Stars*, containing many mass ratios for visual binaries, based on all the available meridian positions; many new orbits, of which we want to mention especially those derived with the help of photographic measures (Strand, van den Bos, Hertzsprung); and several new mass ratios by van de Kamp. For the discussion of the eclipsing binaries, of special importance is Pannekoek's rediscussion of the temperatures of O and B stars, derived from maxima of spectral series. Data for very interesting systems, such as Zeta Aurigae and VV Cephei, have recently been published. Last, but not least, Trumpler's discovery of very massive stars in galactic clusters should be mentioned. These many advances make it promising to assemble and discuss the data now available. . . .

[The author then discusses in turn the contributions from (1) visual binaries, (2) some selected spectroscopic binaries, and (3) Trumpler's massive stars. The final empirical mass-luminosity relation is shown by the graphical presentation of the data in Figure 1.]

40. STELLAR ENERGY AND BETHE'S CARBON CYCLE

By Henry Norris Russell

This lecture should begin by an apology to the audience. The speaker tonight should be the man whose recent and brilliant work has inaugurated a new and very promising stage of astrophysical study—Professor H. A. Bethe of Cornell. The planners of the program can plead but one excuse—his results had not been announced when it was prepared.

They are not yet published in detail; but Professor Bethe has most generously given me a full copy of the proof of his forthcoming paper, and amplified this in personal conversation. What there is of novelty in the present discussion should be credited entirely to him, and not to myself. Indeed my report, apart from this, is mainly upon the work of others in a field in which I did a little pioneering a good many years ago.

A century of precise measurement has furnished astronomers with a large accumulated capital of facts about the stars. One of the most remarkable things that is thus revealed is that the stars differ greatly among themselves in some properties, and relatively little in others. For example, the luminosity—the rate of radiation of energy—ranges (roughly) from a million times the Sun's luminosity to a thousandth part of it, a ratio of a billion to one. The diameters run from at least 1000 times the Sun's down to $\frac{1}{30}$ or less, a range of thirty-thousand-fold. For the masses, the range is roughly from 100 times the Sun to $\frac{1}{10}$ of the Sun, a thousand-fold. But surface temperatures more than ten times the Sun's or less than $\frac{1}{4}$ that of the Sun are practically unheard-of, so that the known range is only forty-fold.

Why should such enormous diversities exist? The limitation of our study is set, on one side, by the obvious fact that we cannot see a star unless it shines—that is, unless it gives out enough light to be perceptible at stellar distances. This obviously sets a limit to our knowledge of *faint* stars, and of *small* stars—which, other things being equal, will shine less brightly because they have fewer square miles of surface, and still more to the study of *cool* stars (of low surface temperature). The heat radiation per square mile varies as the fourth power of the temperature, and a forty-fold increase in the latter changes the former by a factor of about two and a half million. Moreover, at low temperatures, most of the radiant energy is in long infra-red waves, invisible to the eye, so that a star—like a mass of hot iron—ceases to be visible in the dark, long before it stops giving out heat.

But there is no such limitation on the side of great brightness. If we fail to find still brighter stars—though we are fishing for these in very wide waters—it must be because there are no such fish in our sea. Nature herself must in some way set a limit to the brightness attainable by a star—at least as a permanent affair, for we know that the short-lived outbursts of supernovae are enormously more intense. Such a limitation would explain why stars of very large diameter are never found to be very hot—if they were, they would give out too much light; but it has no obvious relation to the mass of a star.

A notable advance was made more than twenty years ago when it was found observationally (first by Halm) that there is a close relation between the mass of a star and its luminosity. All later work has confirmed this, and it now appears that, over almost if not quite the whole available range, the total heat-radiation of a star is nearly proportional to the fourth power of its mass. The most remarkable feature is that, for a given mass, the luminosity does not depend much upon the size of the star—the large ones are cool, and give out less heat per square mile, the small ones hot, and give out more; but the net product is nearly the same.

Eddington's explanation of this, completed fifteen years ago, marks the first, and, even now, the greatest success of modern atomic physics in interpreting the stars.

Since the work of Lane in 1870, it has been realized that, if the familiar laws of perfect gases could be applied to the interior of the Sun, the central temperature must be many millions of degrees. The result may be written

$$T = 11.5 \frac{M}{R} \frac{Y}{X} \mu,$$

where T is the temperature in millions of degrees, M the star's mass and R its radius, μ the mean molecular weight of the material, and Y and X quantities depending on the "density-model," that is, the law in accordance with which the density increases toward the center.

Calculations for different laws of density-distribution show that the "model" has a rather small influence. If the density were everywhere the same, X and Y would be 1 at the star's center. On the model to which Eddington's calculations led, the central density was 54 times the average ($X = 54$), but $Y = 92$, so that the central temperature for the

Sun comes out 19,700,000°. A model recently calculated by Chandrasekhar makes $X = 88$, $Y = 171$, and the temperature 22,300,000°; still another, reported by Bethe, 20,300,000°—all provided that $\mu = 1$. There are good reasons to believe that these "models" give a fairly correct idea of the situation, and therefore that the central temperature of the Sun is close to twenty million degrees—provided that the molecular weight is 1.

But, under ordinary circumstances, μ is 1 for hydrogen, 4 for helium, 16 for oxygen, 56 for iron, and so on, so that the calculated internal temperature of the Sun would depend enormously on its assumed chemical composition.

Fortunately for us theorists, the situation is simpler inside the stars; for the outsides of the atoms—the electrons which surround the nucleus—are pretty well knocked off. A hydrogen atom (with one electron) is broken into 2 particles, a helium atom into 3, one of oxygen into 9, and one of iron into 27 (or perhaps 25, if the two tightest bound electrons stay with the nucleus).

The average mass of a free particle (which is what counts) then comes out ½ for hydrogen, ⅘ for helium and nearly 2 for all heavier atoms. So we can conclude that the central temperature of the Sun is close to 10 million degrees if it is pure hydrogen, 27 million if it is all helium, and 40 million if it is all composed of heavy atoms—and the result for any assumed mixture can easily be computed.

The temperature of the Sun's surface is but a few thousand degrees. Heat will flow outward from the hot interior "down grade" and the surface will adjust its temperature so that it loses by radiation into space just the amount which it gains by transmission from the deep interior. The rate of this transmission thus determines how bright and hot the Sun (or any other star) will be; it takes place mainly by the transfer of radiation from atom to atom through the gas; and the net opacity, which determines the rate of transfer, may be calculated with considerable accuracy, on principles developed by Kramers.

When the net flow of heat out to a star's surface is thus calculated, it is found that it depends relatively little on the "model," and not much upon the radius, but changes rapidly with the mass—at very nearly the rate indicated by observation.

The mass-luminosity relation is thus explained—and with it the fact that stars of less than one-tenth the Sun's have not so far been detected—they are presumably too faint to be seen; and our failure to discover very large masses is connected with the absence of stars of very great luminosity.

Eddington's theory, however, predicts that the luminosity of a star, other things being equal, should change very rapidly with the mean molecular weight of its material (about as its seventh power). A star composed almost entirely of heavy atoms (with $\mu = 2$) will be so hot inside that it will shine brightly, whereas one composed mostly of hydrogen will be cooler inside—and fainter outside by several hundred-fold (the exact amount involving refinements of calculations omitted here).

The observed luminosity of the Sun is about midway between these limits, and may be accounted for, according to Stromgren's calculations, if one-third of the material, by weight, is hydrogen, and the rest heavy atoms; or with more hydrogen if helium is present; for example, 60 percent hydrogen, 36 percent helium, and 4 percent heavy atoms. The mean molecular weight comes out 0.98 and 0.67 in the two cases, and the central temperature 19 million and 13 million degrees. . . .

When first the Sun's heat was measured, it was realized at once that it could hardly be maintained, even for the duration of history, by ordinary chemical combinations—what we now describe as reactions between whole atoms. A far larger source of energy is found in a slow contraction, compressing the gas and turning the gravitational potential energy into heat, and this explanation, due to Helmholtz and Kelvin, was still accepted forty years ago. The energy-supply thus available for the Sun's past history is about thirty million times its present annual expenditure (depending somewhat on the assumed model). But at least half this must still be stored inside the Sun in the form of heat, so that less than fifteen million years of sunshine can be accounted for.

But the Sun has been shining and warming the Earth, very much as now, throughout geological history, which is certainly a hundred times as long as this. There is only one place in the known universe to look for so vast a store of energy—and that is in the minute nuclei of atoms. From these nuclei is liberated the energy of radio-activity—which is of the required order of magnitude, though rather small. But radio-activity itself will not meet our needs, for it takes place at a rate uninfluenced by conditions external to the nucleus—such as the surrounding temperature. To account for the observed variation of luminosity with mass, we would have to assume that in some inscrutable way the amount of active material had been exactly proportional "in the beginning" to the mass of each star, so as to provide the proper heat supply to maintain its radiation; and also that practically none of it had been allowed to enter the Earth or the other planets—for otherwise their surfaces would be red-hot or even white-hot.

It is necessary, therefore, to assume that the great store of energy upon which the stars draw becomes available only under stellar conditions—that is, obviously, at temperatures of millions of degrees. . . .

One other thing was clearly understood, twenty years or so ago; the reactions which provide the stars with their energy must be accompanied by a perceptible loss of mass. The theory of special relativity indicated that mass and energy should be interconvertible, at the rate of c^2 units of energy (ergs) for one unit of mass (gram), where c is the velocity of light. The Sun's rate of loss of energy by radiation (3.8×10^{33} ergs per second) is far too great to be comprehensible in ordinary terms; but, if measured in mass-units, it means that the Sun is actually losing 4,200,000 *tons* of energy every second! The Sun's mass (2×10^{33} g) is so great that even at this enormous rate, it would take 15 billion years to reduce it by one part in a thousand.

The nature of the process—or, at least, of one process—by which this transformation of mass into energy can take place, is now well understood, since it can be made to happen in the laboratory, in many different ways. This is the transformation of atoms of one sort into another, with the appearance of an amount of energy corresponding to the change in total mass; and it is to this that we now look as the source of stellar energy.

The alternative hypothesis, that positively and negatively charged particles (protons and electrons) may utterly annihilate one another, with the appearance of the whole amount of energy corresponding to the sum of their masses, has now been dropped, since no evidence of it has been observed. It does appear to happen for positive and negative electrons; and judgment may well be reserved about "heavy electrons" (mesotrons) till more is known about them.

The masses of the lighter atoms (including their outer electrons) are given in Table 1. They can be very accurately determined—mainly with the mass-spectrograph. The various isotopes—atoms of different weight, but the same chemical properties—are of course listed individually.

From the table it appears that if four hydrogen atoms could in any way be converted into one helium atom, a loss of mass of 0.02866 unit would result. This is 1/141 of the original mass. If, therefore, a thousandth part of the Sun's mass were hydrogen, and could be transmuted into helium, energy enough would be released to keep the Sun shining for 106 million years.

The excess of the tabulated mass above an even value evidently represents the possible value of the given atom as a source of energy. This excess is fairly considerable for most of the lighter atoms; but oxygen,

TABLE I

Masses of Light Atoms *

Name	Symbol	Charge	Mass
Electron	e	−1	0.00055
Neutron	n	0	1.00893
Hydrogen	H	1	1.00813
Deuterium	H^2	1	2.01473
Helium	He^3	2	3.01699
"	He^4	2	4.00386
Lithium	Li^6	3	6.01686
"	Li^7	3	7.01818
Beryllium	Be^9	4	9.01504
Boron	B^{10}	5	10.01631
"	B^{11}	5	11.01292
Carbon	C^{12}	6	12.00398
"	C^{13}	6	13.00761
Nitrogen	N^{14}	7	14.00750
"	N^{15}	7	15.00489
Oxygen	O^{16}	8	16.00000

* These values are from Bethe's computations, with some changes on the basis of his latest work.

carbon and helium are relatively very stable. Some atoms heavier than oxygen are still further "down"; but the differences are small and need not concern us here. Per unit of mass, hydrogen is much the best source of energy.

Two types of nuclear reaction, in the laboratory, give rise to reactions of the sort considered:—the penetration into a heavier nucleus of a neutron, or of a charged particle (proton, deuteron, or alpha-particle). The former are the easiest to produce artificially—since the other nucleus does not repel the neutron. They occur in great variety, and liberate large amounts of energy. For this very reason, it appears that they are not of much astrophysical importance. They happen *too* easily. If there were a lot of neutrons in the interior of a star, they would all collide with atomic nuclei of one sort or another—sometimes building up a heavier atom, sometimes causing the emission of some other particle—and they would be used up in the process, in the twinkling of an eye. There might be an almost explosive outburst of heat (under the altogether unnatural conditions which we have imagined) but anyhow the neutrons would be gone, and similar reactions would occur in future only if "new" neutrons were produced by other nuclear reactions. A careful study by several investigators has shown that this should happen so very rarely that its effects can be neglected.

We are left, then, as a steady source of energy, with the penetration of one charged nucleus into another. The repulsion between the two is always so great that only an exceptionally violent head-on collision between particles moving much faster than the average would be effective. The probability of such a collision has been calculated. It diminishes very rapidly with the charges of the particles, so that a proton is vastly more likely to succeed than an alpha-particle, while, even so, it stands very little chance of getting into a nucleus with a larger charge than oxygen (provided, at least, that the temperature is not more than twenty million degrees).

What happens after the proton gets in depends on the individual peculiarities of the nucleus which it has hit. It may simply go in and stay—producing a new nucleus greater by one in both charge and mass; or a positive electron may be ejected, leaving a nucleus of the same charge as before but of greater mass; or the old nucleus may break up into two or more pieces—one of which is usually an alpha-particle. None of these changes can happen if the hypothetically resulting nucleus is heavier than the reacting particles, for a quite impossible amount of energy would have to be supplied from no known source. If the mass of the product is less than that of the two particles, the excess energy will appear as kinetic energy when the nucleus breaks up, or an electron is ejected; [1] in the first case, it appears as radiation—a gamma-ray.

All three processes occur in the stars.

The astrophysical importance of reactions of this type was first pointed out by Atkinson. They have been discussed by von Weizsäcker, Gamow and Bethe, and enough is now known about the properties of nuclei to give a pretty clear picture of what would happen.

First. The one stable thing in this microcosm of change is the alpha-particle (He^4). Apparently nothing can happen to it. The nuclei which might imaginably be formed by collision with a proton (He^5 or Li^5) are unstable, and do not exist at all; a collision of two alpha-particles would give Be^8, which is slightly unstable, and breaks up again. Three fast-moving alpha-particles, colliding simultaneously, might possibly form C^{12} with liberation of energy; but such an event is so very improbable that it can be neglected, except at much higher temperatures.

If, then, hydrogen is "the fuel of the stars" helium is the ashes.

---

[1] Much of the energy in this case is believed to be carried off by that elusive particle, a neutrino.

Second. The other light nuclei, up to and including boron, are highly susceptible to proton collisions. The different ones go through various transformations, but every sequence ends irrevocably in helium—for example, $Li^7 + H = 2\ He^4$, $B^{11} + H = 3\ He^4$. (In this last case the nucleus breaks into three pieces—all helium.) Hence these elements, if originally present in a star, would be successively exhausted as the reactions went on. If we imagine a mass equal to the Sun's slowly contracting from a large size, with increasing central temperature, then, according to Bethe, the reaction which eats up deuterium would happen fast enough to supply the star's radiation, when the central temperature was 360,000° (more or less, depending on the amount present).

So long as any considerable amount of deuterium remained, the star would not contract further. As it was exhausted a new contraction would begin, to be halted when the central temperature was about 2,000,000° and lithium began to be transformed and used up. Beryllium would have a similar fate at a central temperature of $3\frac{1}{2}$ million degrees, and the two isotopes of boron at five and nine million. Since the Sun's central temperature is much higher than this, there can be practically none of any of those elements left in its interior. At the surface, spectroscopic evidence shows that all three are present, but in very small quantities.

Third. At a temperature somewhat above fifteen million degrees, carbon begins to be attacked, and something quite new happens, which is tabulated by Bethe as follows:

$$C^{12} + H = N^{13} + \gamma \quad \text{2,500,000 years [2]}$$
$$N^{13} = C^{13} + \epsilon^+ \quad \text{9.9 minutes}$$
$$C^{13} + H = N^{14} + \gamma \quad \text{50,000 years}$$
$$N^{14} + H = O^{15} + \gamma \quad \text{4,000,000 years}$$
$$O^{15} = N^{15} + \epsilon^+ \quad \text{2.1 minutes}$$
$$N^{15} + H = C^{12} + He^4 \quad \text{20 years}$$

The proton goes into C^{12} and builds up N^{13} (the spare energy escaping as a γ-ray). This is one of the artificial radio-active elements, which has been produced in the laboratory. It emits a positive electron (ϵ^+) and goes over into the carbon isotope C^{13}. The next proton turns this into ordinary nitrogen N^{14}, and another produces a radio-active oxygen O^{15} which goes over into "heavy" nitrogen N^{15}. A proton collision might build this up into ordinary oxygen O^{16}; but Bethe calculates that

[2] [Subsequent research has considerably modified the original approximate time intervals. The values tabulated above are taken from Ambartsumian's *Theoretical Astrophysics*, J. B. Sykes, trans. (Pergamon, New York, 1958).]

it is ten thousand times more probable that it would split the nucleus into two parts, one of which is helium, and the other the original carbon C^{12}, ready to be used over again! The carbon is not consumed, but acts as a catalyst for the transformation of hydrogen into helium, and we have a regenerative process, which should continue until all the hydrogen has been transformed.

Four of the six steps here described have already been observed in the laboratory. The other two (involving collisions of protons with nitrogen) can reasonably be inferred from known data. The average length of time during which an atom of each kind may be expected to last before its next adventure, is given in the table. The $N^{14} + H$ reaction is the slowest; it will then be the "bottle-neck" for the whole process. Bethe's calculations show that it would supply energy enough for the Sun at a temperature of 18,300,000° (assuming 35 percent of hydrogen and 10 percent of nitrogen). As the rate of reaction increases as the 18th power of the temperature, it would do the work, with 1 percent of nitrogen, at a temperature of 20,800,000°. Now these temperatures, which are calculated from pure nuclear theory, agree excellently with the central temperature of the Sun, as calculated from astrophysical data. An almost equally good agreement is found for Sirius (22 million from theory, 26 from observation), and for the very hot and massive star Y Cygni (30 and 32 million). Moreover, the gradual changes in light, diameter, density and surface temperature for stars of different mass along the whole main sequence are in excellent agreement with the results of the theory.

It appears, then, that one great part of the problem of the source of stellar energy has been fully solved. The main sequence is explained in detail.

[But see the last paragraph below; the proton-proton mechanism is now accepted for fainter main-sequence stars, including the Sun.]

There are still worlds left to conquer. The very rapid energy-liberation in the giants cannot be accounted for in the same way. It might be assumed that they are built on the same sort of model as the main-sequence stars, but contain larger quantities of lithium, beryllium, or boron—but this is a far-fetched idea, and does not explain why they form a sequence in which the reddest stars are the brightest. It is also possible, as Gamow, Öpik, and others have suggested, that they are built on quite a different model, with relatively small and dense cores, which may be as hot as those of the main-sequence stars or hotter. Much more is likely to be done on this problem within a few years.

Reactions involving heavier nuclei are possible, but they would occur at significant rates only at higher temperatures, and so would not have a chance to happen if there were any carbon in the stars—as there assuredly is. If, at the other extreme, we had a star composed originally of pure hydrogen, it might get energy by the following chain of reactions, also according to Bethe:

$$H + H = H^2 + \epsilon^+$$
$$H^2 + H = He^3$$
$$He^3 + He^4 = Be^7$$
$$Be^7 + \epsilon^- = Li^7$$
$$Li^7 + H = 2\ He^4$$

Two protons for a deuteron—a slow reaction; the latter goes over almost at once into He^3; this with an alpha-particle forms a known isotope of beryllium, which turns to lithium and then to helium—the net result being, as usual, the building of four hydrogen atoms into one of helium.

The rate of this process changes much more slowly with the temperature than that of the carbon cycle. Bethe suggests that it may supply the energy in the coolest stars of the main sequence. Present data do not suffice for a decision. . . .

41. TURBULENCE—INTRODUCTION TO A PHYSICAL THEORY OF ASTRONOMICAL INTEREST

By S. Chandrasekhar

May I say at the outset that I have found myself deeply inadequate for the task of giving this third Henry Norris Russell Lecture. I am afraid that I have not discovered or paved a Royal Road that I can describe to you in the manner of Dr. Russell: neither have I the excellence of the material which Dr. Adams presented in his second Henry Norris Russell Lecture. And I am aware that no general interest attaches to matters in which I may claim some degree of competence. I have therefore chosen, after considerable hesitation, to describe to you the recent advances in our understanding of the phenomenon of turbulence, in the belief that these advances are relevant to the progress of astrophysics. Perhaps it is premature to take an occasion like this to describe a physical theory which has yet to establish its relations to astronomical developments. But the history of astronomy and astrophysics shows that major advances in our understanding of astrophysical phenomena have coincided with and depended upon advances in fundamental physical theory. While many examples illustrating this can be given, there is none more conspicuous or notable in recent history than that provided by the work of Henry Norris Russell; thus, during the great period in which the foundations both of atomic spectra and of stellar spectroscopy were laid, Russell was a great exponent of both subjects. As is well known, the main features of the theory of complex spectra emerged for the first time from the pioneering investigations of Drs. Russell and Saunders on the alkaline earths. The main conclusion of these investigations, stated by the authors in the words "both valency electrons may jump at once from outer to inner orbits, while the net energy lost is radiated as a single quantum," has since been incorporated into the analysis of stellar spectra as the "Russell-Saunders" coupling and is one of the keystones of atomic theory. In these early papers of Dr. Russell all the steps preliminary to the formulation of the exclusion principle were taken, and I do not believe that it is a misstatement of history to say that the honor of the discovery of the exclusion principle would have gone to Russell had his concern with the application of the principles of atomic spectra to astrophysical problems been a little less. However that may be, as astronomers we may count ourselves fortunate that Russell's concern with astrophysical problems was as earnest then as it has always been, for otherwise, we should not have had so immediate or so complete an integration of physical and astrophysical theories as was, in fact, achieved when Russell's great work on the quantitative analysis of the solar spectrum and the first determination of the composition of the sun appeared in 1929. I have referred to this example of Russell's work to emphasize the interdependence of physical and astrophysical thought. And, as I have stated, it seems to me probable that the recent advances in the physics of turbulence, due in large measure to G. I. Taylor, von Karman, Kolmogoroff, and Heisenberg, may play an important part in the future developments of astrophysics. But, before I describe the nature of these advances in physical theory, I may perhaps indicate briefly the astrophysical contexts in which they may find their most fruitful applications.

The first person clearly to draw attention to the importance for astrophysics of turbulence with its correct hydrodynamical meaning was Rosseland. In a paper published in 1928, Rosseland pointed out that if different motions—i.e., motions of one part relative to another—occur in cosmical gas masses, then the motions should be turbulent in the sense that we should not expect to describe them in terms of the classical equations of motion of Stokes and Navier. In drawing this inference, Rosseland was guided by the experience in meteorology and oceanography and by the following reasoning.

We are all familiar with the fact that a linear flow of water in a tube can be obtained only for velocities below a certain critical limit and that, when the velocity exceeds this limit, laminar flow ceases and a complex, irregular, and fluctuating motion sets in. More generally than in this context of flow through a tube, it is known that motions governed by the equations of Stokes and Navier change into turbulent motion when a certain nondimensional constant called the "Reynolds number" exceeds a certain value of the order of 1000. This Reynolds number depends upon the linear dimension, L, of the system, the coefficient of viscosity μ, the density ρ, and the velocity v in the following manner:

$$R = \frac{\rho v L}{\mu}.$$

Since R depends directly on the linear dimension of the system, Rosseland argued that motions in the oceans, in terrestrial and planetary atmospheres, and still more in stellar atmospheres, once they occur, must become turbulent in this sense. Rosseland further pointed out that, if turbulence develops, the coefficients of viscosity and heat conduction may be expected to increase a million fold. And the importance of this enhanced efficiency of heat and momentum transport in a turbulent medium cannot be exaggerated.

Stimulated by Rosseland's ideas, McCrea suggested in the same year that the solar chromosphere must be in a state of turbulence and that this turbulence may, in part, contribute to its support against gravity.

About a year later Harold Jeffreys drew attention to a fact which had been ignored until then, namely, that if the generation of energy inside stars is confined to a small region at the center, then the radiation will not be able to dispose of it at a gradient under the adiabatic and that, if a superadiabatic gradient comes into being, vertical currents will be generated which will effectively restore the adiabatic gradient, leaving, however, a slight superadiabatic gradient to make possible the transport of heat. The condition for the occurrence of such convective transport of heat can be written down, and it follows from this condition that even a relatively mild concentration of the energy sources toward the center will lead to its occurrence near the center. Indeed, with the clarification of the source of stellar energy as due to nuclear transformations, it is now generally recognized that all stars must have convective cores in which turbulence prevails. And, as was shown, particularly by Cowling, the existence of turbulence is of primary importance in all considerations relating to the stability of stars.

Returning to the role of turbulence in the atmospheres of the stars, we next observe that the investigations of Struve and Elvey established the occurrence of large-scale motions in the atmospheres of stars like 17 Leporis, ϵ Aurigae, and α Persei. In investigations, which, it may be noted, were also the first to apply the then new method of the curve of growth to the analysis of stellar atmospheres—the method had already been applied to the solar atmosphere by Minnaert—Struve and Elvey showed that the linear portion of the curve of growth, as well as the line profiles themselves, cannot be explained in terms of the Doppler effect due to thermal motions alone and that large-scale motions of a turbulent nature must be postulated. This conclusion has since been confirmed and extended by various other investigators.

That turbulence must play a part also in the solar atmosphere became clear after Unsöld had shown that in the deeper layers of the solar photosphere, where hydrogen begins to get ionized, the radiative gradient must become unstable. Since that time the first view advanced by Siedentopf and Biermann, that the solar granulation must, in some way, be related to this hydrogen convection zone, has been steadily gaining ground.

Again the investigations of Struve and his associates during the past few years have shown that the shells surrounding early-type stars and the gaseous envelopes in which spectroscopic binaries are frequently imbedded must also be turbulent, the turbulence in these contexts arising, in the first instance, from the different parts of the shell or medium rotating with different angular velocities.

And, finally, it would appear that the interstellar clouds must also be in a state of turbulence; for, assuming that a typical cloud is 10 parsecs in diameter and that relative motions to the extent of 10 km/sec occur, we find that the Reynolds number must be of the order of 10^5; and the motions inside the cloud must therefore be turbulent. The even larger question now occurs whether we may not indeed regard the clouds of various dimensions in interstellar space as eddies in a medium occupying the whole of galactic space.

From this brief survey of the various problems in which turbulence may play a role, it would almost appear that, if we are in the mood for it, we may encounter turbulence no matter where we turn. But what is the picture of turbulence in terms of which we wish to interpret such a wide diversity of phenomena? It is that in a turbulent medium there are eddies which spontaneously form and disintegrate; that this process goes on continuously; that each eddy travels a certain average distance with a certain average speed before it loses its identity—a specific enough picture but not one derived from, or justified by, a physical theory. Thus, while the basic concepts of "mean free path" and "root-mean-square velocity" which underlie the picture are plausible enough, it was not known how these quantities were to be related with the physical conditions of the problem. Indeed, from the point of view of a rational physical theory, the situation has been so unsatisfactory that, in a recent conversation, Dr. Russell recalled that E. W. Brown, referring to the frequency with which appeals were being made to the action of a resisting medium to account for this or that anomaly in the motions of celestial bodies, once remarked: "What fifty years ago used to be attributed to the direct intervention of the Deity are now being attributed to a resisting medium." Dr. Russell added that he sometimes felt the same way about the frequency with which turbulence is currently being invoked to account for astrophysical phenomena. Nevertheless, it would seem that the application of the newer developments in the theory of turbulence may help to remove this element of the miraculous in astrophysics.

VIII

THE SPECTRUM–LUMINOSITY
RELATIONS

The plotting of the colors (or spectra) of stars as abscissae against their absolute magnitudes (total luminosities) has become one of the most lucrative adventures in the study of star light. It provides clues to stellar evolution and also helps to estimate distances. It all started by Hertzsprung's pointing out (in a rather obscure paper) that stars of the same color (spectral class) can differ enormously in size and therefore in luminosity, and by Russell's enlarging on this idea and actually plotting and analyzing the available material. Russell's diagram showed the reversed 7 arrangement—that is, a main sequence of stars, with the color index increasing with diminishing brightness, and a horizontal branch of giant stars, centered around absolute magnitude zero, with no particular change of magnitude across the spectral series from blue stars, classes O and B, to red stars of classes M and N.

Hertzsprung writes (1958): "I myself never used the designations 'giants' and 'dwarfs,' as the mass does not vary in an extravagant way, as does the density." On the basis of the great contrast in volume, however, I feel that we are justified in using the common terms Giant and Dwarf when referring to yellowish and reddish stars.

That there exists an important deviation from the horizontal branch of the Hertzsprung-Russell diagram was shown very soon after the publication of Russell's study, which was based only on the neighboring stars. The distant globular star clusters were found, as is shown, for example, in Fig. 3 of selection 43, to reveal a sharply rising luminosity when color progresses from blue to red; but little attention was paid to this indication of a second type of population until W. Baade's noteworthy analysis that is here presented in selection 45.

42. GIANTS AND DWARFS

By Ejnar Hertzsprung

In volume 28 of the "Annals of the Astronomical Observatory of Harvard College" a detailed survey of the spectra is given for northern and southern bright stars by Antonia C. Maury and Annie J. Cannon, respectively.

The first two columns of Table 1 give a short summary of the spectral class designation used by the two authors. In the last two columns are listed characteristic stars along with their spectral types. For a more detailed description of the characteristics used we must refer to the original papers cited above. Here we can find room for only a few words concerning the three sub-classifications b, a, and c. The b stars have broader lines than those of "division" a. The relative intensities of the lines seem, however, to be equal for a- and b-stars "so that there appears to be no decided difference in the constitution of the stars belonging, respectively, to these two divisions." As the most important characteristics of subclass c we can mention, first, that the lines are unusually narrow and sharp; second, that among the "metallic" lines others occur which are not identifiable with any solar lines, and the relative intensities of the remainder do not correspond with the intensities observed in the solar spectrum. "In general, division c is distinguished by the strongly defined character of its lines, and it seems that stars of this division must differ more decidedly in constitution from those of division a than is the case with those of division b." Antonia C. Maury suspects that the a- and b-stars on the one hand and the c-stars on the other, belong to collateral series of development. That is to say not all stars have the same spectral development. What determines such a differentiation (differences in mass and constitution, etc.) is a question that remains unanswered.

The question arises how great the systematic differences of the brightness, reduced to a common distance, of stars of the different groups will be. For this purpose I have used the proper motions of the stars in the following simple manner.

For each group a value was determined above and below which lies,

TABLE I

Spectral Class according to		Number		Mean annual proper motion for $m_H=0$	Mean magnitude reduced to the annual Proper Motion of ".01			Mean deviation from the mean		Mean error of the mean	Number	Star Color		Characteristic Star	
AJC	ACM	in ACM	Adjusted			Number						m_D for $m_H=4.5$	$\dfrac{\Delta m_H}{\Delta m_D}$	Name	Spectrum
I	II	III	IV	V	VI	VII	VIII	IX	X	XI	XII	XIII	XIV	XV	XVI
Oe5B	I	7	3	.067	(4.13)						4	(4.66)		S Monocerotis	Ib
B to B3A	II	14	11	.059	3.84	38	4.37	−3.02	+2.08	±.41	9	4.73	1.36	ε Orionis	IIa
	III	17	11	.069	4.20						13	4.65	1.35	α Virginis	IIIb
	IV	30	9	.069	4.19						26	4.58	1.35	γ Orionis	IVa
	IV'	20	7	.127	5.51						18	4.59	1.17	π₄ Orionis	IV'a
B5A to B9A	V	22	9	.199	6.49	21	7.25	−1.40	+1.22	±.29	19	4.55	1.32	τ Orionis	Vb
	VI	22	10	.411	8.07						19	4.58	1.30	α Leonis	VIb
	VI'	3	2	.520	(8.58)						3	(4.49)		η Aquarii	VI'b
A	VII	32	21	.386	7.93	47	8.05	−1.38	+1.48	±.21	38	4.56	1.27	α Canis maj.	VIIa
	VIII	43	26	.439	8.21						43	4.65	1.34	α Geminorum	VIIIa
	IX	29	16	.547	8.69						26	4.73	1.22	δ Ursae maj.	IXb
AF and F	X	19	9	.646	9.05	34	9.06	−2.03	+1.77	±.33	15	4.82	1.36	α Aquilae	Xb
	XI	13	7	.721	9.29						13	4.86	1.21	δ Aquilae	XIa,b
	XI'	3	2	.568	(8.77)						3	(4.80)		ζ Leonis	XI'ab
FG	XII	30	18	1.197	10.39	30	11.23	−2.25	+2.06	±.40	24	5.03	1.36	α Canis min.	XIIa
	XIII	21	12	3.251	12.56						21	5.02	1.41	χ₁ Orionis	XIIIa
	XIII'	1	1	2.466	(11.96)						1	(5.16)		θ Persei	XIII'a
G and GK	XIV	19	11	.481	8.41	24	7.93	−2.46	+4.01	±.31	16	5.23	1.34	α Aurigae, Sun	XIVa
	XIV'	21	12	.321	7.53						20	5.29	1.53	κ Geminorum	XIV'a
	XV₁	25	21	1.081	10.17						24	5.39	1.46	α Bootis	XV₁a
K	XV	49	21	.745	9.36	74	9.38	−2.06	+1.68	±.22	42	5.44	1.49		XVa
	XV₂	38	32	.658	9.09						36	5.46	1.58	α Cassiopeiae	XV₂a
	XV'	6	5	.908	4.79						6	(5.75)		β Cancri	XV'a
KM	XVI	23	16	.449	8.26	21	7.77	−3.07	+2.01	±.56	22	5.68	1.75	α Tauri	XVIa
Ma	XVII	18	9	.378	7.89	15	8.28	−1.12	+1.15	±.30	15	5.81	1.41	β Andromedae	XVIIa
	XVIII	20	6	.479	8.40						16	5.84	1.45	α Orionis	XVIIIa
	XIX	6	1	.891	(9.75)						5	5.81		ρ Persei	XIXa

respectively, one-half of the proper motions expressed in arc of a great circle, and reduced to magnitude 0. These values are listed in column V of Table 1. In column VI are found the corresponding magnitudes reduced to a proper motion of 1″ in a hundred years. (Reduced to 1″ annual proper motion the stars would be 10 magnitudes brighter.) In column VIII are the mean reduced stellar magnitudes for somewhat large groups, and in the following two columns the values above and below which 15% of the total lies. These values will be, therefore, the mean deviation from the median. Finally there are listed in column XI the mean errors of the medians.

Table 1 contains only stars of subclasses a and b for which I have found proper motions based on the latest determinations of the Fundamental stars (Newcomb precession constants). Also in addition to the c-stars, all stars are omitted which are recognized as variable or the spectra of which were described as "peculiar." The total number of the a and b stars found in Antonia C. Maury's catalogue are given in column III, and in column IV the number of stars remaining after these omissions. I have also attempted to bring together all stars brighter than the 5th magnitude for which spectral class (according to the above-named authors, or to the Draper Catalogue) as well as proper motion could be found, and I come to the same result as that which appears in Table 1. In spite of the small number (308) of stars taken into consideration in Table 1, I consider the picture they give us as more reliable than would be that from a larger number of much more uncertainly classified spectra used in connection with a too great value for the small proper motions (Orion stars).

The radial velocity found for about 60 stars has an approximately typical distribution with a mean deviation from zero of some ±20 km/sec. It is therefore probable that the projection of the absolute proper motions on a randomly chosen direction would also have a typical distribution. We have, however, also considered the projection of the apparent proper motions on a plane at right angles to the line of sight; and we ask which mean deviation in the star magnitudes, reduced to equal apparent proper motions, would uniquely result (corresponding to the assumption that all stars have the same absolute magnitude). The values are about +1.2 and −1.57 magnitudes. Comparing these with those in columns IX and X in Table 1, we find that the stars which were put together in the A-class cannot differ very much among themselves in absolute magnitude. According to this result, combined with the fact that membership in spectral A-class is easily recognized, I have assembled for 100 A-stars of magnitude 4.62–5.00 the proper motions in declination only. If one arranges these according to magnitude, the value $-.''008$ lies in the middle, and respectively 15% of the total is over $+.''0325$ and under $-.''0575$. From this can be calculated the mean deviation $\pm.''0448$ annually, which would correspond to a speed of ± 20 km/sec, or 4 astronomical units per year. According to this, we find for the 100 A-stars of mean magnitude 4.84 the mean parallax of $.''0112$. In Table 1 the magnitudes are reduced to a mean annual proper motion of $.''01$ in arc of a great circle, corresponding to a parallax of some $.''002$. For the 100 A-stars we compute with the parallax the mean stellar magnitude of 8.6, in fair agreement with the value 8.05 from Table 1. . . .

Further I have in column XIII, Table 1, inserted values which can be taken as a sort of color-equivalent and which were derived in the following way from the visual magnitudes taken from the revised Harvard Photometry (H.P.) and the photographic magnitudes (corresponding to G-line light of wave length $.432\mu$) taken from the Draper Catalogue (D.C.). Within each group, for the number of stars in column XII, both magnitudes m_H and m_D were brought together, and, on the approximately correct assumption that a linear relation exists between them, that value of m_D was calculated which corresponds to $M_H = 4.5$. Further we have in column XIV for each group the computed ratios $\Delta m_H : \Delta m_D$. Actually they should be constant with the value 1. That they increase from white through yellow to red may be due to the Purkinje phenomenon. That they all lie appreciably above 1 can be due to the circumstance that the normal intensity scale, which was used for the determination of the D.C. magnitudes through comparison of the spectral darkening in the neighborhood of the G-line ($\lambda = .432\mu$), was established not in pure G-light but by means of the Carcel-lampe. . . .

The minimum shown in column XIII in the neighborhood of the A-group appears to be real. Accordingly the Orion stars would be somewhat yellower than the A-stars. . . .

In any case we may say that the annual proper motion of an average c-star, reduced to magnitude 0, amounts to only a few hundredths of a second. With the relatively large errors of these small values, a dependence on spectral class cannot be recognized. In other words, the c-stars are at least as bright as the Orion stars. In both of the spectroscopic binaries o Andromedae and β Lyrae the brightness of the c-star and of the companion star of the Orion type appear to be of the same order of brightness. The proper motions (not here given) are all small, according to the Auwers-Bradley Catalogue. . . . For the stars in Annie J. Cannon's listing that have narrow sharp lines, I can also find only small proper motions. This result confirms the assumption of Antonia C. Maury that the c-stars are something unique.

When the c- and ac-stars are looked at in summary fashion one sees that with increasing Class number [advancing toward redder spectra] the c-characteristic diminishes, and that these stars stop exactly where the bright K-stars begin.

43. THE SPECTRUM–LUMINOSITY DIAGRAM

By Henry Norris Russell

Investigations into the nature of the stars must necessarily be very largely based upon the average characteristics of groups of stars selected in various ways—as by brightness, proper motion, and the like. The publication within the last few years of a great wealth of accumulated observational material makes the compilation of such data an easy process; but some methods of grouping appear to bring out much more definite and interesting relations than others, and of all the principles of division, that which separates the stars according to their spectral types has revealed the most remarkable differences, and those which most stimulate attempts at a theoretical explanation.

In the present discussion, I shall attempt to review very rapidly the principal results reached by other investigators, and shall then ask your indulgence for an account of certain researches in which I have been engaged during the past few years.

Thanks to the possibility of obtaining with the objective prism photographs of the spectra of hundreds of stars on a single plate, the number of stars whose spectra have been observed and classified now exceeds one hundred thousand, and probably as many more are within the reach of existing instruments. The vast majority of these spectra show only dark lines, indicating that absorption in the outer and least dense layers of the stellar atmospheres is the main cause of their production. Even if we could not identify a single line as arising from some known constituent of these atmospheres, we could nevertheless draw from a study of the spectra, considered merely as line-patterns, a conclusion of fundamental importance.

The spectra of the stars show remarkably few radical differences in type. More than ninety-nine per cent of them fall into one or other of the six great groups which, during the classic work of the Harvard College Observatory, were recognized as of fundamental importance, and received as designations, by the process of "survival of the fittest," the rather arbitrary series of letters B, A, F, G, K, and M. That there should be so very few types is noteworthy; but much more remarkable is the fact that they form a continuous series. Every degree of gradation, for example, between the typical spectra denoted by B and A may be found in different stars, and the same is true to the end of the series, a fact recognized in the familiar decimal classification, in which B5, for example, denotes a spectrum half-way between the typical examples of B and A. This series is not merely continuous; it is *linear*. There exist indeed slight differences between the spectra of different stars of the same spectral class, such as A0; but these relate to minor details, which usually require a trained eye for their detection, while the difference between successive classes, such as A and F, are conspicuous to the novice. Almost all the stars of the small outstanding minority fall into three other classes, denoted by the letters O, N, and R. Of these O undoubtedly precedes B at the head of the series, while R and N, which grade into one another, come probably at its other end, though in this case the transition stages, if they exist, are not yet clearly worked out.

From these facts it may be concluded that the principal differences in stellar spectra, however they may originate, arise in the main from variations in a single physical condition in the stellar atmospheres. This follows at once from the linearity of the series. If the spectra depended, to a comparable degree, on two independently variable conditions, we should expect that we would be obliged to represent their relations, not by points on a line, but by points scattered over an area. The minor differences which are usually described as "peculiarities" may well represent the effects of other physical conditions than the controlling one.

The first great problem of stellar spectroscopy is the identification of this predominant cause of the spectral differences. The hypothesis which suggested itself immediately upon the first studies of stellar spectra was that the differences arose from variations in the chemical composition of the stars. Our knowledge of this composition is now very extensive. Almost every line in the spectra of all the principal classes can be produced in the laboratory, and the evidence so secured regarding the uniformity of nature is probably the most impressive in existence. The lines of certain elements are indeed characteristic of particular spectral classes; those of helium, for instance, appear only in Class B, and form its most distinctive characteristic. But negative conclusions are proverbially unsafe. The integrated spectrum of the Sun shows no evidence whatever of helium, but in that of the chromosphere it is exceedingly conspicuous. Were it not for the fact that we are near this one star of Class G, and can study it in detail, we might have erroneously concluded that helium was confined to the "helium stars." There are

DIAGRAM 255

other cogent arguments against this hypothesis. For example, the members of a star-cluster, which are all moving together, and presumably have a common origin, and even the physically connected components of many double stars, may have spectra of very different types, and it is very hard to see how, in such a case, all the helium and most of the hydrogen could have collected in one star, and practically all the metals in the other. A further argument—and to the [writer] a very convincing one—is that it is almost unbelievable that differences of chemical composition should reduce to a function of a single variable, and give rise to the observed linear series of spectral types.

I need not detain you with the recital of the steps by which astrophysicists have become generally convinced that the main cause of the differences of the spectral classes is difference of temperature of the stellar atmospheres. . . .

I will now ask your attention in greater detail to certain relations which have been the more special objects of my study.

Let us begin with the relations between the spectra and the real brightness of the stars. These have been discussed by many investigators—notably by Kapteyn and Hertzsprung—and many of the facts which will be brought before you are not new; but the observational material here presented is, I believe, much more extensive than has hitherto been assembled. We can only determine the real brightness of a star when we know its distance; but the recent accumulation of direct measures of parallax, and the discovery of several moving clusters of stars whose distances can be determined, put at our disposal far more extensive data than were available a few years ago.

Figure 1 shows graphically the results derived from all the direct measures of parallax available in the spring of 1913 (when the diagram was constructed). The spectral class appears as the horizontal coordinate, while the vertical one is the absolute magnitude, according to Kapteyn's definition,—that is, the visual magnitude which each star would appear to have if it should be brought up to a standard distance, corresponding to a parallax of $0''.1$ (no account being taken of any possible absorption of light in space.) The absolute magnitude -5, at the top of the diagram, corresponds to a luminosity 7500 times that of the Sun, whose absolute magnitude is 4.7. The absolute magnitude 14, at the bottom, corresponds to 1/5000 of the Sun's luminosity. The larger dots denote the stars for which the computed probable error of the parallax is less than 42 per cent of the parallax itself, so that the probable error of the resulting absolute magnitude is less than $\pm 1^m.0$. This is a fairly tolerant criterion for a "good parallax," and the small dots, representing the results derived from the poor parallaxes, should hardly be used as a basis for any argument. The solid black dots represent stars whose parallaxes depend on the mean of two or more determinations; the open circles, those observed but once. In the latter case, only the results of those observers whose work appears to be nearly free from systematic error have been included, and in all cases the observed parallaxes have been corrected for the probable mean parallax of the comparison stars to which they were referred. The large open circles in the upper part of the diagram represent mean results for numerous bright stars of small proper motion (about 120 altogether) whose observed parallaxes hardly exceed their probable errors. In this case the best thing to do is to take means of the observed parallaxes and magnitudes for suitable groups of stars, and then calculate the absolute magnitudes of the typical stars thus defined. These will not exactly correspond to the mean of the individual absolute magnitudes, which we could obtain if we knew all the parallaxes exactly, but they are pretty certainly good enough for our purpose.

Upon studying Figure 1, several things can be observed.

1. All the white stars, of Classes B and A, are bright, far exceeding the Sun; and all the very faint stars,—for example, those less than $\frac{1}{50}$ as bright as the Sun,—are red, and of Classes K and M. We may make this statement more specific by saying, as Hertzsprung does, that there is a certain limit of brightness for each spectral class, below which stars of this class are very rare, if they occur at all. Our diagram shows that this limit varies by rather more than two magnitudes from class to class. The single apparent exception is the faint double companion to o^2 Eridani, concerning whose parallax and brightness there can be no doubt, but whose spectrum, though apparently of Class A, is rendered very difficult of observation by the proximity of its far brighter primary.

2. On the other hand, there are many red stars of great brightness, such as Arcturus, Aldebaran and Antares, and these are as bright, on the average, as the stars of Class A, though probably fainter than those of Class B. Direct measures of parallax are unsuited to furnish even an estimate of the upper limit of brightness to which these stars attain, but it is clear that some stars of all the principal classes must be very bright. The range of actual brightness among the stars of each spectral class therefore increases steadily with increasing redness.

3. But it is further noteworthy that all the stars of Classes K5 and M which appear on our diagram are either very bright or very faint.

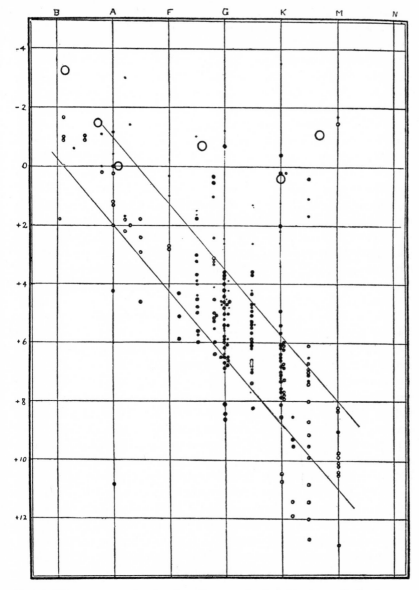

Fig. 1. The spectrum-luminosity diagram for bright stars. Ordinates are absolute magnitudes; abscissae, spectral classes.

There are none comparable with the Sun in brightness. We must be very careful here not to be misled by the results of the methods of selection employed by observers of stellar parallax. They have for the most part observed either the stars which appear brightest to the naked eye or stars of large proper motion. In the first case, the method of selection gives an enormous preference to stars of great luminosity, and, in the second, to the nearest and most rapidly moving stars, without much regard to their actual brightness. It is not surprising, therefore, that the stars picked out in the first way (and represented by the large circles in Figure 1) should be much brighter than those picked out by the second method (and represented by the smaller dots). But if we consider the lower half of the diagram alone, in which all the stars have been picked out for proper-motion, we find that there are no very faint stars of Class G, and no relatively bright ones of Class M. As these stars were selected for observation entirely without consideration of their spectra (most of which were then unknown), it seems clear that this difference, at least, is real, and that there is a real lack of red stars comparable in brightness to the Sun, relatively to the number of those 100 times fainter.

The appearance of Figure 1 therefore suggests the hypothesis that if we could put on it some thousands of stars, instead of the 300 now available, and plot their absolute magnitudes without uncertainty arising from observational error, we would find the points representing them clustered principally close to two lines, one descending sharply along the diagonal, from B to M, the other starting also at B, but running almost horizontally. The individual points, though thickest near the diagonal line, would scatter above and below it to a vertical distance corresponding to at least two magnitudes, and similarly would be thickest near the horizontal line, but scatter above and below it to a distance which cannot so far be definitely specified, so that there would be two fairly broad bands in which most of the points lay. For Classes A and F, these two zones would overlap, while their outliers would still intermingle in Class G, and probably even in Class K. There would however be left a triangular space between the two zones, at the right-hand edge of the diagram, where very few, if any, points appeared; and the lower left-hand corner would be still more nearly vacant.

We may express this hypothesis in another form by saying that there are two great classes of stars,—the one of great brightness (averaging perhaps a hundred times as bright as the Sun), and varying very little in brightness from one class of spectrum to another; the other of smaller brightness, which falls off very rapidly with increasing redness. These

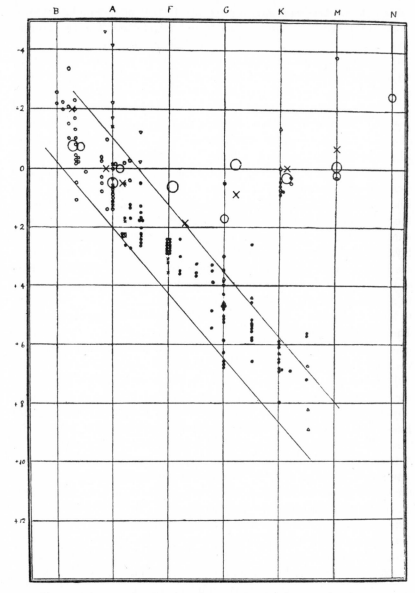

Fig. 2. Spectrum-luminosity diagram for bright groups of stars.

two classes of stars were first noticed by Hertzsprung,[1] who has applied to them the excellent names of *giant* and *dwarf* stars. The two groups, on account of the considerable internal differences in each, are only distinctly separated among the stars of Class K or redder. In Class F they are partially, and in Class A thoroughly intermingled, while the stars of Class B may be regarded equally well as belonging to either series.

In addition to the stars of directly measured parallax, represented in Figure 1, we know with high accuracy the distances and real brightness of about 150 stars which are members of the four moving clusters whose convergent points are known,—namely, the Hyades, the Ursa Major group, the 61 Cygni group, and the large group in Scorpius, discovered independently by Kapteyn, Eddington, and Benjamin Boss, whose motion appears to be almost entirely parallactic. The data for the stars of these four groups are plotted in Figure 2, on the same system as in Figure 1. The solid black dots denote the members of the Hyades; the open circles, those of the group in Scorpius; the crosses the Ursa Major group, and the triangles the 61 Cygni group. Our lists of the members of each group are probably very nearly complete down to a certain limiting (visual) magnitude, but fail at this point, owing to lack of knowledge regarding the proper motions of the fainter stars. The apparently abrupt termination of the Hyades near the absolute magnitude 7.0, and of the Scorpius group at 1.5 arises from this observational limitation.

The large circles and crosses in the upper part of Figure 2 represent the absolute magnitudes calculated from the mean parallaxes and magnitudes of the groups of stars investigated by Kapteyn, Campbell, and Boss. . . . The larger circles represent Boss's results, the smaller circles Kapteyn's, and the large crosses Campbell's.

It is evident that the conclusions previously drawn from Figure 1 are completely corroborated by these new and independent data. Most of the members of these clusters are dwarf stars, and it deserves particular notice that the stars of different clusters, which are presumably of different origin, are similar in absolute magnitude. But there are also a few giant stars, especially of Class K, (among which are the well-known bright stars of this type in the Hyades); and most remarkable of all is Antares, which, though of Class M, shares the proper motion and radial velocity of the adjacent stars of Class B, and is the brightest star in the group, giving out about 2000 times the light of the Sun. It is also clear that the naked-eye stars, studied by Boss, Campbell and Kapteyn, are for the most part giants.

[1] *Zeitschrift für Wissenschaftliche Photographie 3,* 442 (1905).

Color Class

Limits of Photovisual Magnitude	BO	BO to B5	B5 to AO	AO to A5	A5 to FO	FO to F5	F5 to GO	GO to G5	G5 to KO	KO to K5	K5 to MO	MO to M5	M5	All Colors
10.20–10.39													1	1
10.40–10.59														
10.60–10.79													1	1
10.80–10.99						1			1				3	4
11.00–11.19											3	3		7
11.20–11.39					1					2	2	6		9
11.40–11.59									1	2	2	1		5
11.60–11.79								2	2	5	1			5
11.80–11.99							1	3	3		1	1		9
12.00–12.19						1		6	8	2				5
12.20–12.39							1	4	5	2	2			14
12.40–12.59						1	3	4	15	2				13
12.60–12.79		1				1	6	9	3	1				25
12.80–12.99					1	8	6	6	3	1				11
13.00–13.19			1	1	1	12	28	16	1	1				10
13.20–13.39	1	1	2	1	4	34	59	7						20
13.40–13.59		1	11		17	39	12							23
13.60–13.79			19	40	61	39	3							57
13.80–13.99			3	26	20	2								106
14.00–14.19														71
14.20–14.39	1													156
14.40–14.59														68
14.60–14.79														3
Totals	1	3	36	68	105	138	123	57	45	17	10	12	8	623

Fig. 3. The color-luminosity diagram for the bright stars in the globular cluster Messier 22.

[The first evidence of a stellar population other than that depicted here by Russell came soon, in 1915, through the determination of the color-magnitude arrays for the brighter stars in Messier 13 (Mt. Wilson Contr. No. 116, Table XII), and later for other globular clusters. To illustrate the difference from the Russell "reversed seven" distribution, the array for the giant stars in the globular cluster Messier 22 is reproduced in Figure 3 (from "Star Clusters," pp. 29, 205 [1930]) ; the array shows no trace of Russell's giant-star branch, which depends on the stars of the solar neighborhood, and it does reveal a steep rise from the blue stars to the red giants.

[These early globular cluster observations and the steep rise were also discussed by ten Brugencate in his book, *Sternhaufen* (Berlin, 1927). Since 1950 the color-magnitude arrays for a few globular clusters have been much extended, with important bearing on stellar evolution problems, by Sandage, Johnson, Arp, and Baum, working with the 200-inch telescope on Mount Palomar.]

44. SPECTRAL TYPES IN OPEN CLUSTERS

By Robert J. Trumpler

Among the variety of objects that are included in the category of star clusters there is only one class that stands out distinctly from the rest: the extremely rich and highly condensed globular cluster. All the remaining objects are generally called *open* clusters; since they lie scattered along the path of the Milky Way, they might also be designated as *galactic* clusters. They are less distant than the globular clusters and contain in most cases stars bright enough for spectroscopic observations.

In a cluster we can consider all its members to be practically at the same distance from us. The absolute magnitudes of stars constituting a cluster then differ from their apparent magnitudes only by a constant, and this constant is a direct function of the distance. Star clusters thus offer an exceptionally favorable opportunity for studying the relation between spectral type and absolute magnitude. Trigonometric parallaxes as well as proper motions give us quite reliable data about the distances and luminosities of the dwarf stars of classes A to M; but so far this information is insufficient for B type stars which are of great luminosity and great average distance so that both parallaxes and proper motions are small and cannot be determined accurately. In many star clusters we find B type stars associated with dwarf stars of classes A, F or G. For these latter stars we can adopt the average luminosities of stars in general, as obtained from parallax and proper motion measures, while spectroscopic and photometric observations of such a cluster tell us directly how many magnitudes a certain subdivision of class B is more luminous than those dwarf stars. The difference between the observed magnitudes of cluster stars and the mean absolute magnitudes adopted for those spectral types further furnish the distance of the cluster.

In so far as clusters seem to be systems with common origin, and in so far as we can consider the spectral types to represent different stages of evolution, a study of spectral types in star clusters should give valuable help toward solving the problem of stellar evolution. The fact that in most clusters the stars cover a wide range of spectral types is capable of two interpretations. Either we may assume that the stars of a cluster do not originate all at the same time but vary in age, or we may follow the view that the origin of all cluster members falls in a limited period of time, that they vary, however, in original mass and on this account run through their evolutionary course with unequal speed and perhaps also following somewhat different courses of evolution.

With these problems in view an extensive program of spectroscopic observations of open clusters was undertaken. Two instruments are used for this purpose: the slitless quartz spectrograph attached to the Crossley reflector and the one-prism slit spectrograph of the 36-inch refractor. The first of these two instruments, designed by Mr. Wright, is described in *L.O.Bulletin, 9, 52*. Operating somewhat like an objective prism it photographs the spectra of all stars within a field of 20′ diameter on the same plate. The dispersion of the two prisms is such as to separate the Hβ and Hϵ lines by 7.2 mm. Because of the dense crowding of stars in the richer clusters the spectra of several stars are often overlapping; in order to avoid this condition as much as possible, the spectra are widened very little. Although this makes the classification more difficult, the spectra can in general be estimated to a few tenths of a class. For each cluster one or two long exposures of 3^h to 7^h are secured, giving useful spectra of stars up to the photographic magnitudes 13 – 14; if necessary shorter exposures are also made for the brighter stars. The field of the spectrograms includes the smaller clusters entirely, while it covers only the central parts of the larger ones. For a few of the largest clusters several plates with different centers are taken so as to obtain a more complete representation of these objects.

Twelve to fifteen of the brightest stars in each cluster are also being observed with the one-prism slit spectrograph for more accurate classification as well as for radial velocity. This part of the program is progressing more slowly because each star has to be observed separately. Both instruments combined so far furnish data for 52 clusters. While a complete evaluation and publication of these observations will yet take considerable time, a preliminary examination of the plates already gives some definite results which seem of sufficient interest to be communicated immediately.

For a statistical discussion of the estimated spectral classes it is best to relate them to the magnitudes of the stars; this is done by the construction of a *magnitude-spectral class diagram*. Each star of the cluster is plotted as a dot at the point which has the star's spectral class as abscissa and its apparent visual magnitude as ordinate. An example of such a diagram is found in Fig. 1 for the cluster Messier 34 (N.G.C. 1039). Graff's measures were used for the visual magnitudes and the spectral classes were estimated on three slitless spectrograms of 20^m, 3^h, and $5\frac{1}{2}^h$ exposure. Six of the brightest stars had also been observed with the slit spectrograph.

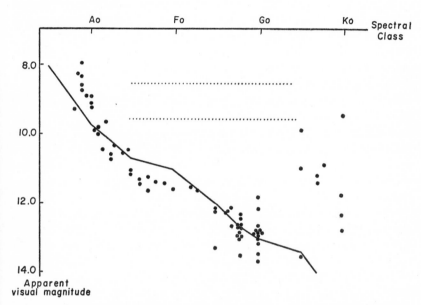

Fig. 1. Magnitude-spectral class diagram for the cluster Messier 34.

The broken inclined line around which the stars are evidently scattered represents the dwarf branch. This line was constructed with Lundmark's values of the mean absolute magnitudes of dwarf stars, adding to them a constant so determined as to fit the line closely to the plotted cluster stars. This constant is the apparent magnitude of a cluster star of absolute magnitude 0; its value was found to be $8^m.56$ corresponding to a parallax of $".0019$ or a distance of 515 parsecs (1700 light years) for Messier 34. The branch of giant stars lies between the two dotted horizontal lines at absolute magnitude 0 and $+1^m$. No clustering of dots is noticeable in this space; our cluster does not contain any yellow or red giant stars except perhaps one or two. A number of scattered stars fit neither into the giant nor the dwarf branch. This is not astonishing when we consider that all stars within 10′ from the center of the cluster were plotted and that among these there must be some stars not physically connected with the cluster but accidentally appearing projected on that part of the sky, while in reality situated before or behind the cluster. We should expect these background stars to be more numerous among the fainter stars and this is indeed the case in our diagram of Messier 34. . . .

As a first result of our investigation the statement can be made that the members of each cluster exhibit a close relationship between spectral class and magnitude and that the magnitude-spectral class diagrams of clusters show a marked crowding of points along the lines which can be identified with parts of the Hertzsprung-Russell diagram of giant and dwarf stars. Among the 52 objects so far investigated there is only one, N.G.C. 6885, for which this statement does not hold; this being a loose clustering of very little concentration is probably no physical system at all, but an accidental apparent conglomeration of stars.

After the cluster members are identified as giants or dwarfs, the A to G dwarf stars or the K type giants can be used for a determination of the distance of the cluster, as described above. The results thus obtained place the open clusters observed at distances ranging between 40 and 3000 parsecs, but most of them lie between 500 and 2000 parsecs. Some of the faintest clusters which are beyond the reach of the spectrograph may be situated at still larger distances, but the number of such objects so far discovered is relatively small. . . .

Disregarding for the moment the end of the dwarf series, missing on account of instrumental limitations, we find that the open clusters represent the Hertzsprung-Russell diagram in different ways, emphasizing some parts more strongly, while neglecting others. It seems promising then to try a classification of open clusters according to the character of the magnitude-spectral class diagram. Although at the present stage of this investigation such an attempt is necessarily tentative, some of its features are definite enough to be given here.

The distribution of spectral types in open clusters varies essentially along two lines. On the one hand we find great differences in the proportion of cluster stars that fall on the giant branch (F – M). In class *1* we shall include clusters in which the giant branch is entirely missing or cases in which there are so few scattered stars falling within its limits that it must remain doubtful whether they are physical members or background stars. Class *2* comprises the clusters which show a marked crowding of stars along the giant branch although their number may still be small compared with that of the dwarf stars. There is not quite sufficient evidence yet for the establishment of a third class in which the yellow and red giant stars are very numerous and form the most important constituents of the cluster.

The dwarf series, as defined above, on the other hand, while always present does not always extend equally far in the direction of the hotter spectral types. We shall designate with b those clusters in which the dwarf branch reaches up to the spectral classes B_0–B_5. Clusters of class a contain no types of higher temperature than B_8, while those of class f are entirely composed of stars with spectral types F_0–M. A priori we might expect the existence of all possible combinations of 1 and 2 with b, a, f; so far, however, only the following four were found to be represented:

1b: The brightest cluster stars are of type B_0–B_5, including occasionally an O type star. The fainter stars are closely crowding around the dwarf branch, tracing it continuously down to the instrumental limit. Giant stars A_5–M are entirely missing or very rare. Typical examples are h and χ *Persei,* Messier 36, Messier 35, the *Pleiades.*

1a: The brightest cluster stars are of class B_8–A_5; the fainter stars follow the dwarf branch closely. No stars of class B_0–B_5 are present and no or very few giant stars. Typical examples are Messier 34 (see Fig. 1), N.G.C. 1647, Messier 39.

2a: The brighter cluster stars are scattered along the giant branch from B_8 to K, the fainter stars follow the dwarf branch from B_8 toward F and G. In general the giant branch is not uniformly represented but shows a concentration at G_5–K_0; around class A, where the giant and dwarf branches meet the stars are numerous and pretty much scattered. Often there is even a marked gap in the giant branch between A and G making the G_5–K_0 giants appear as an isolated group. The dwarf branch is continuous. Typical examples are: Melotte 210, Messier 37, Messier 11, *Praesepe, Taurus cluster, Come Berenices.*

2f: No stars of class B or A are present. The brighter stars follow the giant branch from K to F, which turns somewhat down to reach the F dwarf stage; the points are rather widely scattered at this point, and the fainter stars seem to follow the dwarf branch from F_0 to G. N.G.C. 752 is the only representative of this class so far met with, but N.G.C. 6811 is intermediate between *2a* and *2f* containing no stars of hotter types than A_5. . . .

The 52 clusters examined are distributed among the four types in the following manner:

Type *1b* : 24

1a : 6

2a : 20

2f : 1

No physical system   1

Types *1b* and *2a* are the most frequent and the former seems characteristic for the many small cluster knots scattered throughout the Milky Way. . . .

As a whole, the spectroscopic observations are in good agreement with the hypothesis that all stars of a cluster originate at the same time or within a limited period, that the stars differ in original mass and run through their evolutionary course with different speed. If these views are correct we are led to the conclusion that the open clusters are already of considerable age; otherwise we would not find the dwarf branch so well formed in all cases, nor could we explain the general scarcity or total absence of yellow and red giant stars. As class *1b* and the immediately adjoining stages are by far the most frequent, we might even say that the open clusters seem to be of similar age.

45. TYPES OF STAR POPULATION

By Walter Baade

In contrast to the majority of the neublae [1] within the local group of galaxies [1] which are easily resolved into stars on photographs with our present instruments, the two companions of the Andromeda nebula—Messier 32 and NGC 205—and the central region of the Andromeda nebula itself have always represented an entirely nebulous appearance. Since there is no reason to doubt the stellar composition of these unresolved nebulae—the high frequency with which novae occur in the central region of the Andromeda nebula could hardly be explained otherwise—we must conclude that the luminosities of their brightest stars are abnormally low, of the order of $M_{pg} = -1$ or less compared with $M_{pg} = -5$ to -6 for the brightest stars in our own galaxy and for the resolved members of the local group. Although these data contain the first clear indication that in dealing with galaxies we have to distinguish two different types of stellar populations, the peculiar characteristics of the stars in unresolved nebulae remained, in view of the vague data available, a matter of speculation; and, since all former attempts to force a resolution of these nebulae had ended in failure, the problem was considered one of those which had to be put aside until the new 200-inch telescope should come into operation.

It was therefore quite a surprise when plates of the Andromeda nebula, taken at the 100-inch reflector in the fall of 1942, revealed for the first time unmistakable signs of incipient resolution in the hitherto apparently amorphous central region—signs which left no doubt that a comparatively small additional gain in limiting magnitude, of perhaps 0.3–0.5 mag., would bring out the brightest stars in large numbers.

How to obtain these few additional tenths in limiting magnitude was another question. Certainly there was little hope for any further gain from the blue-sensitive plates hitherto used, because the limit set by the sky-fog, even under the most favorable conditions, had been reached. However, the possibility of success with red-sensitive plates remained.

---

[1] [The words "galaxy" and "nebula" are used interchangeably in this article. "Early type" nebulae are the ellipsoidal galaxies, E0 to E7 in Hubble's classification.]

From data accumulated in recent years it is known that the limiting red magnitude which can be reached on ammoniated red-sensitive plates at the 100-inch in reasonable exposure times is close to $m_{pr} = 20.0$, the limiting photographic magnitude being $m_{pg} = 21.0$. These figures make it clear at once that stars beyond the reach of the blue-sensitive plates can be recorded in the red only if their color indices are larger than +1.0 mag.—the larger, the better. Now there are good reasons to believe that the brightest stars in the unresolved early-type galaxies actually have large color indices. When a few years ago the Sculptor and Fornax systems were discovered at the Harvard Observatory, Shapley introduced these members of the local group of galaxies as stellar systems of a new kind.[2] Shortly afterward, however, Hubble and the writer pointed out that in all essential characteristics, particularly the absence of highly luminous O- and B-type stars, these systems are closely related to the unresolved members of the local group.[3] It was therefore suggested that in dealing with the Sculptor and Fornax systems "we are now observing extragalactic systems which lack supergiants and are yet close enough to be resolved." Since the brightest stars in the Sculptor system, according to later observations by the present writer, have large color indices (suggesting spectral type K), it appeared probable that this would hold true for the brightest stars in the unresolved members of the Andromeda group. Altogether there was good reason to expect that the resolution of these systems could be achieved with the 100-inch reflector on fast red-sensitive plates if every precaution were taken to utilize to the fullest extent the small margin available in the present circumstances.

Since success depended so much upon a careful use of the available light-intensities, it may be surprising that the final tests were made in the light of the narrow band $\lambda\lambda6300$–6700 (on ammoniated Eastman 103E plates behind a Schott RG 2 filter). The reason is the following: It is quite true that nearly twice the speed in the red could have been obtained if a yellow filter, transmitting wave lengths $>\lambda5000$, had been used instead of the red filter. But experience has shown that the benefits to be derived from the larger range of wave lengths are of doubtful value, particularly in long exposures, because the larger range includes two of the strongest emission lines of the night sky—the green aurora line at $\lambda5577$ and the red [OI] doublet $\lambda6300$, $\lambda6364$.

The red doublet at $\lambda6300$, $\lambda6364$ has proved to be especially troublesome for astronomical photography, partly because it falls into the re-

[2] *Nature 142,* 715 (1938); *Proc. Nat. Acad. 25,* 565 (1939).
[3] *Pub. A.S.P. 51,* 40 (1939).

gion of maximum sensitivity of the E plates, partly because it displays erratic intensity changes from night to night and even in the same night. These changes are large, and it is well known that not infrequently, partly at the times of sunspot maxima, the intensity of the red doublet surpasses that of the strong green line by a factor of 2 or more. Consequently, it is impossible to predict whether on a given night the exposure time for the range $\lambda\lambda 5000$–6700 has to be restricted to 1 hour or can be safely extended to several hours. To avoid any difficulties resulting from uncontrolled sky fog, which are especially serious for objects near the plate limit, it was decided to use the narrower range of wave lengths cut out by the RG 2 filter. Although this filter transmits about 24 per cent of the red doublet, no difficulties have thus far been encountered even with exposure times up to 9 hours. It may be remarked here that the plates to be discussed later are practically free from sky fog.

The minimum exposure times required with the RG 2 filter turned out to be 4 hours. Exposures of this length with a large reflector present a number of problems if critical definition is the prime requisite. That only nights with exceptionally fine definition, together with a practically perfect state of the mirror, would do hardly needs mention. Fortunately, these conditions are easily met on Mount Wilson during the fall months when the Andromeda region is in opposition. But real difficulties were presented by changes of focus during the relatively long exposures. On account of the normal drop in temperature during the night these changes are quite large under average conditions; hence repeated refocusing with the knife edge—usually once every hour—is necessary as the exposure proceeds. Although a special, precision-built plateholder arrangement is available for such purposes, its manipulation is always somewhat risky because the change from the field to a suitable focus star and back has to be made in complete darkness. Even if such repeated manipulations are performed without mishap during a prolonged exposure, the method remains a makeshift, since between two settings the plate will gradually move out of focus. To avoid both difficulties it seemed best to use only nights on which the focus-changes at the 100-inch are very small if not entirely negligible. Such conditions are not infrequently met on Mount Wilson during the fall, when, owing to a temperature inversion, the temperature stays practically constant all night. Neither was it difficult in the present case to select the proper nights. Since in the fall the Andromeda region culminates around midnight, a careful watch of the state of the mirror and of the temperature in the early evening hours permits a fair prediction of the focus-changes during the latter part of the night. Eventual small changes in focus during the exposure can then be inferred from changes in the coma of the guiding star. Although this method has fallen into disrepute because of some bad experiences of earlier observers, the writer has found it as good as the knife-edge test if the following conditions are fulfilled: (1) a nearly perfect figure of the mirror; (2) steady and crisp images; and (3) such an adjustment of the guiding eyepiece that small focus-changes produce marked changes in the coma pattern of the guiding star. All exposures discussed in the following pages have been made in this manner. As a control of the correct handling of the focus-changes, the focus was checked with the knife edge at the end of each exposure. In every case the difference between the last actually used focus and the knife-edge setting was well below 0.1 mm.

The plates of the Andromeda nebula, of Messier 32, and of NGC 205, taken in this manner at the 100-inch reflector during the fall months of 1943, led to the expected results. All three systems were resolved into stars. A description of the plates thus far obtained follows. Since the preparation of adequate reproductions would involve time-consuming experiments impossible under present conditions, illustrations will be published later. The plate of NGC 185 in the following *Contribution* will give the reader an idea how far the resolution of the hitherto unresolved systems of the local group has been successful.

I. *Messier 32, the brighter, round companion of the Andromeda nebula (ammoniated 103E plate behind Schott RG 2 filter, λλ6300–6700; exposure 3^{h}30^m; August 25, 1943)*.—The plate was obtained under ideal conditions: a perfect mirror, seeing 5–6, and no change in focus during the whole exposure (which was cut short by the oncoming twilight). As a result, the smallest stellar images on the plate have diameters of less than 0.7″ of arc.

The central part of Messier 32 is completely burned out, but the outer parts have disintegrated into an unbelievable mass of the faintest stellar images. The plate is of special interest because it shows in an instructive manner which features are the first signs of resolution in systems of this type. They are star chains, formed by accidental groupings of some of the brightest members of the system. Clearly resolved into stars on the red exposure, they were indicated on the best blue-sensitive plates taken previously, where they appear as very weak, ill-defined filaments in the otherwise amorphous structure of the nebula.

The extent of Messier 32—i.e., the distance to which its members can be traced—is difficult to ascertain, since a spiral arm of the Andromeda nebula sweeps over the field in such a way that at greater distances from the center of Messier 32 the members of both systems are hope-lessly mixed. But there are indications that the situation is even more complicated. To gain more intensity, another 4-hour exposure of Messier 32 was made on August 26, 1943, this time behind a Schott GG 11 filter, so that the plate covered the range from λ5000 to λ6700. It so hap-pened that the sky began to brighten up after the exposure was started —probably on account of a diffuse aurora—with the result that the plate fog became rather dense. The plate is interesting, however, be-cause it shows that up to a distance of 17′ south of Messier 32 the field is covered with a stratum of extremely faint stars. Obviously these stars belong to the Andromeda nebula, since their slowly decreasing density in a southward direction follows the contour lines of the nebula. There seems to be little doubt that this mass of faint stars, in luminosity and color index similar to the brightest stars in Messier 32, is identical with the faint extension of the Andromeda nebula first recorded photo-electrically by Stebbins and Whitford.[4] On the plate just mentioned the stars can be traced along the minor axis of the Andromeda nebula to a distance of 32′ from the center, corresponding to the isophote 25.4 mag. per square second of arc (Stebbins and Whitford). Prop-erly centered plates may well shift the limit farther out to lower iso-photes.

II. *NGC 205, the fainter elliptical companion of the Andromeda neb-ula (ammoniated 103E plate behind Schott RG 2 filter, λλ6300–6700; exposure 4 hours, September 29, 1943).*—During the 4-hour exposure thin haze occasionally drifted over the field, probably reducing the ef-fective exposure time to 3½ hours. The plate was taken under excellent seeing conditions, but with a fast-deteriorating figure of mirror caused by rising temperatures. As a result the otherwise small and crisp images show an irregular flare which may have reduced both resolving-power and limiting magnitude. In spite of these shortcomings, NGC 205 is beautifully resolved up to the very nucleus. It is a much looser aggre-gration of stars than Messier 32, as was to be expected from its lower surface brightness.

In order to test how far the faint stars revealed on the red exposures are reproduced from one plate to another, a second plate of NGC 205, of only 90 minutes exposure, was obtained in the larger range λλ5000–6700 on December 23, 1943. This shorter exposure registers stars as faint as the earlier 4-hour exposure behind the RG 2 filter. The inter-comparison of the two plates in the blink comparator showed that the pattern of resolution is identical on both plates, each configuration of

[4] *Proc. Nat. Acad. 20*, 93 (1934).

faint stellar images on one plate being reproduced on the other. Undoubtedly, a small percentage of the images are still unresolved doubles and accidental groupings of stars, but the majority are certainly single stars. Intercomparison of the two plates led to the discovery of 3 faint variable stars which are undoubtedly members of NGC 205.

Nebulae of the globular type like NGC 205 have always presented the difficulty that their dimensions, as inferred from the extent of the nebulosity, were rather indeterminate. With the resolution of NGC 205 it has become possible to use a definition of the radius which has proved both significant and practical for globular clusters. The radius is defined as the maximum distance from the center up to which the members can be traced. The dimensions of NGC 205 derived in this manner are $2a = 15'.8$, $2b = 9'.1$ The only comparable value is that published by Reynolds,[5] who derived $2a = 12'$ from photometric measures on a plate taken with the Helwan reflector. Reynolds' value should be considered as a lower limit, since his plate was exposed for only 30 minutes.

Because the resolution of NGC 205 proved so easy in red light, a corresponding test on a fast blue-sensitive plate seemed to be of special interest. The nebula was therefore photographed at the 100-inch on the remarkably fast Eastman 103a-O emulsion. The exposure time was 90 minutes, which represents about the practical limit for plates of this type. The plate reveals incipient resolution of NGC 205 quite unmistakably; but the prevailing pattern is still very soft, and the smallest elements are not yet stars but small-scale fluctuations in the stellar distribution. The resulting impression is very irritating to the eye. The nebulosity has lost its amorphous character, but nothing definite has yet emerged.

III. *The inner amorphous region of the Andromeda nebula (ammoniated 103E plate behind a Schott RG 2 filter, λλ6300–6700; exposure 4 hours, September 28, 1943).*—The plate was taken under excellent conditions: a perfect mirror, seeing 3–6, focus-changes during the whole 4-hour exposure amounting to less than 0.1 mm. Since it was to be expected that the nuclear region of the nebula would be burned out in a 4-hour exposure, the plate was centered on a point of the preceding major axis, 11' distant from the nucleus. It shows the hitherto amorphous nebulosity disintegrated into a dense sheet of extremely faint stars, all close to the limit of the plate. As expected, the resolution decreases somewhat in the denser parts of the nebulosity but is easily traced to a point 3'.5 from the nucleus where the burnt-out area sets in. Altogether,

[5] *Mon. Not. Roy. Ast. Soc. 94,* 519 (1933–1934).

there is not the slightest doubt that with the proper optical means the Andromeda nebula is resolvable into stars right up to the very nucleus.

The main facts presented in the preceding descriptions can be summarized in the following four statements:

1. By using red-sensitive plates we have recorded the brightest stars in the hitherto unresolved members of the local group of galaxies.

2. The apparent magnitudes of the brightest stars are closely the same in all three systems, a result which was to be expected because the three nebulae form a triple system.

3. At the upper limit of stellar luminosity, stars appear at once in great numbers in these systems. (In what have been termed the resolvable systems, the brightest stars increase very slowly in numbers for the first 1.0–1.5 mag. below the upper limit of luminosity.)

4. With our present instruments early-type nebulae can be resolved on red-sensitive plates if their distance modulus does not exceed that of the Andromeda group. . . .

With these data at hand we are in the position to draw an important conclusion regarding the Hertzsprung-Russell diagram of the stars in early-type nebulae. As pointed out earlier, it has been known for some time that the highly luminous stars of the main branch (O- and B-type stars), together with the supergiants of types F–M, are absent in these systems; in fact, their absence was the reason why up to now the early-type nebulae have proved to be unresolvable. But neither are the brightest stars which we find in them the common giants of the ordinary H–R diagram, because as a group they are nearly 3 mag. brighter (the average early K-type giant of the H–R diagram has the absolute photographic magnitude $M_{pg} = +1.7$, compared with $M_{pg} = -1.1$ for the mean absolute magnitude of the brightest stars in the early-type nebulae).

It is significant that the same situation is known to exist in the globular clusters. Table 1 serves to illustrate this point. It gives M_{25}—the mean absolute photographic magnitude of the 25 brightest stars in a globular cluster—as a function of M_t, the total brightness (stellar content) of the cluster. Only clusters with distance moduli determined from cluster-type variables have been used. Table 2 shows that for the richest globular clusters M_{25} is -1.3, compared with $M_{pg} = -1.1$ for the brightest stars in NGC 205. Now M_{25} in globular clusters and our mean value for the brightest stars in NGC 205 should be closely comparable; for although the value for NGC 205 refers to the several hundred of its brightest stars, it should define nearly the same group of stars as M_{25} in the clusters, because the population of NGC 205, according to its luminosity, exceeds that of the richest globular clusters by a factor 10 to 20. The agreement of the values quoted above is therefore as good as one could expect.

TABLE 1. Data for Globular Clusters

M_t	M_{25}
−8.12	−1.32
−7.70	−1.21
−7.22	−1.14
−6.70	−1.03
−5.65	−0.55

Similarly, there is perfect agreement in the color indices of the brightest stars in early-type nebulae and globular clusters. We derived $CI = +1.3$ mag. for the brightest stars in NGC 205, a value identical with that found by H. Shapley in globular clusters.[6]

We conclude, therefore, that, within the present uncertainties, absolute magnitude and color index of the brightest stars in early-type nebulae are the same as those of the brightest stars in globular clusters. However, the similarity of the stellar populations of early-type nebulae and globular clusters does not end here; for there are strong indications that another, even more unique feature of the H–R diagram of the globular clusters is shared by the stars of the early-type nebulae.

Figure 1 represents schematically the H–R diagrams of the stars in the neighborhood of the sun (*shaded*) and of those in globular clusters (*hatched*).[7] To conform with the usual practice, photovisual magnitudes have been used for the absolute magnitudes; hence the brightest stars in globular clusters appear now as stars of $M_{pv} = -2.4$. Both the dispersion and the frequency of the stars have been roughly indicated to convey an idea of the distribution of the two groups of stars in the H–R plane.

As already remarked, the H–R diagram for globular clusters begins with early K-type stars of $M_{pv} = -2.4$. On its downward slope the giant branch soon splits into two separate branches, the one continuing more or less in the original direction, the other proceeding nearly horizontally from spectral type G through F and A into the early B's. For our following argument we are concerned with this horizontal branch of the cluster diagram, which is remarkable for two reasons: (1) it sweeps

---

[6] *Star Clusters* (Harvard Observatory Monographs No. 2; McGraw-Hill, New York, 1930), p. 29.

[7] [Compare Fig. 3 in selection 43 above.]

through the well-known Hertzsprung gap of the ordinary H–R diagram; or, to put it differently, stellar states which seem to be excluded in the ordinary H–R diagram for galactic stars in our neighborhood are quite frequent in the H–R diagram of the globular clusters; (2) the short-period Cepheids, which are such a characteristic feature of the globular clusters, are located along this horizontal branch of the cluster diagram.

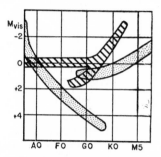

Fig. 1. The spectrum (color)-luminosity diagram: for stars in the solar neighborhood (shaded) and in globular clusters (hatched). Ordinates are provisional absolute magnitudes.

In a very interesting paper M. Schwarzschild [8] has recently shown that, if the mean absolute magnitudes and the mean color indices of the short-period Cepheids in a cluster are used as co-ordinates, their domain is restricted to a well-defined, exceedingly narrow strip within the horizontal branch. More than that, Schwarzschild produces excellent evidence that any cluster star located within this strip is actually a cluster-type variable. This suggests the following interpretation: Since the short-period Cepheids are localized in a well-defined, narrow strip of the H–R plane, they can be expected in considerable numbers only in stellar populations which possess a high density in this particular region of the H–R plane. This condition is fulfilled by the H–R distribution of the stars in globular clusters. It is not fulfilled by the stars in the solar neighborhood (the slow-moving stars) because their distribution exhibits the Hertzsprung gap. . . .

We thus have two strong arguments which indicate that the H–R diagrams of globular clusters and of early-type nebulae are similar, if not identical:

1. In both populations the brightest stars are K-type stars of $M_{pg} \sim -1.1$

[8] *Harv. Circ.*, No. 438 (1940).

2. In both populations the distribution in the H–R plane is characterized by high density in the Hertzsprung gap, with the resulting appearance of cluster-type variables.

But we can advance a third argument which explains at the same time why the globular clusters happen to be the prototypes of this peculiar type of stellar population which we will call type II in distinction from populations defined by the ordinary H–R diagram—type I. This is the fact that, as far as the present evidence goes, globular clusters are always associated with stellar populations of type II. A good example is our own galaxy, where the globular clusters clearly have the same spatial distribution as the cluster-type variables which are representative of the stars of the second type. It is also significant that among the nebulae composed solely of stars of type II even the absolutely faintest usually have one or two globular clusters. Examples are NGC 205, the Fornax system, and the two faint globular nebulae NGC 147 and NGC 185, discussed in the following paper. This association suggests that globular clusters are properly regarded as condensations in stellar populations of the second type. Under these circumstances it is hardly surprising that their H–R diagram should be essentially identical with that of the larger populations of which they are members.[9]

Although the evidence presented in the preceding discussion is still very fragmentary, there can be no doubt that, in dealing with galaxies, we have to distinguish two types of stellar populations, one which is represented by the ordinary H–R diagram (type I), the other by the H–R diagram of the globular clusters (type II) (see Fig. 1). Characteristic of the first type are highly luminous O- and B-type stars and open clusters; of the second, globular clusters and short-period Cepheids. Early-type nebulae (E–Sa) seem to have populations of pure type II. Both types co-exist, although differentiated by their spatial arrangement, in the intermediate spirals like the Andromeda nebula and our own galaxy.[10] In the late-type spirals and in most of the ir-

---

[9] Similarly, we should regard the open clusters as condensations in populations of type I, an interpretation which hardly needs comment in view of the intimate association of open clusters and slow-moving stars in our own galaxy. It is the more acceptable because it would ascribe the curious variations in the composition of open clusters (Trumpler's types) to the large-scale variations in the composition of populations of type I which have been noted not only in our own galaxy but also in several of the nearer extragalactic systems.

[10] The strong concentration of both globular clusters and short-period Cepheids toward the center of our galaxy indicates that the main mass of the stars of type II is located in this region, which, in turn, suggests a structure of our galaxy very similar to that of the Andromeda nebula.

regular nebulae the highly luminous stars of type I are the most conspicuous feature. It would probably be wrong, however, to conclude that we are dealing with populations of pure type I, because the occurrence of globular clusters in these late-type systems, for instance, in the Magellanic Clouds, indicates that a population of type II is present too. Altogether it seems that, whereas stars of the second type may occur alone in a galaxy, those of type I occur only in association with type II.

IX

INTERSTELLAR PHENOMENA

What material, if any, is between the stars? The macrouniverse of stars, like the microuniverse of atoms, is mostly space. But how empty? One constituent of open space, both between stars and between galaxies, is, of course, the "dying" photon. Through every cubic inch of open space the weak radiation of billions of stars is continually flowing. Is space perhaps made up of such radiation?

The problem of interstellar material has interested astronomers (and philosophers) for centuries. Of late, methods of investigating space have increased in number and power and have led to the present situation where a large proportion of current astronomical research is devoted to interstellar phenomena.

Ever since photography became a sensitive astronomical tool in the 1870's, we have known that shining nebulosities, like the Orion Nebula, are common in many star fields; and gradually, through the efforts of E. E. Barnard, Father J. W. J. A. Stein, S.J., Max Wolf, and others, we have learned that nonshining nebulosity is also widely distributed. The investigation of nonluminous interstellar material got its real start when J. F. Hartmann interpreted the spectrum of Delta Orionis as that of a typical star shining through an absorbing cloud of calcium gas (selection 46). The interstellar calcium alone held our attention until Mary Heger, using much the same spectroscopic technique, demonstrated the existence also of interstellar sodium. The faster spectroscopes and improved analyses then began to find other gases outside the stars. We quote from a summary entitled "The Material of Interstellar Space," by C. S. Beals of the Victoria Observatory (*Popular Astronomy 52*, 216 [1944]).

For many years the H and K lines of calcium discovered by Hartmann, and the two sodium doublets at λ5893 and λ3302 discovered by Miss Heger were the only lines appearing in interstellar spectra which were known to be due to interstellar matter. Recent discoveries by Dunham and Adams, and Dunham have added lines due to Ti II, Ca I, and K I, while in a paper which has just appeared Dunham and Adams have reported the definite identification of lines due to interstellar Fe I.

The physical nature of interstellar hydrogen has been analyzed by Bengt Strömgren in an important article which unfortunately is too technical for inclusion here.

Considerations of interstellar matter have now become inextricably involved in the census of stars in the Milky Way (selection 47) and with the birth of stars (see Bok and Spitzer in part **X**). The research has been extended by Sinclair Smith and especially by Fritz Zwicky to the question of intergalactic matter.

The polarization of star light by interstellar dust particles is also a new field for exploration; it is preliminarily described by W. A. Hiltner and J. S. Hall in selection 49. An early interpretation of interstellar polarization was advanced by Lyman Spitzer, and John W. Tukey in *Science 109*, 461 (1949).

The most remarkable advance in interstellar astronomy, however, has been van de Hulst's prediction, and the subsequent discovery by Harold Ewen and E. M. Purcell, of neutral hydrogen atoms throughout all space— hydrogen radiating in such a way that its presence can be readily detected with radio telescopes of special design. Much has resulted therefrom. The detailed structure of the spiral arms of our Galaxy as revealed by the Leiden and Sydney observers and the hydrogen gas enveloping the Magellanic Clouds as reported by Frank Kerr and associates, are but two of the important products of this checking up by radio on the "emptiness" of space. These radio telescope activities have occurred since 1950, however, and are not described in this volume.

46. THE DISCOVERY OF INTERSTELLAR CALCIUM

By J. F. Hartmann

Delta Orionis belongs to the type of *Orion* stars (1*b*), whose spectrum shows, besides the hydrogen lines, chiefly the lines of helium, all of which in this case are exceedingly diffuse and dim, so that their measurement is very difficult and uncertain. On account of the slight intensity of the lines, all defects of the film are very disturbing, and, in consequence of irregular distribution of the silver grains, the lines often appear crooked and unsymmetrical, sometimes indeed double. I have convinced myself by a special investigation that the indications of duplicity and unsymmetrical broadening cannot be caused by lines belonging to a second component of the stellar system; but I do not hold it to be impossible that the form of the lines is subject to small real changes, perhaps in consequence of violent motions in the gaseous envelope of the star.

Although we must, accordingly, regard δ *Orionis* as a binary system having one of its components "dark," in the customary phrase, I would nevertheless point out that by "dark" we here must understand only a relatively small difference of brightness. A difference of only about one magnitude would be sufficient to bring the spectrum of the fainter component to almost complete disappearance, and a difference of two magnitudes would make it impossible for even a trace of the fainter spectrum to be visible on the plate. The slight difference of magnitude necessary for the extinction of the fainter spectrum explains the fact that among the numerous spectroscopic binaries so far discovered there are only very few which show the lines of the second component in the spectrum. . . .

The calcium line at λ3934 exhibits a very peculiar behavior. It is distinguished from all the other lines of this spectrum, first by the fact that it always appears extraordinarily weak, but almost perfectly sharp; and it therefore attracted my attention that in computing the wave lengths . . . for this particular line, the agreement between the results from the different plates was decidedly less than for the other, much less sharp lines. Closer study on this point now led me to the quite surprising result *that the calcium line at λ3934 does not share in the periodic displacements of the lines caused by the orbital motion of the star.* . . .

The fact being thus fully established that a single line of the spectrum does not participate in the oscillatory motion of the other lines, the question arises as to how it can be explained. The character of the absorption corresponding to the line makes it highly improbable that it should have originated in the Earth's atmosphere. In that case, moreover, the line in question would have to appear in every stellar spectrum, and the velocities computed from its position would come into worse agreement on applying the correction for reduction to the Sun. But the case is quite the opposite—the value does not become constant until after applying the reduction to the Sun, and thus the cosmical origin of the line is proven. . . .

We are thus led to the assumption that at some point in space in the line of sight between the Sun and δ *Orionis* there is a cloud which produces that absorption, and which recedes with a velocity of 16 km, in case we admit the further assumption, very probable from the nature of the observed line, that the cloud consists of calcium vapor. This reasoning finds a distinct support in a quite similar phenomenon exhibited by the spectrum of *Nova Persei* in 1901. While the lines of hydrogen and other elements in that spectrum led us, by their enormous broadening and displacement and the continuous changing of their form, to conclude that stormy processes were going on within the gaseous envelope of the star, the two calcium lines at λ3934 and λ3969, as well as the D lines, were observed as perfectly sharp absorption lines, which yielded the constant velocity of +7 km during the whole duration of the phenomenon. I then expressed the opinion that these sharp lines probably did not have their origin in the *Nova* itself, but in a nebulous mass lying in the line of sight—a view which only gained in probability on the later discovery of the nebula in the neighborhood of the *Nova*. In the case of δ *Orionis* also it is not unlikely that the cloud stands in some relation to the extensive nebulous masses shown by Barnard to be present in the neighborhood. The second calcium line at λ3969 is concealed in the spectrum of δ *Orionis* by the broad hydrogen line *H*ε, and therefore cannot be observed.

47. WOLF'S METHOD OF MEASURING DARK NEBULAE

By Bart J. Bok

In 1923 Wolf illustrated, with a striking diagram, a paper on the distribution of the stars for a field in the region of obscuration near the Veil Nebula in Cygnus. Similar diagrams have been published in several of Wolf's more recent papers. The advantages of this form of presentation are so obvious that it is not surprising that most of the papers relating to dark nebulae, published during the last ten years, are illustrated with "Wolf diagrams" for the regions under investigation. Figure 1 shows a Wolf diagram for the region near the North America Nebula. The abscissae are apparent photographic magnitudes; and the ordinates are the values of $\log A(m)$, $A(m)dm$ being the number of stars per square degree having apparent magnitudes between m and $m + dm$. The full-drawn curve gives the counts for the unobscured comparison field, and the dotted curve the data for the region of the dark nebula. Wolf concludes from this curve that the counts for the obscured region indicate the presence of two dark nebulae, one at a distance corresponding to that of the average star of the eighth or ninth magnitude, the other at a distance of an average star of the twelfth magnitude. Using the secular parallaxes of Kapteyn and van Rhijn, Wolf places the first nebula at a distance of between 100 and 200 parsecs, the second at a distance of 600–700 parsecs. The absorbing power of the first nebula would amount to 0.5 mag., while that of the second nebula is estimated to be as high as 3.5 mag.

The very extensive use which has been made in recent years of Wolf diagrams, similar to Figure 1, renders it necessary to examine critically the conclusions that may be drawn from a mere inspection of the curves in such a diagram. Unfortunately, it appears that the results obtained in this fashion are frequently open to criticism. The first source of error may be in drawing the continuous curve through the points in the diagram. All too frequently investigators have disregarded the fundamental fact that the natural uncertainty equals the square root of the average counted number. Before attempting to draw the smooth curve, the investigator should know exactly the amount of the fluctuations in the value of log $A(m)$ to be expected because of purely random fluctuations. As an illustration of the procedure we shall consider a dark nebula with a total area of 0.5 square degree in which 50 stars have been counted between magnitude limits 13.5 and 14.5. The natural uncertainty in this counted number is equal to ±7. Reduced to an area of one square degree, the counted number becomes equal to 100, with a natural uncertainty of ±14. In our particular case the value of log

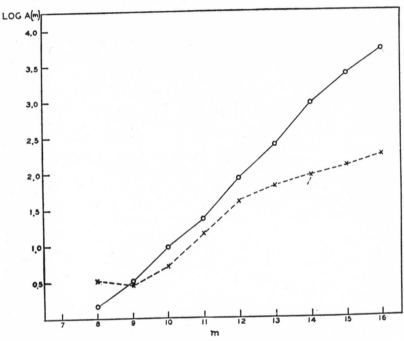

Fig. 1. A Wolf diagram for the region of the North American Nebula.

$A'(14)$ should be taken equal to 2.00±0.06; the influence of random fluctuations may be best represented by drawing a vertical arrow of appropriate length. The only justifiable procedure for obtaining the Wolf curves is to draw through the observed points the smoothest curves which represent the observed values of log $A(m)$ within the limits of accuracy set by the random fluctuations. The difference between the observed and the smoothed values of log $A(m)$ should exceed the natural uncertainty in one out of three cases, because of random fluctuations. In the drawing of the smooth curves the influence of the purely accidental errors in the magnitudes and in the counts should, of course, be considered.

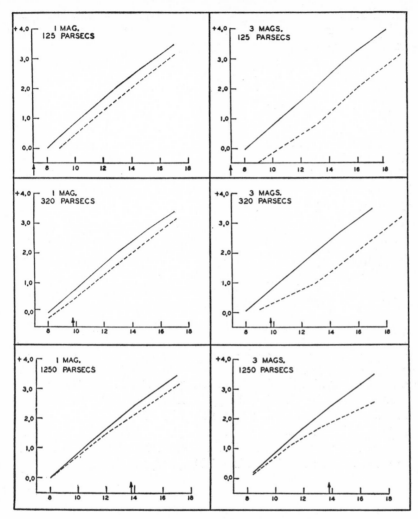

Fig. 2. Imaginary Wolf diagrams, constructed by Freeman D. Miller.

There has been a regrettable tendency among the users of Wolf diagrams to assert that every single discontinuity in the curve for the obscured region is indicative of the presence of another dark nebula. In hardly any case is it justifiable to postulate, on the basis of general star counts alone, the existence of more than a single dark nebula in a field. The influence of the large spread in the general luminosity function has not been sufficiently appreciated by most workers in this field.

Pannekoek's theoretical curves for the field in Taurus show how gradually, for an infinitely thin nebula, the divergence between the curves for the free and for the obscured regions grows with the apparent magnitude. The point is further illustrated in Figure 2, where a series of imaginary Wolf diagrams, constructed by Freeman D. Miller for use in his unpublished dissertation, is reproduced. The upper curve in each diagram represents the value of log $A(m)$ computed for the average unobscured region in the zone of galactic latitude 0°–20°, from Seares's counts. Using van Rhijn's general luminosity function, the density distribution for the average field was found by a method of trial and error. With these densities Miller computed the log $A(m)$ curves which would have been observed if a dark nebula absorbing either one or three magnitudes were interposed at distances of 125, 320, or 1250 parsecs. I doubt very much whether anyone who might be presented with a series of counts similar to those of the last diagram of Figure 2 would have suspected, without further analysis, that the difference between the free and the obscured regions could be explained by the presence of a single dark nebula absorbing 3.0 mag. at a distance of 1250 parsecs! It is significant that some observed Wolf curves bear a marked resemblance to Miller's theoretical curves; the Wolf diagram for the obscuration near the North America Nebula (Fig. 1) is similar to the last diagram of Figure 2. A single dark nebula at a distance of the order of 1000 parsecs, with an absorption of at least 2.5 mag., will explain all the significant features of the Wolf diagram in Figure 1. The large spread in the general luminosity function is responsible for the fact that the influence of this particular dark nebula may be already traced among the ninth magnitude stars.

[For Wolf's method also see "Ueber den dunkeln Nebel N.G.C. 6960" by M. Wolf, *Astronomische Nachrichten, 219,* 109–116 (1923).]

48. PREDICTION OF 21–CENTIMETER INTERSTELLAR RADIO SIGNALS

By H. C. van de Hulst

Astronomical Importance.—Although the existence of radio waves of extraterrestrial origin has been known for about ten years, astronomers have not yet paid much attention to them. This is partly due to the incomplete data at hand; not much more than an order of magnitude of the intensity and a rough dependence on direction of the radiation has been established. Hence little is to be expected from a careful discussion of observational material.

Also the existence of these radio waves is not very interesting from the purely theoretical point of view. The production of radio waves is by no means an essential feature of the physical condition of the interstellar matter. The amounts of energy which are transformed into radio radiation are so small that they are negligible in the large energy balance which starts with the ionization of interstellar atoms by the light of the stars. The radio waves may *not* form a negligible item in the energy balance of sources like the one in Cygnus, where no comparable output of visual light has been found. We cannot expect any new insight regarding the physical condition of the interstellar gas from purely theoretical considerations of the origin of these radio waves; the condition of the interstellar gas is mainly characterized by its density, degree of ionization, and the distribution of electron velocities.

However, the possibility of direct observation of these waves has now made the subject attractive. For twenty centuries astronomers have obtained all their knowledge from observations in the rather narrow energy range around the visual frequencies. They built themselves powerful instruments and did not shrink from any trouble. Now we know that the earth's atmosphere leaves open another frequency range, near the radio frequencies. The first observations have been made, but the technique of observing is still in its infancy. Bakker has shown that it can be improved greatly with presently available means.

The long wavelengths involve one difficulty. Without telescopes of enormous apertures the radio waves will never yield a detailed picture of the sky. For the time being we shall have to be satisfied to reach a resolving power of about 1°. The sun, the Milky Way, and the brightest extra galactic nebulae would then be measurable objects. In the Milky Way the run of the intensity with latitude and longitude, and especially the separation into H I and H II regions of the interstellar gas, could be investigated. Moreover it is possible that the very distant extra-galactic nebulae would constitute a diffuse background, which would be of special cosmological interest.

The Spectrum of the Milky Way.—We shall try to derive, purely theoretically, what the spectrum of the Milky Way will be at these frequencies. We extend our considerations over the entire width of the 'radio windows' in the earth's atmosphere, that is, wavelengths from 20 m to 1 cm. The observations at wavelengths 14, 16, and 1.86 meters yield intensities of equal order of magnitude. Although a continuous spectrum is probable, we do not want to exclude the possibility of discrete spectral lines. (No restriction is introduced by speaking about the Milky Way; the Milky Way is what we observe of our stellar system with its threefold population of stars, interstellar smoke, and interstellar gas.) The maximum intensity of the radio waves has been observed in the direction of the constellation of Sagittarius. We know indeed from many other data that this direction is the one towards the galactic center, in which we look through the deepest and densest layers of our stellar system. As a working model we shall schematize it greatly by assuming a homogeneous layer of constitution equal to what we know near the sun, and a depth of 16,000 parsecs $= 5 \times 10^{22}$ cm.

First we try to establish in which group of the population of our stellar system the observed radio waves chiefly originate. We use the law of radiation of a black body, which in this region, since $h\nu << kT$, is

$$j_{\text{black}} = 2\nu^2 kT/c^2. \tag{1}$$

a) *Stars.* These probably radiate like black bodies with a mean temperature of T $\approx 5,000°$K. . . .

It does not seem impossible that in the future the intensity of the sun in the region of the decimeter waves would be measurable. Meanwhile Southworth (*Jour. Franklin Inst., 239,* 285ff [1945]) has succeeded in measuring the solar radiation in the wavelength region 1 to 10 centimeters.

Reber's method was not yet accurate enough to observe the sun. It is certain, moreover, that the stars together do not give a sky background of sufficient intensity; for the stellar discs certainly cover less than one part in 10^{10} of the area of the sky.

It is now known that the sun's radiation in meter waves is enhanced to an equivalent temperature of 10^6 degrees and, in occasional bursts, to 10^{11} degrees. The suggestion has been made that similar enhancements in other stars may account for all of the observed galactic rediation. However, the observed intensity would be reached

(1) if all stars are continuously enhanced to 10^{17} degrees, or

(2) if all stars are enhanced 10^{21} degrees for 10^{-4} of the time, or

(3) if some peculiar stars (10^{-4} of all) are continuously at 10^{21} degrees.

None of these assumptions seems very likely (cf. Greenstein, Henyey and Keenan, *Nature, 157,* 805, 1946).

b) *Interstellar Smoke.* These solid particles of interstellar matter have a temperature of about 3°K. Even if they would cover the whole sky, their contribution could be neglected. This conclusion is not changed by the greater details which Whipple and Greenstein (*Proc. Nat, Ac. Sc., 23,* 177–181, 1937) take into account; viz. (1) the argument that too small a particle cannot radiate at all in this frequency range because it has no 'eigen-frequency,' (2) the possibility that the smoke particles near the galactic center may have a higher temperature, as high as 30°K, because of the greater energy density.

c) *Interstellar Gas.* The interstellar gas has a kinetic temperature of around 10,000°K and, different from the stars, it radiates from the whole area of the sky. At present our knowledge of the physical status of the interstellar gas is much more confused than it seemed in 1944. There are strong density fluctuations (cloudiness) and possibly also temperature fluctuations. The temperature cannot be measured directly. Theoretical estimates show that T may be as low as 200°K in H I regions and possibly as high as 100,000°K in H II regions (cf. L. Spitzer, *Ap. J., 107,* 6 ff 1948). The question is only if the layers are optically thick enough to yield black radiation. And this question cannot be answered without going to some detail into the mechanism of emission. . . .

[The author then presents a technical discussion of the radiation of a homogeneous layer of gas, and of the continuous spectrum associated with interstellar hydrogen.]

Are there any Separate Spectral Lines? We have stated that the bound-bound transitions contribute, on the average, only a negligible amount to the continuum. However, the energy liberated in these transitions is emitted in separate lines, and it is feasible that within some rather narrow lines the intensity would be appreciably higher than in the continuum. . . .

All these lines, like the free-free continuum, would be formed only in the H II regions. But quite a different possibility is left. The ground level of hydrogen is split by hyperfine structure into two states of distance 0.047 cm^{-1}. In the one state the spins of the electron and the proton are parallel, in the other anti-parallel. At the spontaneous reversal of the spin a quantum of 21.2 cm would be emitted. . . .

49. THE POLARIZATION OF STAR LIGHT BY INTERSTELLAR PARTICLES

I. By W. A. Hiltner
II. By John S. Hall

I

In the course of photoelectric observations made last summer with the 82-inch telescope of the McDonald Observatory (University of Texas) the writer found that the light from distant galactic stars is polarized.[1] Polarizations as high as 12 percent were found. The plane of polarization appears to be close to the galactic plane in the cases examined. More recently control measures were made at the Lick Observatory, thanks to the courtesy of Director Shane and Dr. G. Kron; and during December the work at the McDonald Observatory was extended to different regions of the Milky Way.

In view of the unexpected nature of this result the circumstances leading to its discovery are recorded. Photometric observations for the detection of partially polarized radiation from eclipsing binary stars have been in progress at the Yerkes Observatory for several years with a view to establishing observationally the effect pointed out by Chandrasekhar that the continuous radiation of early-type stars should be polarized.[2,3] On the assumption that the opacity of early-type stars is due to scattering by electrons, the continuous radiation emerging from a star should be polarized with a maximum of polarization of 11 percent at the limb. Since the presence of this polarization can be detected only when the early-type star is partially eclipsed by a larger-type companion of the system, the effect is masked by radiation from this companion so that the expected maximum observable effect was only of the order of 1.2 percent in one case investigated (RY Persei).

At this stage Dr. John Hall, of Amherst College, proposed to the writer a program of collaboration whereby Dr. Hall would construct a

---

[1] [An early interpretation of interstellar polarization was advanced by Lyman Spitzer, Jr., and John W. Tukey in *Science 109*, 461 (1949).]

[2] Chandrasekhar, S., *Astroph. Jour. 103*, 365 (1946).

[3] Hiltner, W. A., *Astroph. Jour. 106*, 231 (1947).

"flicker" photometer which was to be tested jointly at the McDonald Observatory. Independently the writer was developing his own equipment which used polaroids. Dr. Hall's equipment was tested in August 1947, during a short session at the McDonald Observatory, but no dependable results were obtained and it was found that the equipment had to be remodeled. Unfortunately, Dr. Hall was unable to come for a second trial period, scheduled for August 1948.

Meanwhile the writer's own equipment was completed and put to use during the summer of 1948 and was found satisfactory. Certain Wolf-Rayet stars which were known or suspected to be eclipsing binaries were examined for polarization. Fairly large polarizations were found, but *they did not appear to depend on the phase of the binary motion.* The possibility of instrumental polarizaton was considered, of course, but ruled out by control measures on check stars. The Wolf-Rayet stars give the following results:

Star	*Polarization*	
	%	Position angle
CQ Cep	10.0	62°
BD 55°2721	8.0	44
WN Anon [4]	12.5	44

The control stars had similar color and brightness, but showed no polarization except for one object, BD + 55°2723, which gave 3 percent. This star, however, is a giant and more distant than the other control stars. Similar observations made on a group of Wolf-Rayet stars in Cygnus showed no appreciable polarization, while two stars in Scutum gave positive results. Other regions, such as the double cluster in Perseus, also show polarization with values ranging up to 12 percent.

We conclude from the positive and negative results quoted that the measured polarization does not arise in the atmospheres of these stars but must have been introduced by the intervening interstellar medium. If this conclusion is accepted, a new factor in the study of interstellar clouds is introduced. Further observations are in progress for relating this phenomenon with other observable characteristics of the interstellar medium. As has been stated, the results already at hand indicate that the plane of polarization approximates the plane of the galaxy.

II

Photoelectric observations of the polarization of starlight made during the period November 1948 to January 1949 with the 40-inch re-

[4] Coordinates: 22^{h}08^m, + 57°26′ (1945); 12.5 magnitude.

flector at Washington substantiate the hypothesis of W. A. Hiltner [5]
that this effect is produced by interstellar matter. Furthermore, the
percentage of polarization appears to be independent of wavelength;

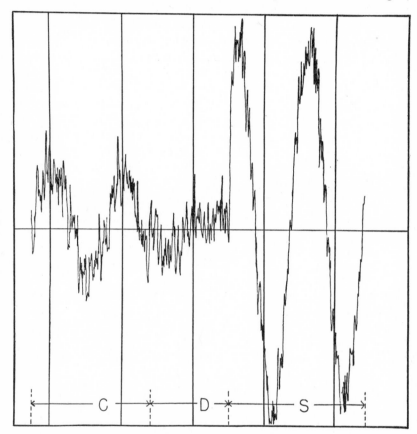

Fig. 1. Record of a large percentage of polarized light.

and the plane of polarization (plane containing the magnetic vector
and the line of sight) appears to have no one preferential orientation.

The observations were obtained with a photoelectric polarizing photometer [6] built at Amherst College in 1946 with the aid of a grant from
the Research Corporation of New York. The light from a star is col-
limated and directed through a cover glass, which serves as a calibrat-
ing device, and then through a Glan-Thompson prism rotated at 15

[5] Hiltner, W. A., *Science 109,* 165 (1949).
[6] Hall, John S., *Astron. Jour. 54,* 39 (1948).

cycles per second to a 1P21 photomultiplier. The 30-cycle voltage developed by the polarized component of the light is selectively amplified and mixed with a phasing voltage in such a way that the d-c output can be impressed as a sine wave on a Brown recorder. The amplitude of this wave is proportional to the intensity of the polarized light, and the phase of maximum defines the plane of polarization. Records of two

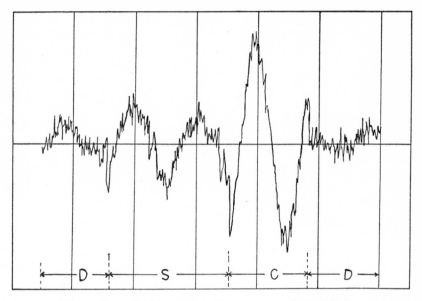

Fig. 2. Record of a small percentage of polarized light.

stars showing large and small percentages of polarized light are shown in Figs. 1 and 2. The vertical lines represent two-minute intervals. The trace during interval S is produced by polarized light from the star. During interval D a quartz depolarizer is placed in the light path, and C is the result when the cover glass is tilted 20° about an axis whose position angle is arbitrarily set at 94.° The starlight was already depolarized during the interval C. The plane of polarization is defined by the direction of the light and the axis about which the glass is tilted. A 20° tilt corresponds to 1.4% polarization.

The percentages of polarization of the light from 27 early-type stars are shown in Fig. 3 as a function of the color excesses determined by Stebbins and Huffer.[7] A strong correlation is obvious; the scatter, however, is much greater than the accidental errors of the observations.

[7] Stebbins, J., and Huffer, C. M., *Publ. Washburn Observ. 15*, 5 (1934).

The dependence of polarization on color was determined on three nights from the observations of ζ Persei using Schott filters UGI and BG14 for the ultraviolet region and the RG1 or a Wratten yellow filter for the red region. The effective wavelengths of the two spectral regions were near 3,700 A and 6,200 A. The observed percentages with the ultraviolet filter were 2.0, 1.6, and 1.8; and with the red filter, 1.6, 1.0, and 2.2. The average value obtained when no filter was used is

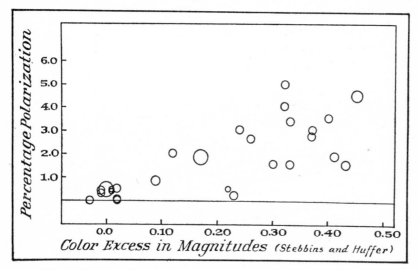

Fig. 3. Polarization percentages (27 stars) and color excesses.

1.8 percent. A second star, HD 33,203, was observed on one night, the result being 1.8 in the ultraviolet, 2.2 in the red, and 1.8 with no filter. No definite variation of the orientation of the plane of polarization with color is indicated by these observations.

Fig. 4 shows the observed planes of polarization for 28 early-type stars. The amount of polarization and the orientation of the plane of each is indicated by the length and direction of the line, whose midpoint represents the position of a star. The group of seven stars near the middle of the diagram exhibits a remarkable similarity in percentages of polarization and orientations of the planes, which may be a consequence of the relative homogeneity of the obscuring material in the direction from which their light comes.

I have obtained these preliminary results from a project initiated in collaboration with W. A. Hiltner. My grateful appreciation is expressed to Dr. Hiltner and to the Yerkes and McDonald Observatories

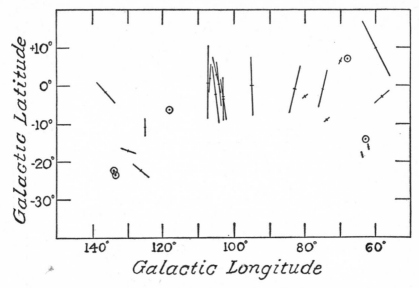

Fig. 4. Planes of polarization for 28 early-type stars.

for the use of the 82-inch reflector for a period of two weeks during the summer of 1947. Despite very unfavorable weather conditions and some difficulty with a new type of photometer, we obtained some evidence of polarization in the light from one star, CQ Cephei. Accordingly it was planned to make a second trial at McDonald during the summer of 1948 with improved equipment, but a second trial could not be made because of other obligations incurred by my transfer from Amherst to Washington on September 1, 1948. With the improved photometer, however, it was easily possible to detect polarization in the light from CQ Cephei with the 18-inch refractor at the Amherst College Observatory. Furthermore, these observations, made during the summer of 1948, showed little if any change in the amount of polarization with the phase of this eclipsing binary star.

X

STELLAR EVOLUTION

The two papers in this part deal directly with stellar evolution; but indirectly various evolutionary evidences and processes have been presented in other parts. W. Baade's study of populations touches on the ages of various types of stars, and therefore on growth and decay; H. N. Russell's spectrum-magnitude array, V. A. Ambartsumian's expanding stellar associations, and W. J. Luyten's degenerate stars are all evolutionary products, since they involve irreversible changes. Part VI (variable stars) and much of part XII (relativity and cosmogony) deal indirectly with evolution. In other words, evolution is a central theme of stellar astronomy. The simple fact that stars do shine emphasizes the nonstatic nature of the universe, for the radiation carries away mass, as the Einstein formula indicates (selection 58), and a loss of mass means a change in density or in volume or both. Eventually a change in energy output occurs along with a change in the atomic make-up of the stars. It is this evolution of the chemical composition that controls, sometimes gradually, sometimes explosively, the evolution of stars.

50. CONCERNING PROTOSTARS

By Bart J. Bok

The present symposium [1] on interstellar matter states as a central theme that in some cosmic clouds we are now witnessing the operation of the process of star formation. It is then our function to provide the theoretician with information on the dimensions and masses of the observed varieties of cosmic clouds. Special attention must be paid to those cosmic clouds which, from their general appearance, seem to be fairly homogeneous single dynamical units; these represent presumably the closest observed approach to the prestellar stage.

In our survey, we have concentrated on roundish dark nebulae, from the smallest dark "holes," seen in projection against some diffuse nebulae, or occasionally against a very rich stellar background, to the larger forms, such as the southern Coalsack.

We shall summarize first some recent work on small dark nebulae. Then we shall turn our attention to the Coalsack variety of dark nebulae and show that this type represents in all probability the largest observed single dynamical unit. Probable dimensions and minimum masses will then be derived and the results of some preliminary calculations about the effects of radiation pressure in the equilibrium of the cosmic clouds will be presented. In the final section we shall consider briefly the future development of the observed varieties of cosmic clouds.

1. THE GLOBULES

Some diffuse nebulae (Messier 8 for example) serve as a background for the delineation of small and roundish dark nebulae, which, for convenience, we have called *globules*. Similar dark objects are seen in projection against star-rich fields like those in Sagittarius, Ophiuchus and Scutum. Barnard's list of dark objects yields some 20 distinct globules.

---

[1] [In the same conference (Centennial Symposia) were presented important papers, closely related, by Hendrik C. van de Hulst, Lyman Spitzer, Jr., and Fred L. Whipple. See *Centennial Smyposia* (Harvard Observatory Monographs No. 7; Harvard College Observatory, Cambridge, 1948).]

In a recent joint note with Edith F. Reilly we have suggested that as many as 50 globules can be recognized along the Milky Way. This number may have to be revised downward, since not all objects listed by us as globules seen projected against the luminous background of

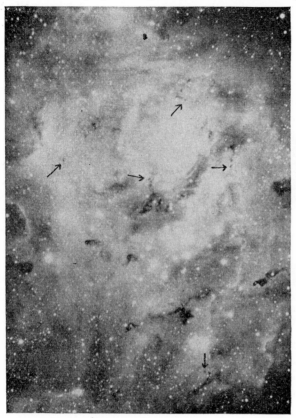

Fig. 1. The diffuse nebula Messier 8, which provides a bright back-drop for dark globules.

Messier 8 are as round and sharply delineated as we suspected them to be. During the summer of 1947 Dr. Walter Baade made available to us several original photographs of Messier 8, taken with the 100-inch reflector at the Mount Wilson Observatory, and on these plates a number of objects identified originally as globules appear to show marked structure of a filamentary nature. This applies especially to the larger objects of our list. Several of the smaller objects are globule-like in appearance and we may consider them tentatively as real physical units.

The diameters of the smallest recognizable globules are of the order of 12,000 A.U. Their estimated minimum total absorptions are two to five magnitudes; they may be totally opaque, of course.

Most of the globules from Barnard's list are larger and definitely more transparent than the globules of Messier 8. The smallest have diameters comparable to those of the globules of Messier 8, but the larger ones (like Barnard 34 in Auriga) may measure 100,000 A.U. or more across. In some cases rough total absorptions may be derived from star counts. For Barnard 34 the total photographic absorption is about one magnitude.

The globules are apparently not distributed uniformly over the sky. They abound in the Sagittarius-Ophiuchus-Scutum section of the Milky Way, where—it must be admitted—the conditions for their detection are especially favorable. In the southern Milky Way, the nebula near Eta Carinae shows a number of globules superposed on the diffuse nebula. They are, however, relatively rare in the anticenter region. None, for example, is seen projected against the Great Nebula in Orion, or against the California Nebula in Perseus. The few globules observed in the anticenter region are large and relatively transparent. Their average diameters are of the order of 100,000 A.U. and their total absorption for photographic light is only about one magnitude. Barnard 34 is a typical example of this larger variety of globule.

2. THE SOUTHERN COALSACK

The distance from our sun to the southern Coalsack is probably of the order of 150 parsecs. The linear diameter of the southern Coalsack is, therefore, probably of the order of 8 parsecs.

Starcounts indicate that the total photographic absorption for stars seen shining through the Coalsack dark nebula varies between one mag. and 3 mags., with 1.5 mags. representing a fair average.

The southern Coalsack differs from the larger globules mostly through its size. Its diameter is 8 parsecs as opposed to 100,000 A.U., or 0.5 parsecs, for the larger globules. The *total* absorption produced by the Coalsack dark nebula is only twice that of the largest globules.

An inspection of available Milky Way photographs shows that the Coalsack-type of dark nebula is in all probability the largest observed variety of unit dark nebula. There are larger systems of dark nebulae, such as the complex in Taurus-Auriga and the nebulae in Ophiuchus. These dark clouds have diameters of the order of 30 to 40 parsecs and total absorptions of about 1.5 mags. They lack, however, the compactness and regular outlines that one expects in a dynamically stable cosmic cloud. They may well be groupings of smaller dark nebulae. The prevalence of canal-like dark lanes and of thin dark streaks suggests the presence of large-scale turbulence as opposed to the quiescence that is in all probability required in the prestellar stage.

3. Minimum Estimated Masses

From the observed total photographic absorptions and dimensions of the known obscuring clouds, minimum values may be derived for the total masses and densities. These are obtained on the assumption that the observed absorptions of the clouds are produced wholly by particles of optimum size for scattering and absorption. Particles that are either larger or smaller than this optimum size are supposed to be non-existent and the contributions from gaseous atoms or molecules are ignored.

We shall suppose that all particles have identical radii, $a = 10^{-5}$ cm, and that they are either of a metallic or of a dielectric nature. According to Greenstein, we have, for dielectric particles with $N = \frac{4}{3}$, $\log K_{pg}(a) = +4.2$ for $a = 10^{-5}$ cm, where $K_{pg}(a) =$ total photographic absorption per cm that is produced if the density of particles of radius a is 1 gram per cc. It seems likely that the interstellar particles are preponderantly dielectric rather than metallic particles, but it may be of interest to note that for iron and for $a = 10^{-5}$ cm, Greenstein finds $\log K_{pg}(a) = +4.4$. We are, therefore, on relatively safe ground if we say that, for $a = 10^{-5}$ cm, we may assume that

$$K_{pg}(a) = 2 \times 10^4. \qquad (1)$$

For a particle density of $\delta(a)$ grams per cc, the expression for A_{pg}, the absorption for photographic light in magnitudes per kiloparsec, becomes:

$$A_{pg} = 3.36 \times 10^{21} \times K_{pg}(a) \times \delta(a). \qquad (2)$$

For each type of cosmic cloud mentioned above, we have given approximate values for the true dimensions and for the total photographic absorptions. From these we may obtain the approximate value of A_{pg} and then, with the aid of (2), the corresponding value for $\delta(a)$. Since the true diameter for each object has also been estimated, we can readily compute its corresponding total mass.

The results of these computations are in Table 1, where, for comparison purposes we are presenting side by side the values for the smaller

TABLE I

	Globule I	Globule II	Coalsack	Large Cloud
Diameter (parsecs)	0.06	0.5	8	40
Total Pg. Absorption	5^m	$1^m.5$	$1^m.5$	$1^m.4$
A_{pg}	8×10^4	3×10^3	200	35
$\delta(a)(g/cc)$	1.3×10^{-21}	5×10^{-23}	3×10^{-24}	5×10^{-25}
$\delta(a)(\odot/pc^3)$	20	0.8	0.05	0.01
Min. Mass ($\odot$)	0.002	0.05	13	300

(I) and larger (II) variety of globule, for the southern Coalsack and for a large obscuring cloud.

In connection with the data of Table 1, we make the following comments:

a. The last column of Table 1 has been included for comparison purposes only. We have already noted that we do not feel inclined to consider the "Large Clouds" as single dynamical units. The derived low density would make it almost impossible for an average large cloud to remain as a unit in a part of the galactic system where the average density of interstellar gas and dust combined exceeds that for the cloud by a factor of six.

b. The calculations on stability of dark nebulae by Klauder suggest that an object like the obscuring cloud of the last column of Table 1 is not sufficiently dense to withstand the disruptive effects of the galactic tidal forces, whereas the globules and the Coalsack are sufficiently dense to remain intact.

c. It must be kept in mind that all densities and masses in Table 1 are minimum values. An admixture of particles of different sizes, and perhaps of some gas atoms or molecules, might well lead to true masses and densities that exceed those given here by a factor of ten or more. There is little possibility that the masses and densities given here are in excess of the true values.

d. The most uncertain quantity derived from observation is the total photographic absorption of Globule I. For all we know, the smaller globules may be totally opaque. The true density and mass of Globule I may therefore be considerably greater than the values given here. . . .

6. GROWTH AND EVOLUTION

In sections 4 and 5 we have been concerned with globules in a state of equilibrium. This is not a permanent condition, since, through slightly inelastic collisions, there will always be some gradual loss of energy. An isolated globule will therefore, in the course of time, contract under the combined influence of its own gravitational attraction and of radiation pressure exerted from outside.

The globules and small dark nebulae are, however, not isolated in our galaxy. They occur in regions where there is considerable interstellar gas and dust. Some of this dust and gas will "fall" into the globule, attracted to it by the globule's gravitational attraction and by the effect of external radiation pressure. The globule may further gain in mass through the sweeping up of cosmic dust in its motion through the general medium.

The conditions for growth are best in the regions of our galaxy where there exists a dense substratum of dust and gas. We list three arguments in support of this statement.

(1) Inside a large dark cloud the density of cosmic dust exceeds that for average interstellar space from 10 to 100 times. This gain is offset only in part by the reduction in the general radiation pressure to one-half or one-fifth of the average for the dust-free regions of space.

(2) The interior parts of large dark nebulae are mostly H I regions, where—according to Spitzer—the kinetic temperature may be as low as 100°K.

(3) A large dark cloud serves as a protecting atmosphere against collisional evaporation. Oort and van de Hulst find that collisional evaporation is effective mostly at the periphery of two colliding large clouds. It should have little effect on an imbedded globule. . . .

At present nothing is known about the rates at which the globules and the Coalsack move through the general interstellar substratum, but we should keep in mind that these objects may grow in mass by the simple process of sweeping up cosmic grains, *i.e.*, by accretion.

For the case of a small globule, the gravitational attraction is so small and the resultant modification in the external radiation pressure so slight that the curvature in the orbits of the captured grains is negligible; the capture radius of the globule is very nearly equal to its true radius. This is not so for the Coalsack variety of dark nebulae. Approximate calculations indicate that the capture radius is at least 50 percent greater than the true radius.

The computed rates of increase in the mass depend entirely upon the assumed value for the velocity of the objects with respect to the surrounding general substratum. The rate of growth through accretion will equal that from the falling in of cosmic grains, provided the speeds of the nebulae are of the order of 1 km/sec. These calculated rates of accretion are not unreasonable, if we bear in mind that turbulent velocities of the order of several kilometers per second are found in diffuse nebulae.

We conclude that there exist in our galactic system several varieties of small dark nebulae in the prestellar stage. These objects will tend to contract as a consequence of their own gravitational attraction and the pressure of radiation exerted by the general galactic field. It is further suggested that during the process of contraction the small dark nebulae will grow in part through matter being accelerated toward the object, and also through matter being swept up as the object travels through the interstellar medium. Approximate calculations show that the prestellar stage may last for times of the order of 10^7 to 10^8 years, depending somewhat on the initial mass and dimensions of the condensing dark nebula.

51. THE FORMATION OF STARS

By Lyman Spitzer, Jr.

If research in astronomy had stopped in 1913, our knowledge of stellar evolution today would be in a satisfactory state. At that time astronomers had a plausible theory of a star's life cycle. Einstein's theory of relativity, advanced only a few years before, showed that mass and energy were interchangeable. It was therefore natural for astronomers to assume that stars were formed as large massive bodies which through successive century after century continued to radiate away matter. Ultimately most of the matter in a star, according to this picture, would be radiated away as light and heat. In this way all the stars, despite their large differences in mass, formed part of the same evolutionary sequence.

Unfortunately, this simple, sweeping, and satisfying picture became discredited by additional information, both astronomical and physical. On the astronomical side, evidence began to accumulate that the universe has not lasted long enough for most stars to radiate away much of their matter. The expansion of the universe, the presence of uranium on the earth, the existence of certain relatively transitory clusters of stars, all indicate that something happened about three billion years ago.[1] If the universe was not created then, it was certainly very extensively reorganized; some sort of cosmic explosion apparently took place at that time. Since the sun, a fairly typical star, would require many hundreds of billions of years to radiate an appreciable fraction of its mass, its total mass has obviously not changed appreciably within the past two billion years.

On the physical side, nuclear physicists have learned a great deal about the specific processes by which matter can be converted into energy. The only known process of importance which can liberate energy inside a star is the combination of four hydrogen atoms to form a helium atom. Calculations carried out by the nuclear physicist Professor Hans Bethe show that in the stars this process occurs through the

[1] [Because of subsequent revision of the extragalactic distance scales, the author would now double this number.]

catalytic action of carbon and nitrogen nuclei. Since four hydrogen atoms weigh 0.7 percent more than one helium atom, the additional mass is released as energy and can be radiated by the star. Even if a star is originally all hydrogen, the total mass radiated can evidently not exceed a very small fraction of the mass of the star.

As a result of these findings we now know that the universe has apparently not lasted long enough in its present form for stars to radiate much of their mass, and in any case there seems to be no physical process by which a star could radiate away most of its matter even if there were time enough. We are forced to conclude that the present variety of stars in the sky is the result of the original method of star formation rather than of any evolutionary process. And the formation of stars in general is still a closed book, since the explosion of the universe a few billion years ago has so far defied any attempts at detailed analysis. It is even possible that the basic laws of nature may have been quite different at that time. Thus our research in the direction of general stellar evolution reminds one of Browning's philosopher, who had

> ". . . written three books on the soul,
> Proving absurd all written hitherto
> And putting us to ignorance again."

Supergiant Stars

While the origin of the universe is still beyond our understanding, some progress has been made in explaining the origin of a certain class of stars, which may have been created relatively recently. A supergiant star is one which radiates light and heat some ten thousand times as strongly as our own sun. There are not many of these stars, but in a galaxy of many billions of lesser stars they stand out in the same way that a searchlight stands out from a swarm of fireflies. These stars are burning their candle at both ends and they cannot last very long, astronomically speaking. Within a mere hundred million years, such a star must burn all its hydrogen into helium. There is no known way in which a star can remain dark for a long period of time and then suddenly start shining. We conclude that these supergiant stars have formed within the last hundred million years—less than a tenth of the age of the universe.

Of course, it is possible that nuclear physicists have overlooked some important process by which a star can radiate a much larger fraction of its mass than the hydrogen-into-helium process liberates. This does not seem very likely, since the energies with which the atoms hit each other inside a star average only a few thousand electron volts—a small fraction of the energies developed in such atom-busting devices as the cyclotron and synchrotron—and since the nuclear reactions produced at low energies have been fairly well explored in the laboratory.

If it is assumed that these stars have in fact been formed within the last hundred million years, the mechanism for this formation is a problem which astronomers may hope to investigate with some hope of success. Within this interval, conditions in the universe have apparently not changed very much and an examination of the universe about us may actually indicate how supergiant stars have formed in the past, and may even be forming at the present time.

CLOUDS—THE CLUE

The clouds of matter which float about between the stars are an obvious source of material for star formation. Recent investigations show that these clouds are in fact so closely associated with supergiant stars that a physical connection between them seems very likely.

In brief, the observations indicate that supergiant stars are found only in those aggregations of stars where interstellar clouds of matter are also present. More specifically, observations of stellar galaxies, each one a million or so light years away and each, like our own galaxy, containing many billions of stars, show that the supergiant stars are found only in spiral galaxies. These spiral systems, like the huge galaxy in which our sun is located, are flattened, disk-shaped systems some hundred thousand light years in diameter, each one rotating about an axis perpendicular to the plane of its disk. A typical spiral galaxy is shown in the accompanying figure. The characteristic feature of these systems, after which they are named, is the presence of a pair of arms which apparently come out of the central nucleus and wind around the system.

In the elliptical galaxies—which are not rotating so rapidly, are not so flattened, and show no spiral structure—no supergiant stars are found. In fact, long-exposure plates at the Mount Wilson Observatory have shown that the stars in these systems have a sharp upper limit on their brightness; no star greater than the critical brightness can be found, while below this critical brightness myriads of stars appear on the photographic plate. This result is in marked contrast to the observed brightness of the stars in spiral galaxies, where there are always one or two brightest supergiant stars, a number of less bright supergiants, and a gradually increasing number of fainter and fainter stars. This sharp

Fig. 1. The southern spiral galaxy Messier 83.

upper limit on the brightness of stars in elliptical galaxies is just what
one would expect if no new stars had been formed since the beginning
of the universe, and if the brightest ones had burned up all their fuel
and gone out.

Detailed examination of galaxies also indicates that clouds of matter
between the stars are found only in spiral systems. In elliptical galaxies
the vast stretches between the stars are very nearly empty, but in flattened spiral galaxies like our own there is about as much matter between the stars as there is inside the stars. This association between obscuring clouds and supergiant stars is strengthened by the fact that in
the closest galaxy, the great nebula in Andromeda, supergiant stars are
observed to occur in exactly those regions where the obscuring clouds

are most prominent. Thus the observational evidence indicating a phys- ical connection between clouds and supergiant stars is very strong.

Before we can accept the hypothesis that supergiant stars have in fact formed from these clouds we must investigate whether or not there is some process which could cause interstellar matter to condense into stars. In this way we are led to consider the physical nature of the stuff between the stars, and the forces which operate on it. Thirty-five years ago the very existence of interstellar matter was not fully realized but recently extensive information on this topic has been obtained.

Atoms in Space

The dominant constituents of interstellar matter are believed to be individual atoms. These atoms absorb or emit light of particular wave- lengths, which can be measured accurately by use of the spectroscope. In some regions, where the gas is at a high temperature, bright emission lines of hydrogen, oxygen, and nitrogen are observed. Measurements of the intensities of these lines show that the density of the interstellar gas is about one hydrogen atom in each cubic centimeter, with other ele- ments present as slight impurities. The interstellar medium is a much better vacuum than is ever obtained in a terrestrial laboratory. If a fly were to breathe a single breath into a vacuum chamber as big as the Empire State Building, the resulting density of the air would still be much greater than the density of the interstellar gas.

In other regions of space the interstellar gas is cool, and no emission lines are produced. Instead, the atoms absorb the light from distant stars, producing absorption lines at particular wave-lengths. The ab- sorption lines of the abundant gases, hydrogen, helium, nitrogen, oxy- gen, etc., when these are cool, lie far out in the ultraviolet where they cannot be detected. Interstellar absorption lines of sodium, calcium, titanium, and iron lie within the observable spectrum and have been ob- served in the spectra of bright stars a few thousand light years away. These lines are very sharp, and can usually be distinguished from the lines produced by the atoms in a stellar atmosphere, where the high temperature and pressure give wide lines.

Recent work has been concerned with the detailed distribution of in- terstellar gas. Measurement of the strongest absorption lines, with the most powerful spectrographs available at the 100-inch telescope of the Mount Wilson Observatory, shows that a single line is frequently made up of several components. Each separate component is produced by ab- sorption in a single cloud of gas, the different components being separated in wave-length by the difference in Doppler effect produced by the different cloud velocities. These clouds, each one about twenty light years across, are moving through space at speeds of some ten to twenty miles a second. A more detailed understanding of the nature of these clouds is desirable before one can discuss in detail how interstellar matter can form new stars. Further work along these lines is now in progress.

SOLID PARTICLES

In addition to the separate atoms drifting about in space, small solid particles, or grains, are also present. Each grain is about one hundred-thousandth of an inch in diameter; ten thousand placed end to end would make a line about as long as a period on this page. Since the size of these grains is just about equal to the wave-length of visible light, these particles are of the size which is most effective in absorbing and scattering light waves. These particles are responsible for the general obscuration produced by the clouds shown in the accompanying figure. Particles of smaller size are presumably also present, but these do not produce such a noticeable effect, and can therefore not be detected.

The properties of these particles have been determined from accurate measurements of the obscuration which they produce in light of different wave-lengths. This obscuration is greater for blue light than for red light, which proves that the particles cannot be much *larger* in size than the wave-length of light. On the other hand, the obscuration varies inversely only as the first power of the wave-length, instead of as the fourth power which is observed for scattering by the molecules of the atmosphere. From this one can conclude that the grains are not very much *smaller* in size than the wave-length of light. In this way a particle size of about the wave-length of light has been determined. From the fact that the grains seem to scatter more than they absorb it seems likely that they are dielectric rather than metallic in composition. If, as seems likely, these grains were produced by the sticking together of individual atoms, the enormous abundance of hydrogen relative to other elements would be expected to produce solid hydrogen compounds, in particular, ordinary ice. However, impurities of all other elements would also be present.

Studies of the distribution of these grains have indicated that the clouds in which these grains are concentrated are apparently identical with the gaseous clouds already described. Thus whatever pushes atoms into clouds also pushes the grains together.

Forces in Interstellar Space

To discuss in detail how stuff in space can condense to form new stars we must determine the physical conditions of matter in space. In particular, we must combine the observational evidence described above with our knowledge of basic physical principles to investigate the different forces that are at work on the different particles. Only in this way can we predict how the interstellar medium will behave under various widely different conditions.

In the immense vacuum between the stars, an interstellar particle spends most of its time moving in a straight line without interruption. Occasionally, one of two things may happen to it: an encounter with another interstellar particle, or an encounter with a light wave, or photon. The information which physicists have obtained on such processes is not so complete as astronomers would like, but is sufficient for an approximate evaluation of the effects which these various collisions will produce.

The collisions of the interstellar atoms and grains with each other help determine the temperature of matter in space. In most cases, the collisions are elastic and the kinetic energy of the different particles is exchanged back and forth; as a result, the distribution of velocities corresponds to that in thermal equilibrium at some particular temperature. Photoemission of energetic electrons from hydrogen atoms and grains, on absorption of photons, tends to keep the temperature high, but inelastic collisions between atoms and grains tend to give a low temperature. Near a very hot and very bright star the gas will be heated up to about 10,000°K, but in other regions a temperature of about 100°K seems likely. This difference of temperature between different regions is believed to produce cosmic currents, or winds, in the same way as the winds on earth are produced.

In some cases the interstellar particles stick to each other on collision. Thus atoms stick together to form molecules, molecules stick together to form larger molecules, and grains grow by slow accretion. This process was analyzed during the war by a number of Dutch astronomers, who were able to show that the interstellar grains have probably been formed by this evolutionary process within the last few billion years. More accurate physical information on collisions between particles at low energies is required to make this theory more quantitative.

Collisions between grains and photons are important in star building. It is well known that light exerts pressure. Since starlight in a galaxy comes from all directions in the galactic plane, a single grain will be knocked this way or that by photon collisions, without any net motion resulting. However, when several grains are present, the shadow of each one on the other unbalances the radiative force, and photons striking from the opposite sides push the grains toward each other. As a result, there is an effective force of attraction between grains which is several thousand times as great as the gravitational force between them.

STAR FORMATION

The further we go away from observational data the more uncertain our theories become. The mechanism of star formation, which is the ultimate objective of much of the work described above, is still in a rather speculative state. However, putting all the above information together does provide a reasonable preliminary picture for the process by which stars can be formed from interstellar matter.

The process may be assumed to start with an interstellar gas, formed at the same time as the rest of the universe. The first step in the process is then the slow condensation of interstellar particles from the gas. After these particles have reached a certain size, the radiative attraction between them forces them together and they drift toward each other, forming an obscuring cloud in a time of about ten million years. In a cloud, where the density of grains is high, the temperature tends to be low. In the surrounding region the high temperature produces high pressure, and the low-temperature, low-pressure cloud therefore becomes compressed. In this way the density of gas within a cloud will be increased, corresponding to the observed result that a cloud of grains is also a cloud of gas.

Currents produced by differences of temperature and also by the general rotation of the galaxy will tend to tear some of these clouds to pieces. On the other hand, the forces of condensation will pull them together, and some clouds may be expected to go on contracting. The radiative force becomes ineffective when the clouds become so opaque that light does not penetrate into them very far. At this point gravitation takes over and tends to produce a further contraction. In this stage a cloud has a diameter of a light year or less. Small opaque clouds of this type, called globules, have been known for some time, and are shown in the accompanying figure [in the preceding chapter].

One of the chief problems concerns the angular momentum of this prestellar globule, or protostar. According to Newton's laws of motion, the angular momentum, which is proportional to the product of the radius and the rotational velocity, remains constant; as the radius decreases the rotational speed increases. Since the radius of a typical cloud is some ten million times the radius of a supergiant star, this increase in rotational speed can be quite impressive. Unless some way can be found to dispose of the angular momentum, a protostar would hurl itself to pieces by centrifugal force. The possibility that turbulent motions in the gas may carry the angular momentum away has been explored by several German astronomers. However, the turbulent velocities involved would exceed the velocity of sound in the interstellar gas, and physical information about this type of turbulence is virtually non-existent. In this country the possibility has been advanced that a galactic magnetic field might produce electrical eddy currents in a rotating protostar, which would then damp out the angular momentum.

An interesting variant of this star-building picture has been proposed by Dr. Fred Whipple, one of the astronomers who has contributed most to this theory of star building. He suggests that a condensing cloud may have produced our solar system. In view of the widespread general interest in the formation of the solar system, such a bold extrapolation of these theoretical concepts back to conditions several billion years ago is naturally of much significance.

It is evident that the picture of star formation which has been described here is still in a formative stage. The work in progress is being carried out cooperatively by a number of astronomers all over the world. Perhaps when the 200-inch telescope probes further into the secrets of space, and when further progress in experimental and theoretical physics increases our understanding of the processes at work between the stars, we may then outline with more assurance the detailed steps by which supergiant stars may be forming almost before our very eyes.

XI

GALAXIES

If at the proper time a Source Book of Astronomy is prepared for the interval 1950 to 1975, or 1950 to 2000, the galaxies will receive a much greater share of attention than that accorded to them in this volume. The reason is that several large telescopes, both optical and radio, have come into operation since 1950, and others soon will be built. Also the metagalactic picture is growing in interest, and the measures thereof in accuracy. In fact, this section on galaxies might have been considerably enlarged through further selections from the many contributions published by Mount Wilson, Lick, and Harvard astronomers.

The Magellanic Clouds, which must be rated as the most fruitful of all external galaxies, are not here discussed as such, although the Harvard Observatory has published scores of papers on this pair of exciting systems. The next Source Book will be more generous to the Clouds, for not only Harvard, but a half-dozen other well-equipped observatories will give the Clouds serious attention; they will turn on them the analytical power of radio, fast spectroscopes, polarimeters, and electronic photometers. Indirectly the Clouds make their debut in this Source Book through Miss Leavitt's classic investigations of the Small Cloud's cepheids (selection 32).

Among other significant contributions not included in this part on galaxies is the work of Sinclair Smith on the mass of the Virgo Cloud, and of K. Lundmark and his associates at the Lund Observatory on the classification and distribution of galaxies. The identification of spiral nebulae as star-composed galaxies was the work of many from Immanuel Kant in 1755 to H. D. Curtis, K. Lundmark, E. P. Hubble, and others. No specific account is given here of this activity. Also not included are the Spitzer-Baade speculations on the collisions of galaxies (their paper was published in 1951); the census-taking by Max Wolf and by C. W. Wirtz in Germany; the early descriptive photographic studies by Curtis and visual surveys by Bigourdan; the work of Shapley and colleagues on metagalactic population gradients; the examination by Carpenter, Page, Holmberg, and Zwicky of double galaxies and their connectors; the classification scheme by Hubble; the beginning of large-scale survey work by Zwicky and colleagues with the Palomar schmidt telescopes and by Shane at the Lick Observatory with the Carnegie telescope; and the early radial velocity work by Slipher at Flagstaff and the later programs by Mayall on Mount Hamilton. The practical reasons for not including some of these have been noted in the general introduction to this volume. Certainly the next Source Book of Astronomy, which may be more technical and specialized, will have much to say about galaxies.

The articles which appear in this part deal with the discovery of the center of the galaxy, galactic rotation, and the expanding metagalaxy. Since 1950, additional work has also been done on these three problems, but, in general, the conclusions here presented stand, notwithstanding some numerical improvements.

In the three papers from Mount Wilson the term "extragalactic nebula" is used instead of the now more common term "galaxy."

52. FROM HELIOCENTRIC TO GALACTOCENTRIC

By Harlow Shapley

The attempt to find the extent of the Galaxy from measures of its star clusters [1] succeeds in the gross but not in detail. The clusters dimly outline the size but show little of the structure. We may confidently expect that analyses of star motions, star clouds, and individual stellar distances will in time reveal with more than present clarity the significance of our galactic system in the total material universe.

MEMBERSHIP IN THE GALAXY

By the term "Galaxy," or "the galactic system," is meant the aggregate of stars and nebulae for which the distributions appear to be organized with respect to the galactic plane. Globular clusters are therefore included, with galactic stars and galactic clusters and nebulae; but the Magellanic Clouds and the extragalactic nebulae (spiral nebula family) are outside the organization.

Possibly, however, some of the remote globular clusters (e.g., N.G.C. 7006, N.G.C. 4147, and Messier 75) are actually independent, being either fugitives from the Galaxy or chancing for the moment (cosmically speaking) to be moving in this part of space. Some of the high-velocity stars of the sun's neighborhood also may eventually escape from galactic control. We need more information on speeds and masses and on the phenomena of galactic rotation before we can pass judgment on these questions of membership.

The complete freedom of the Magellanic Clouds from our Galaxy is but a surmise based on relative masses, present positions, and radial velocities. Without accurate information concerning their proper motions, we cannot safely assume from radial components that they are receding from the Galaxy; [2] in fact, the increasing evidence that the Galaxy is in rapid rotation argues for the affiliation of the Magellanic

---

[1] [This account summarizes several articles published as Mount Wilson Contributions in 1918.]
[2] Luyten, W. J., *Harv. Circ.*, 326, 327 (1928).

Clouds with our Galaxy, or, at least, with a local group of galaxies that would also include the three Andromeda nebulae, Messier 33, and some others.

In measuring the Galaxy, we should at the start admit indefinite limits, and also striking irregularities, not only in the interior but probably also at the edges. The dimensions discussed below are therefore not to be taken too literally as marking the boundaries or even as giving sharp limits of star density. At best, we measure or estimate the distances of the remotest attainable stars or group of stars which yield to present methods and which appear to be members of the Galaxy.

The Higher System of Globular Clusters

To illustrate the space distribution of the 93 known globular clusters of the galactic system, the following rectangular coordinates have been computed for all clusters:

$$X = R \cos (\lambda - 327°) \cos \beta$$
$$Y = R \sin (\lambda - 327°) \cos \beta$$
$$Z = R \sin \beta$$

where R is the distance in kiloparsecs, β is the galactic latitude, and $(\lambda - 327°)$ is the galactic longitude measured from the direction to the center of the cluster system.

Eccentric Position of the Solar System

A diagram of the distribution of the globular clusters in the plane of the Galaxy (XY plane) is shown in Figure 1. Crosses indicate clusters lying on the north of the galactic plane, and dots those on the south; the smaller the symbol, the more distant is the object from the plane. The equality of the division by the galactic plane is remarkable—47 clusters are on the north, 46 on the south.

The direction to the center of the system, derived from the apparent positions of globular clusters, is seen to agree with the direction on the basis of space coordinates; in Figure 1 there are 46 positive values of Y and 46 negative values, with $Y = 0$ for one cluster. (The remote system N.G.C. 7006, with coordinates $X = +22.4$, $Y = +48.0$, falls outside the limits of the diagram.)

The origin of coordinates for Figure 1—that is, the position of the observer—is on the border of the globular cluster system. The center of gravity of the system of clusters, indicated by an open square, has the

coordinates $X = +16.4$, $Y = -0.3$ (or $Y = +0.2$ if the remote and isolated N.G.C. 7006 is included).

Probably 10 or 20 globular clusters, within the limits of space represented by the diagram, await discovery. Obscuring nebulosity possibly

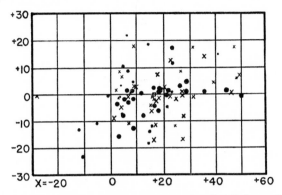

Fig. 1. Distribution of the globular clusters in the Galaxy (XY plane).

conceals most of these systems, of which the existence is intimated by
the scarcity of observed points in the right half of the figure. Of course,
we need not assume high regularity in distribution, or even approximate circularity in the projected array; but it is probably observational
difficulties caused by nebulosity and by the small dimensions and faint

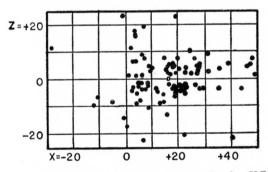

Fig. 2. Distribution of the globular clusters in the XZ plane.

magnitudes of remote clusters, and not inherent irregularities, that
have produced the apparent incompleteness for X greater than $+30$
kiloparsecs. On the left side of the diagram, from $X = 0$ to $X = -20$,
where the survey may be considered sufficiently exhaustive, the complete absence of clusters from one quadrant is even more striking and

structurally significant than the scarcity of values of X greater than +30.

The cluster at the extreme left is N.G.C. 2419—an object in a region far from other clusters, found through studies of the Lowell Observatory photographs.

The "Region of Avoidance"

The same asymmetrical position of the sun with respect to the super-system of globular clusters is shown in Figure 2, where all 93 systems

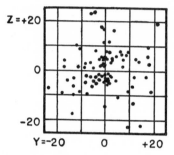

Fig. 3. Distribution of the globular clusters in the *YZ* plane.

are plotted on the *XZ* plane. The center of gravity—that is, the algebraic mean values of X and Z, indicated by an open square—is at $X= +16.4$, $Z = +0.4$. N.G.C. 2419 again stands out on the extreme left.

The most interesting feature of Figure 2, which represents a section perpendicular to the galactic plane, is the "region of avoidance." The scarcity of globular clusters in low galactic latitudes is again in evidence. Here is a central section 2.5 kiloparsecs (8,000 light years) in diameter, in which no globular cluster has been found. On the other hand, there is only one galactic cluster out of the 249 listed [in an appendix] that does not fall well within this mid-galactic segment, and that cluster, N.G.C. 2243, is of doubtful nature and uncertain distance. Practically all known stars and nebulae also fall within this "region of avoidance."

Projection on the YZ Plane

Figure 3 shows the distribution of globular clusters on the YZ plane, which is perpendicular to the line joining the sun and the center of the system at galactic longitude 327°, galactic latitude 0°. The "region of avoidance" is again clearly shown, and also the essential symmetry of the globular cluster system, for the numbers of clusters in the four quadrants are 23, 24, 22, and 24. Again N.G.C. 7006 is outside the diagram, with coordinates $Y = +48.0$, $Z = -20.4$.

THE DISTANCE TO THE GALACTIC CENTER

It appears to be a tenable hypothesis that the supersystem of globular clusters is coextensive with the Galaxy itself. Researches on variable stars in the Milky Way will eventually afford an instructive check on this hypothesis. Until we have such direct measures we can only assume that the galactic system is at least as large as the system of globular clusters, and note that we have shown rather convincingly that the globular clusters are galactic members or associates.

The algebraic mean of the values of $R \cos (\lambda - 327°) \cos \beta$ gives a satisfactory indication of the distance to the center of the cluster system and provisionally, therefore, of the distance to the center of the Galaxy. In so far as it depends on the globular clusters now known, the uncertainty of the distance does not exceed ten percent. Further research on faint globular clusters, especially if new ones be found, will be more likely to extend the system than to reduce it; on the other hand, the distances of the more remote clusters are the least certainly determined and must be given low weight. We shall adopt as the distance to the center $R_g = 16$ kiloparsecs.[3]

The galactic clusters and the ordinary individual stars are too near the sun to contribute effectively to the determination of the distance to the center; but the direction to the center, as is well known, is confirmed by the counts of faint stars [by Nort and Seares] and through the recent studies of galactic rotation by Oort, J. S. Plaskett, Lindblad, Schilt, and others; it is shown, though less definitely, by the distribution of Milky Way star clouds, planetary nebulae, Class O stars, and other objects of high luminosity. Most galactic objects, however, with the possible exception of novae and cepheid variable stars, are too faint absolutely and too infrequent in number to contribute in current surveys of galactic regions at and beyond the center of the cluster system.

Studies of the cluster-type cepheids and the long-period variable stars in the star clouds of the southern Milky Way have led the writer

[3] [This value, derived in 1918, is considerably diminished when allowance is made for space absorption. The current (1959) value lies between 8 and 12 kiloparsecs. Notwithstanding the uncertainty in the distance, the direction of the center has not been appreciably altered by the various investigations since 1918.]

and Miss Swope to a value of the distance of the centrally located star clouds that is very much like the value given above. The variable star investigations of several observers at Harvard tend to support this suggestion that the heavy star clouds in Ophiuchus, Sagittarius, Scorpio, and neighboring constellations are parts of a massive stellar nucleus of the galactic system. The distribution of the cluster-type cepheids in these regions suggests that the nucleus extends perhaps halfway from the center to the sun. The speed of rotation about this nucleus is approximately 300 kilometers a second at the sun's distance from the center.

One important feature of the galactic central region is that the center itself lies behind heavily obscuring nebulosity. The dark clouds are apparently but a part of those causing the rift in the Milky Way that extends from Cygnus southward to Centaurus; they seem to be largely responsible for the apparent avoidance of the mid-galactic regions by globular clusters.

53. CONFIRMING LINDBLAD'S HYPOTHESIS OF THE ROTATION OF THE GALAXY

By Jan H. Oort

INTRODUCTION

It is well known that the motions of the globular clusters and RR Lyrae variables differ considerably from those of the brighter stars in our neighborhood. The former give evidence of a systematic drift of some 200 or 300 km/sec with respect to the bright stars, while their peculiar velocity averages about 80 km/sec in one component, which is nearly six times higher than the average velocity of the bright stars.

Because the globular clusters and the bright stars seem to possess rather accurately the same plane of symmetry, we are easily led to the assumption that there exists a connection between the two. But what is the nature of the connection?

It is clear that we must not arrange the hypothetical universe in such a way that it is very far from dynamical equilibrium. Following Kapteyn and Jeans let us for a moment suppose that the bulk of the stars are arranged in an ellipsoidal space whose dimensions are small compared to those of the system of globular clusters as outlined by Shapley. From the observed motions of the stars we can then obtain an estimate of the gravitational force and of the velocity of escape. An arrangement as supposed by Kapteyn and Jeans, which ensures a state of dynamical equilibrium for the bright stars, implies, however, that the velocities of the clusters and RR Lyrae variables are very much too high. A majority of these would be escaping from the system. As we do not notice the consequent velocity of recession it seems that this arrangement fails to represent the facts.

As a possible way out of the difficulty we might suppose that the brighter stars around us are members of a local cloud which is moving at fairly high speed inside a larger galactic system, of dimensions comparable to those of the globular cluster system. We must then postulate the existence of a number of similar clouds, in order to provide a gravitational potential which is sufficiently large to keep the globular clusters from dispersing into space too rapidly. The argument that we cannot observe these large masses outside the Kapteyn-system is not at all conclusive against the supposition. There are indications that enough dark matter exists to blot out all galactic starclouds beyond the limits of the Kapteyn-system.

Lindblad has recently put forward an extremely suggestive hypothesis, giving a beautiful explanation of the general character of the systematic motions of the stars of high velocity. He supposes that the greater galactic system as outlined above may be divided up into subsystems, each of which is symmetrical around the axis of symmetry of the greater system and each of which is approximately in a state of dynamical equilibrium. The sub-systems rotate around their common axis, but each one has a different speed of rotation. One of these subsystems is defined by the globular clusters, for instance; this one has a very low speed of rotation. The stars of low velocity observed in our neighborhood form part of another sub-system. As the rotational velocity of the slow moving stars is about 300 km/sec and the average random velocity only 30 km/sec, these stars can be considered as moving very nearly in circular orbits around the centre.

We may now apply an analysis similar to that used by Jeans in his discussion of the motions of the stars in a "Kapteyn-universe," the only difference being that in the present analysis we do not introduce a second system rotating in the opposite direction. Adopting some probable formula for the gravitational potential we can derive the rotational velocities for each of the sub-systems from our knowledge of the distribution of the peculiar velocities (defined as the velocities remaining after correction for the effects of rotation). The higher the average peculiar velocity in a certain sub-system the slower its rotation will be, and the less flattened it will appear in a direction perpendicular to the galactic plane. If we refer our motions to the centre of the slow moving stars in our neighborhood, the members of a sub-system with higher internal velocities will appear to lag behind, and Lindblad has shown that in this way we can arrive at a connection between average peculiar velocity and systematic motion of the same form as that computed from observation.

Lindblad's hypothesis conforms beautifully with the well-established fact that the average direction of the systematic motion of the high velocity stars is perpendicular to the direction in which the globular clusters are concentrated (galactic longitude 325°, latitude 0°). At first sight it might be hard to imagine how such a mixture of sub-systems of different angular speeds could ever come into existence; but the *possi-*

bility cannot be denied, as is apparent from a comparison with spiral nebulae.

If somewhere there existed a rapidly rotating system of stars and by some cause the internal velocities in this star-system were increased, an asymmetry in the stellar motions would necessarily result in the long run. It must be admitted, however, that the part played by the globular clusters cannot be so easily understood.

The following paper is an attempt to verify in a direct way the fundamental hypothesis underlying Lindblad's theory, namely that of the rotation of the galactic system around a point near the centre of the system of globular clusters.

[Then follows the mathematical development of the theory of galactic rotation and a consideration of the relevant observations of the radial velocities and proper motions of stars.]

CONCLUDING REMARKS

It has been shown from radial velocities that for all distant galactic objects there exist systematic motions varying with the galactic longitudes of the stars considered. The relative systematic motions are always of the same nature and they increase roughly proportional with the distance of the objects. Probably the simplest explanation is that of non-uniform rotation of the galactic system around a very distant centre. This explanation is capable of representing all the observed systematic motions within their range of uncertainty (except perhaps in the case of the B stars). If with this supposition we compute the position of the centre from the radial velocities, we find that it lies in the galactic plane, either at 323° longitude or at the opposite point. The first direction is in remarkably close agreement with the longitude of the centre of the system of globular clusters (325°). The observations would therefore seem to confirm Lindblad's hypothesis of a rotation of the entire galactic system around the latter centre.

The proper motions corroborate the above interpretation, at least qualitatively. They were used mainly to determine the character of the non-uniformity of the rotation. This character corresponds to a gravitational force which can sufficiently well be represented by the following formula: $K = \frac{c_1}{R^2} + c_2 R$, if R is the distance of the centre. A provisional solution gave: $\frac{c_2}{c_1} = \frac{0.11}{R^3}$

Such a force would for example result if $9/10$th of the total force came from mass concentrated near the centre and $1/10$th from an ellipsoid of constant density large enough to contain the sun within its borders. The true character of the force will of course be more complicated.

We can derive a numerical result for R as soon as the circular velocity, V, is known. An estimate of this circular velocity may be made from the radial velocities of the globular clusters. According to Strömberg these clusters possess a systematic motion nearly perpendicular to the direction of their centre and equal to 286 $km/sec \pm 67$ (m.e.) relatively to the sun, or 272 km/sec relatively to the centre of the slow moving stars. This would give us an estimate of the circular velocity if we were sure that the system of globular clusters had no rotation. From the ellipsoidal arrangement in space Lindblad derives a rotation for these objects, such that the circular velocity would be increased to 426 km/sec. As, however, the apparent positions in the sky give no indication whatever of the system of globular clusters being flattened towards the galactic plane, it seems better to assume that they possess no rotation. In fact it seems more probable from dynamical considerations that the true circular velocity is below the value found by Strömberg than above it. Assuming $V = 272 \ km/sec$ we find $R = 5{,}900$ parsecs. As the longitude of the centre of rotation agreed with that of the system of globular clusters, it is probable that the distance will agree as well. The distance of the centre of the globular cluster system is very uncertain, however, on the one hand by an uncertainty in the scale of the cluster distances and on the other hand by our incomplete knowledge of the more distant clusters. Shapley gives estimates varying from 13,000 to 25,000 parsecs. The value found above is considerably smaller. Even if we made the extreme supposition that $K_2 = 0$, or that all the mass of the galactic system were concentrated near the centre, the distance of R would only be increased by 6,600 parsecs.

In order to explain the rotation there must be near the centre an attracting mass of at least 8×10^{10} times the mass of the sun. There remains the difficulty why we do not observe this large mass. Near 6,000 parsecs Kapteyn and van Rhijn find an almost negligible density, whereas it *should* be very much greater than in our neighborhood. Part of the discrepancy may have resulted from the approximative character of their solution, in which all galactic longitudes were combined. Discussing various galactic regions separately Kreiken finds indications of a centre near 314° longitude, at a distance of 2,270 parsecs which is in the right direction, but certainly at too small a distance and too little defined. The most probable explanation is that the decrease of density in the galactic plane indicated for larger distances is mainly due to obscuration by dark matter. Such a hypothesis receives considerable support from the marked avoidance of the ·galactic plane by the globular clusters, a phenomenon for which up to the present time no other well defensible explanation has been put forward.

Lindblad suggests that the starstreaming is an indirect consequence of the rotation. In so far as his computations result in a starstreaming to and from the centre, in the same way as originally suggested by Turner, the present data do not entirely confirm the hypothesis. The true vertices according to Eddington lie at 166° and 346° longitude; it does not seem possible to admit an error of over 20°, which would be required to make vertex II coincide with the direction to the centre as derived in the present paper.

It may be remarked that the rotation offers a means of determining average distances from radial velocities, the relative accuracy of the method increasing with the distance and being independent of possible absorption of light in space. We have thus been able to derive a value for the average distance of the most distant planetary nebulae.

The question naturally arises which would be the most valuable observations that could be made in order to check the present results. Probably the most promising results could be derived from the radial velocities of some very faint c-stars. If one could get down to the 8th or 9th magnitude the semi-amplitude of the rotation term might become as large as 50 km/sec; a small number of stars would then suffice to give reliable results. There is another class of stars which deserves mentioning, *viz.* very faint δ Cephei variables. Several variables are known whose estimated distances are of the order of 5000 parsecs and larger, thus bringing us quite near the hypothetical centre of the galactic system. If rough values were known for the velocities of a few stars of this type favourably situated for the purpose, it might well become possible to derive reliable absolute values of the distance to this centre as well as of the circular velocity in our neighborhood.

54. A RELATION BETWEEN DISTANCE AND RADIAL VELOCITY AMONG EXTRA–GALACTIC NEBULAE

By Edwin Hubble

Determinations of the motion of the sun with respect to the extragalactic nebulae have involved a K term of several hundred kilometers which appears to be variable. Explanations of this paradox have been sought in a correlation between apparent radial velocities and distances, but so far the results have not been convincing. The present paper is a re-examination of the question, based on only those nebular distances which are believed to be fairly reliable.

Distances of extra-galactic nebulae depend ultimately upon the application of absolute-luminosity criteria to involved stars whose types can be recognized. These include, among others, Cepheid variables, novae, and blue stars involved in emission nebulosity. Numerical values depend upon the zero point of the period-luminosity relation among Cepheids, the other criteria merely check the order of the distances. This method is restricted to the few nebulae which are well resolved by existing instruments. A study of these nebulae, together with those in which any stars at all can be recognized, indicates the probability of an approximately uniform upper limit to the absolute luminosity of stars, in the late-type spirals and irregular nebulae at least, of the order of M(photographic) $= -6.3$. The· apparent luminosities of the brightest stars in such nebulae are thus criteria which, although rough and to be applied with caution, furnish reasonable estimates of the distances of all extra-galactic systems in which even a few stars can be detected.

Finally, the nebulae themselves appear to be of a definite order of absolute luminosity, exhibiting a range of four or five magnitudes about an average value of M(visual) $= -15.2$. The application of this statistical average to individual cases can rarely be used to advantage, but where considerable numbers are involved, and especially in the various clusters of nebulae, mean apparent luminosities of the nebulae themselves offer reliable estimates of the mean distances.

Radial velocities of 46 extra-galactic nebulae are now available, but individual distances are estimated for only 24. For one other, N.G.C.

3521, an estimate could probably be made, but no photographs are available at Mount Wilson. The data are given in table 1. The first

TABLE I. Nebulae Whose Distances Have Been Estimated from Stars Involved or from Mean Luminosities in a Cluster

Object	m_s	r	v	m_t	M_t
S. Mag.	—	0.032	+170	1.5	−16.0
L. Mag.	—	0.034	+290	0.5	17.2
NGC 6822	—	0.214	−130	9.0	12.7
598	—	0.263	−70	7.0	15.1
221	—	0.275	−185	8.8	13.4
224	—	0.275	−220	5.0	17.2
5457	17.0	0.45	+200	9.9	13.3
4736	17.3	0.5	+290	8.4	15.1
5194	17.3	0.5	+270	7.4	16.1
4449	17.8	0.63	+200	9.5	14.5
4214	18.3	0.8	+300	11.3	13.2
3031	18.5	0.9	−30	8.3	16.4
3627	18.5	0.9	+650	9.1	15.7
4826	18.5	0.9	+150	9.0	15.7
5236	18.5	0.9	+500	10.4	14.4
1068	18.7	1.0	+920	9.1	15.9
5055	19.0	1.1	+450	9.6	15.6
7331	19.0	1.1	+500	10.4	14.8
4258	19.5	1.4	+500	8.7	17.0
4151	20.0	1.7	+960	12.0	14.2
4382	—	2.0	+500	10.0	16.5
4472	—	2.0	+850	8.8	17.7
4486	—	2.0	+800	9.7	16.8
4649	—	2.0	+1090	9.5	17.0
Mean					−15.5

m_s = photographic magnitude of brightest stars involved.
r = distance in units of 10^6 parsecs. The first two are Shapley's values.
v = measured velocities in km./sec. N.G.C. 6822, 221, 224 and 5457 are recent determinations by Humason.
m_t = Holetschek's visual magnitude as corrected by Hopmann. The first three objects were not measured by Holetschek, and the values of m_t represent estimates by the author based upon such data as are available.
M_t = total visual absolute magnitude computed from m_t and r.

seven distances are the most reliable, depending, except for M 32, the companion of M 31, upon extensive investigations of many stars involved. The next thirteen distances, depending upon the criterion of a uniform upper limit of stellar luminosity, are subject to considerable probable errors but are believed to be the most reasonable values at present available. The last four objects appear to be in the Virgo Cluster. The distance assigned to the cluster, 2×10^6 parsecs, is derived from the distribution of nebular luminosities, together with luminosities of stars in some of the later-type spirals, and differs somewhat from the Harvard estimate of ten million light years.

The data in the table indicate a linear correlation between distances and velocities, whether the latter are used directly or are corrected for solar motion, according to the older solutions. This suggests a new solution for the solar motion in which the distances are introduced as coefficients of the K term, i.e., the velocities are assumed to vary directly with the distances, and hence K represents the velocity at unit distance due to this effect. The equations of condition then take the form

$$rK + X \cos a \cos \delta + Y \sin a \cos \delta + Z \sin \delta = v.$$

Two solutions have been made, one using the 24 nebulae individually, the other combining them into 9 groups according to proximity in direction and in distance. The results are

	24 objects	*9 groups*
X	-65 ± 50	$+3 \pm 70$
Y	$+226 \pm 95$	$+230 \pm 120$
Z	-195 ± 40	-133 ± 70
K	$+465 \pm 50$	$+513 \pm 60$ km./sec. per 10^6 parsecs.
A	$286°$	$269°$
D	$+40°$	$+33°$
V_o	306 km./sec.	247 km./sec.

For such scanty material, so poorly distributed, the results are fairly definite. Differences between the two solutions are due largely to the four Virgo nebulae, which being the most distant objects and all sharing the peculiar motion of the cluster, unduly influence the value of K and hence of V_o. New data on more distant objects will be required to reduce the effect of such peculiar motion. Meanwhile round numbers, intermediate between the two solutions, will represent the probable order of the values. For instance, let $A = 277°$, $D = +36°$ (Gal. long. $= 32°$, lat. $= +18°$), $V_o = 280$ km./sec., $K = +500$ km./sec. per million parsecs. Mr. Strömberg has very kindly checked the general order of these values by independent solutions for different groupings of the data.

A constant term, introduced into the equations, was found to be small and negative. This seems to dispose of the necessity for the old constant

K term. Solutions of this sort have been published by Lundmark, who replaced the old K by $k + lr + mr^2$. His favored solution gave $k = 513$, as against the former value of the order of 700, and hence offered little advantage. . . .

The results establish a roughly linear relation between velocities and distances among nebulae for which velocities have been previously pub-

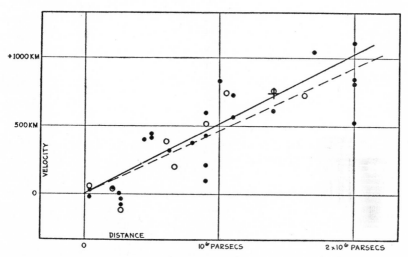

Fig. 1. Velocity-distance relation among extra-galactic nebulae. Radial velocities, corrected for solar motion, are plotted against distances estimated from involved stars and mean luminosities of nebulae in a cluster. The black discs and full line represent the solution for solar motion using the nebulae individually; the circles and broken line represent the solution combining the nebulae into groups; the cross represents the mean velocity corresponding to the mean distance of 22 nebulae whose distances could not be estimated individually.

lished, and the relation appears to dominate the distribution of velocities. In order to investigate the matter on a much larger scale, Mr. Humason at Mount Wilson has initiated a program of determining velocities of the most distant nebulae that can be observed with confidence. These, naturally, are the brightest nebulae in clusters of nebulae. The first definite result, $v = +3779$ km./sec. for N.G.C. 7619, is thoroughly consistent with the present conclusions. Corrected for the solar motion, this velocity is $+3910$, which, with $K = 500$, corresponds to a distance of 7.8×10^6 parsecs. Since the apparent magnitude is 11.8, the absolute magnitude at such a distance is -17.65, which is of the right order for the brightest nebulae in a cluster. A preliminary distance, derived independently from the cluster of which this nebula appears to be a member, is of the order of 7×10^6 parseces.

New data to be expected in the near future may modify the significance of the present investigation or, if confirmatory will lead to a solution having many times the weight. For this reason it is thought premature to discuss in detail the obvious consequences of the present results. For example, if the solar motion with respect to the clusters represents the rotation of the galactic system, this motion could be subtracted from the results for the nebulae and the remainder would represent the motion of the galactic system with respect to the extra-galactic nebulae.

The outstanding feature, however, is the possibility that the velocity-distance relation may represent the de Sitter effect [expanding universe], and hence that numerical data may be introduced into discussions of the general curvature of space. In the de Sitter cosmology, displacements of the spectra arise from two sources, an apparent slowing down to atomic vibrations and a general tendency of material particles to scatter. The latter involves an acceleration and hence introduces the element of time. The relative importance of these two effects should determine the form of the relation between distances and observed velocities; and in this connection it may be emphasized that the linear relation found in the present discussion is a first approximation representing a restricted range in distance.

55. THE VELOCITY–DISTANCE RELATION AMONG EXTRA–GALACTIC NEBULAE

By Edwin Hubble and Milton L. Humason

Abstract.—Methods of determining distances of extra-galactic nebulae are discussed, and the mean absolute magnitude is revised on the basis of (1) Shapley's revision of the zero-point of the period-luminosity curve for Cepheids, and (2) more extensive observations of stars involved in nebulae. The revised value is M(vis) $= -14.9$.

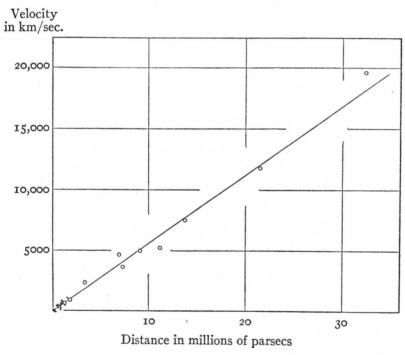

Velocity in km/sec.

Distance in millions of parsecs

Fig. 1. The velocity-distance relation. The circles represent mean values for clusters or groups of nebulae. The dots near the origin represent individual nebulae, which, together with the groups indicated by the lowest two circles, were used in the first formulation of the velocity-distance relation.

The mean color-index of the nearer extra-galactic nebulae appears to be of the order of +1.1 mag., hence $M(pg) = -13.8$. A color-excess is suggested which is independent of distance but shows some relation to galactic latitude.

The velocity-distance relation is re-examined with the aid of 40 new velocities, 26 of which refer to nebulae in 8 clusters or groups. Distances of the clusters, ranging out to about 32 million parsecs, have been derived from the most frequent apparent magnitudes. The velocity displacements reduce the apparent magnitudes by amounts which become appreciable for the more distant clusters.

The new data extend out to about eighteen times the distance available in the first formulation of the velocity-distance relation, but the form of the relation remains unchanged except for the revision of the unit of distance. The relation is Vel. = Dist. (parsecs)/1790, and the uncertainty is estimated to be of the order of 10 per cent.

56. THE LARGE APPARENT VELOCITIES
OF EXTRA–GALACTIC NEBULAE

By Milton L. Humason

It is now well established that the lines in the spectra of faint nebulae [1] are shifted to the red by large amounts. As ordinarily interpreted,
this would indicate that the nebulae are receding from the earth at very
great speeds, the greatest that have been found among celestial bodies.
It appears, also, that the more distant a nebula is, the faster it is receding.

Extra-galactic nebulae, that is, nebulae outside our own stellar system, are very faint and difficult to observe with a spectrograph. Up to
three years ago radial velocities of only forty-three had been measured,
most of them by Dr. V. M. Slipher of the Lowell Observatory. About
that time Dr. Edwin Hubble at Mount Wilson showed that the velocities were linear functions of the distance, and that the velocities increased at the rate of about 100 miles per second per million light-
years of distance. The relation was established only to about 6 million
light-years, since spectra of the fainter, more distant nebulae were not
available. In view of its fundamental significance, however, it was desirable to test the relation over as great a range in distance as could be
observed with existing instruments, and the investigation was made
with the aid of the 100-inch reflector.

The many thousands of extra-galactic nebulae appear to be scattered over the sky, more or less at random, but a few actual clusters or
related groups are known. The cluster in Leo at a distance of about 105
million light-years is one of these. This cluster was discovered about
one year ago by W. H. Christie on direct photographs taken with the
60-inch reflector and over 300 nebulae have been counted within it; the
faintest ones are probably beyond the limit of the plates available. . . .
The nebulae are small and extremely faint. For two reasons, these remote clusters offer the best test of the curious relation found by Hubble.
Reliable distances can be derived from the mean apparent brightness

---

[1] [The terms "nebula" and "extra-galactic nebula" are here used as equivalent to the
more common term "galaxy."]

of the many nebulae in a given cluster, and velocities for various nebulae in one cluster should be approximately equal if they are at the same distance from the earth. In the last two years forty-six new velocities have been determined at Mount Wilson, the majority referring to nebulae in seven clusters. Practically all of them are velocities of recession, that is, the lines in the spectra are shifted towards the red, indicating motions away from the earth. The results confirm the velocity-distance relation and extend the observed range out to more than a hundred million light-years. Distances of isolated nebulae cannot be determined. They may be formed into groups, however, and a mean distance of a group may be obtained from the mean apparent faintness of the nebulae which make up the group. The relationship is now so well established, however, that it affords a means of obtaining the distances of isolated nebulae from their velocities.

The clusters observed, each named for the constellation in which it appears, together with distances in millions of light-years and velocities of recession in miles per second are as follows:

Cluster	Distance*	Velocity	Number
Virgo..............	6.0	560	7
Pegasus............	23.5	2,400	4
Pisces.............	24.0	2,900	4
Cancer.............	29.5	3,000	2
Perseus............	36.0	3,200	4
Coma Berenices......	45.0	4,700	8
Ursa Major.........	72.0	7,300	1
Leo................	105.0	12,000	1

* [On the basis of the older value of the zero point of the period-luminosity relation for cepheid variables.]

The single velocities in the last two clusters refer to the brightest nebulae in each. That in the Leo cluster is the faintest and most distant object for which a measurable spectrum has been photographed. The visual magnitude is about 15.5, or 6,300 times fainter than the faintest star seen with the naked eye.

The increase in velocities with distance is evident in the table, and the diagram (selection 55) shows the linear relation in a striking way. The boxed-off section at the lower left-hand, represents the material from which Hubble's preliminary correlation was derived, and the rest of the diagram represents the extension effected in the last two years. The enormous extension of the distances has not changed the shape of the correlation "curve" as originally determined.

The number of velocities obtained during the past two years is almost entirely due to a new high-speed spectrograph lens designed by Dr. W. B. Rayton of the Bausch and Lomb Optical Company. Before the Rayton lens was available some of the exposures were as long as 60 hours, while with it exposures vary from 5 to 40 hours, depending on the brightness of the nebulae to be photographed. During the longer exposures the plate is protected from daylight and left in the spectrograph from one night to the next. The velocities from both the Ursa Major and Leo clusters could not have been obtained except with the use of this exceedingly high-speed Rayton lens.

To understand better the difficulties with which the astronomer contends in making his interpretations of the messages brought to us by the camera from these remote depths of space, it is perhaps interesting to read that the size of the photographic plate is but ⅝ by 1½ inches, and that the total length of the spectrum obtained is from one twenty-fifth to one twelfth of an inch, depending upon the prism used.

The results of the investigation are observational facts. From them a close relation between apparent brightness of nebulae and red shifts in their spectra has been found. The former are confidently interpreted as distances but the interpretation of the red shifts as velocities of recession is controversial. For the present we prefer to speak of these velocities as *apparent*. In this sense the velocity-distance relation is one of the two established characteristics of the observable region of the universe as a whole. The other is the approximately uniform distribution of the nebulae. The two together form the observational basis for theories concerning the structure of the universe.

XII

RELATIVITY AND COSMOGONY

Except for the report by W. W. Campbell and R. Trumpler on the observational test of one of the predictions from the general theory of relativity, and except for the presentation of some generally accepted cosmogonic theory by Henri Poincaré and Albert Einstein, this whole part is riddled with speculation. It is, however, both very interesting and rather profitable speculation. "The Beginning of the World" and "The End of the World" are brave advances in human thought; and although both the primeval atom and the perfect cosmological principle are now considered by many to be dead-end excursions, they have served and will continue to serve in leading to more acceptable speculations about a most tremendous theme—the origin of the universe. George Gamow's thoughts in the area of explosive beginnings, and Fred Hoyle's writings on the continuous creation hypotheses, have helped to keep our attention turned to the "big bang" and "steady state" versions of the universe. It now appears that the origin of the universe which we presently observe, and the origin of the planet on which we live and speculate, will long remain unsettled questions. Currently our best contributions seem to be those researches that show what is not true.

Again attention is drawn to the essays by Poincaré and W. de Sitter—both masters of their technical subjects, both artists in their presentations. On the other hand, quite unreadable except by experts is the derivation by Einstein of the most momentous formula, dangerous and utopian, of the atomic age.

The brief paper by Sir James Jeans slightly antedated the appearance of Einstein's formulation; it contained the important suggestion that available energy sources lie within matter.

57. ON THE TRANSFORMATION OF MATTER INTO ENERGY

By J. H. Jeans

I am venturing, in the present note, to add another to the already large number of suggestions as to the meaning of the phenomenon of radio-activity.

It seems to be well established that the apparent instability of the atoms of radio-active substances is not to any great extent dependent on the temperature of the mass; the instability, therefore, is not the outcome of intermolecular collisions. Neither does it seem to arise from an excess of the internal energy of the molecule. For the internal agitation of the molecule, so far as is known, shows itself in the emission of light, and this is associated with high mass-temperature. There is, of course, the possibility, suggested by Prof. J. J. Thomson, that there are internal degrees of freedom not represented in the spectrum of the gas, and that it is the energy of these which forms the starting point of the radio-active process. On the other hand, it is possible that the atomic instability, not being the result of the agitation of the molecules or of the component material parts (ions or corpuscles) of which the molecules are composed, must be traced to the agitation of the ultimate constituents of these ions or corpuscles. If, for instance, we take a definite mechanical illustration, and imagine our universe constructed on the model suggested by Prof. Osborne Reynolds, the source of instability must be looked for in the agitation of the "grains" of which he supposes the ether to be constituted. The velocities of these grains follow Maxwell's law of distribution, so that very high velocities, although rare, are not impossible. It is at least thinkable that a grain moving with exceptionally high velocity may succeed in breaking down the normal piling in its immediate neighborhood when this is possible (*i.e.*, probably, when in the immediate proximity of matter), and may therefore effect a rearrangement of the adjacent ether structure. A process of this kind would be independent of the mass-temperature; it would, so to speak, depend solely on the ether temperature, which is supposed, on Prof. Reynolds' hypothesis, to be constant throughout space. It seems probable that the rearrangement would consist of the combination and mutual annihilation of two ether strains of opposite kinds, *i.e.* in the coalescence of a positive and negative ion, and would therefore result in the disappearance of a certain amount of mass. There would, therefore, be conservation neither of mass nor of material energy; the process of radio-activity would consist in an increase of material energy at the expense of the destruction of a certain amount of matter. . . .

[In a monographic treatise "Astronomy and Cosmogony (1929, Chapter IV), Jeans argues the inadequacy of terrestrially-known sources to account for the sun's continuing radiation, and enlarges on the foregoing idea of the annihilation of matter as a source of stellar energy. He also quotes this remarkable foresighted statement from Newton's Opticks (1704): "Are not gross bodies and light convertible into one another; and may not bodies receive much of their activity from the particles of light which enter into their composition?"

[For H. N. Russell's discussion of the sources of stellar energy, see selection 40, above.]

58. THE $E = Mc^2$ EQUATION

By Albert Einstein

The results of my recently published electrodynamic investigation lead to a very interesting conclusion, which is here to be deduced.

I based that investigation on the Maxwell-Hertz equations for empty space, together with the Maxwellian expression for the electromagnetic energy of space, and in addition the principle that:—

The laws by which the states of physical systems change do not depend on which of two systems of co-ordinates, in uniform motion of parallel translation relatively to each other, these alterations of state are referred to (principle of relativity).

With these principles [1] as my basis I deduced *inter alia* the following result:—

Let a system of plane waves of light, referred to the system of co-ordinates (x, y, z), possess the energy l; let the direction of the ray (the wave-normal) make an angle ϕ with the axis of x of the system. If we introduce a new system of co-ordinates (ξ, η, ζ) moving in uniform parallel translation with respect to the system (x, y, z), and its origin of co-ordinates moving along the axis of x with the velocity v, then this quantity of light—measured in the system (ξ, η, ζ)—possesses the energy

$$l^* = l \frac{1 - \dfrac{v}{c} \cos \phi}{\sqrt{1 - v^2/c^2}}$$

where c denotes the velocity of light. We shall make use of this result in what follows.

Let there be a stationary body in the system (x, y, z), and let its energy—referred to the system (x, y, z)—be E_0. Let the energy of the body relative to the system (ξ, η, ζ), moving as above with the velocity v, be H_0.

Let this body send out, in a direction making an angle ϕ with the axis

---

[1] The principle of the constancy of the velocity of light is of course contained in Maxwell's equations.

of x, plane waves of light, of energy $\frac{1}{2}L$ (measured relatively to (x, y, z)), and simultaneously an equal quantity of light in the opposite direction. Meanwhile the body remains at rest with respect to the system (x, y, z). The principle of energy must apply to this process, and in fact (by the principle of relativity) with respect to both systems of co-ordinates. If we call the energy of the body after the emission of light E_1 or H_1 respectively, measured relatively to the system (x, y, z) or (ξ, η, ζ) respectively, then by employing the relation given above we obtain

$$E_0 = E_1 + \tfrac{1}{2}L + \tfrac{1}{2}L,$$

$$H_0 = H_1 + \frac{1}{2}\,L\,\frac{1 - \dfrac{v}{c}\cos\phi}{\sqrt{1 - v^2/c^2}} + \frac{1}{2}\,L\,\frac{1 + \dfrac{v}{c}\cos\phi}{\sqrt{1 - v^2/c^2}}$$

$$= H + \frac{L}{\sqrt{1 - v^2/c^2}}.$$

By subtraction we obtain from these equations

$$H_0 - E_0 - (H_1 - E_1) = L\left\{\frac{1}{\sqrt{1 - v^2/c^2}} - 1\right\}.$$

The two differences of the form $H - E$ occurring in this expression have simple physical meanings. H and E are energy values of the same body referred to two systems of co-ordinates which are in motion relatively to each other, the body being at rest in one of the two systems (namely, (x, y, z)). Thus it is clear that the difference $H - E$ can differ from the kinetic energy K of the body, with respect to the other system (ξ, η, ζ), only by an additive constant C, which depends on the choice of the arbitrary additive constants of the energies H and E. Thus we may place

$$H_0 - E_0 = K_0 + C,$$
$$H_1 - E_1 = K_1 + C,$$

since C does not change during the emission of light. So we have

$$K_0 - K_1 = L\left\{\frac{1}{\sqrt{1 - v^2/c^2}} - 1\right\}.$$

The kinetic energy of the body with respect to (ξ, η, ζ) diminishes as a result of the emission of light, and the amount of diminution is independent of the properties of the body. Moreover, the difference $K_0 - K_1$, like the kinetic energy of the electron, depends on the velocity.

Gibt ein Körper die Energie L in Form von Strahlung ab, so verkleinert sich seine Masse um L/V^2. Hierbei ist es offenbar unwesentlich, daß die dem Körper entzogene Energie gerade in Energie der Strahlung übergeht, so daß wir zu der allgemeineren Folgerung geführt werden:

Die Masse eines Körpers ist ein Maß für dessen Energie-inhalt; ändert sich die Energie um L, so ändert sich die Masse in demselben Sinne um $L/9.10^{20}$, wenn die Energie in Erg und die Masse in Grammen gemessen wird.

Es ist nicht ausgeschlossen, daß bei Körpern, deren Energieinhalt in hohem Maße veränderlich ist (z. B. bei den Radiumsalzen), eine Prüfung der Theorie gelingen wird.

Wenn die Theorie den Tatsachen entspricht, so überträgt die Strahlung Trägheit zwischen den emittierenden und absor-bierenden Körpern.

Bern, September 1905.

Fig. 1. Einstein's original statement of the mass-energy equivalence.

Neglecting magnitudes of fourth and higher orders we may place

$$K_0 - K_1 = \frac{1}{2}\frac{L}{c^2}\,v^2.$$

From this equation it directly follows that:—

If a body gives off the energy L in the form of radiation, its mass diminishes by L/c². The fact that the energy withdrawn from the body becomes energy of radiation evidently makes no difference so that we are led to the more general conclusion that:

The mass of a body is a measure of its energy-content; if the energy changes by L, the mass changes in the same sense by $L/9 \times 10^{20}$, the energy being measured in ergs, and the mass in grams.

It is not impossible that with bodies whose energy-content is variable to a high degree (e.g. with radium salts) the theory may be success-fully put to the test.

59. COSMOGONICAL HYPOTHESES

By Henri Poincaré

The question of the origin of the world has always preoccupied the thoughts of everyone who reflects, for it is impossible to contemplate the spectacle of the starry universe without asking how it has been formed. We should perhaps put off looking for a solution until we have patiently assembled the data for it, and until we have acquired in this way some serious hope of finding a solution. But if we had always been so reasonable, so curious without impatience, it is probable that we should never have created Science and that we should have always contented ourselves to live out our little lives without vision.

Our minds have indeed imperiously demanded a solution, even before the problem was ripe, and when we had only vague glimpses that permitted guessing rather than attainment. And it is because of that demand that the cosmological hypotheses are so numerous, so varied; one is born every new day, all equally uncertain, but all as plausible as the older theories, in the midst of which they take their place without causing the others to be forgotten.

One could assume that the universe has always been what it is today, that the small phenomena that play on the surface of the stars are temporary while the stars themselves do not change, and that the stars follow gloriously their eternal lives without concerning themselves about their miserable and ephemeral parasites. But there are two reasons for rejecting this [static] way of looking at things.

The solar system presents to us a spectacle of perfect harmony; the orbits of the planets are all nearly circular, all nearly in the same plane, all circulating in the same direction. This cannot be the result of chance. We could suppose that an infinite intelligence has established this order from the beginning, once for all and forever; and formerly everybody was content with that explanation. Today we are no longer satisfied with this cheap solution.

Certainly there are still many who hold that a creative God is a necessary hypothesis, but they would no longer construe divine intervention as would their forefathers. Their God is less of an architect and more of a mechanician; and it remains for them to explain by what mechanism order has been extracted from chaos. If the order which we recognize is not due to chance, and if we decline to attribute it to some directly administrative divine decree, then order must have developed out of chaos; it follows therefore that the stars must have changed. It is thus that Laplace reasoned.

On the other hand, the second principle of Thermodynamics, the principle of Carnot, informs us that the World tends toward a final state. Energy "dissipates"; that is, friction continually tends to transform motion into heat, and temperature tends to equalize everywhere. Consequently the final state of the World is a state of uniformity. This state that it is destined to attain has not yet been reached; therefore the universe now changes, and likewise it has always changed.

And therein lies the open field for hypotheses. That of Laplace is the oldest, but its old age is vigorous, and considering its age it does not have too many wrinkles. In spite of the objections that have been raised against it, in spite of the discoveries that astronomers have made and which would have astonished Laplace, it always stands up and is still the hypothesis that best accounts for the facts. It still responds the best to the question that the author posed to himself: Why does order reign in the solar system if that order is not due to change? From time to time a breach has opened in the old edifice; but it was promptly repaired and the edifice did not fall.

We know in what this [Laplacian] hypothesis consists. The solar system is developed from a nebula that at one time extended out to the orbit of Neptune. This nebula was animated with a motion of uniform rotation. It could not be homogeneous; it was condensed, even strongly condensed toward the center. A relatively dense nucleus formed which has become the Sun, surrounded by an extremely tenuous atmosphere which has given birth to the planets. It contracted through cooling, abandoning from time to time at its equator nebulous rings. These rings were unstable, or promptly became so; they had to break up and finally re-assemble in a single spheroidal mass.

At the moment when the system began to form, there was already a beginning of order; the internal movements of the nebula were not capricious or disorderly; they restored a uniform rotation. It is this initial harmony which has produced the final harmony that we admire. This initial harmony is easy to explain. The internal friction of a mass should promptly destroy the irregularities of the internal movements and permit the persistence only of a perfectly regular rotation of the whole. Promptly? That depends upon the meaning attached to the word. The inequalities disappear promptly if we regard some milliards of years as a very short interval. If we desire to make the calculation, attributing to the matter of the nebula the viscosity of the gases with which we are acquainted, we arrive at fantastic numbers. And that is not all: the cooling, and the contraction which results therefrom, tends to trouble this harmony, so slowly acquired. To preserve it, the contraction and entire evolution of the system must also be prodigiously slow, the more so because we have established that it will require hundreds of millions of years for the various parts of a given ring, moving separately according to Keplerian laws, to arrive at the point of collisions and adhesions—phenomena which can be regarded only as brief episodes in the general evolution.

These numbers should not frighten us. They are in disagreement with the age that other theories attribute to the Sun and to stars; but these other theories on their own part have great difficulties. One thought always presents itself: other systems similar to ours should have undergone in the same time the same evolution. Each of them has occupied a considerable space, extending indeed to the realm of our Sun. If this evolution has continued very long, we are obliged to reckon with the probability of collisions destroying everything before its completion. . . .

[The author then presents in general and critical terms the hypotheses of Herve Faye, R. du Ligondes, T. J. J. See, and G. H. Darwin, concerning the origin of the planetary system, and discusses some of the relevant ideas of J. C. Maxwell, Lord Kelvin, Hermann von Helmholtz, Norman Lockyer, E. Belot, and Svante Arrhenius, and relates their work to the Kant-Laplacian theory of an original contracting nebula.]

All these hypotheses, so divergent in other respects, have one characteristic in common: they are theories of Rational Mechanics, of Mathematical Astronomy. They borrow little from physical science and for that reason are incomplete. The physicists, whose intervention was as inevitable as it was desirable, have above all interested themselves in the origin of the solar heat. From precise measures we have shown the astonishing expenditure of heat that the Sun emits each second. What resources has it that permit such prodigality? Where has it been able to store the energy sufficient for millions of years? And what can be the origin of this store? One might think first that this energy is of chemical origin—the Sun burning like a large chunk of carbon. But this hypothesis is not tenable; for in such a case the Sun would have been only an ephemeral bonfire, scarcely sufficient to illuminate mankind during its historical past. . . .

After this exposition one doubtless expects from me a conclusion, and that is what embarasses me. The more one studies the origin of the stars, the less one is in a hurry to draw conclusions. Each of the theories proposed is attractive in certain respects. Some give, in a very nice fashion, an explanation of a certain number of facts; others embrace more facts, but the explanations lose in precision what they gain in compass; or, on the contrary, they give us too great precision, but one that is only illusory and which seems grossly artificial.

If there were only the solar system, I would not hesitate to prefer the old Laplacian hypothesis; there is very little to do to renovate it. But the variety of stellar systems oblige us to enlarge our frame in such a way that the hypothesis of Laplace, if it should not be abandoned entirely, should be modified in a fashion to be other than a theory adapted specially to the solar system, to become rather a more general hypothesis that would suit the total universe, and which would explain to us at the same time the diverse destinies of the stars and how each of them takes its place in the grand total. . . .

One fact that strikes everybody is the spiral form of certain nebulae. That form is met much too often for one to consider it due to chance. We recognize how incomplete is every cosmogonic theory that disregards it. None of them now take it into account in a satisfactory manner, and the explanation that once I myself gave rather incidentally is worth no more than the others. We can only conclude with an interrogation point.

60. TESTING THE THEORY OF RELATIVITY

By W. W. Campbell and R. Trumpler

The W. H. Crocker Eclipse Expedition to Wallal, Western Australia, included in its program observations on the eclipse test of Einstein's Generalized Theory of Relativity. According to this theory light rays, when passing near the Sun are bent in such a way that a star seen at the edge of the Sun appears displaced from its normal position by $1''.745$, while for other stars the displacements are inversely proportional to their angular distances from the Sun's center. Two twin cameras of fifteen feet and five feet focal length, respectively, were used in connection with this problem. As far as the photographs obtained with the larger instrument are concerned, the measures and reductions have just been completed and will form the subject of a Lick Observatory Bulletin, now in press.

The problem of observing the predicted light deflections is solved by photographing during the eclipse the stars surrounding the eclipsed Sun and comparing these photographs, by accurate measures, with other photographs of the same stars taken at night several months before or after the eclipse, when the Sun is in another part of the sky and has no disturbing influence on the positions of these stars. Such comparison photographs were secured by Trumpler in May 1922, on the island of Tahiti, as nearly as possible under the same conditions (hour-angle, zenith-distance, temperature) as were expected to prevail during the eclipse. As a check on any instrumental sources of error that might affect such delicate measures an auxiliary star group was selected, following the eclipse star field by 6^h15^m in right ascension but at the same declination. At both observing stations these check stars were photographed on the same plates and at the same hour angle as the eclipse stars, and at both observing dates this check field was over $90°$ distant from the Sun and therefore unaffected by any light deflections.

At Tahiti the comparison plates were given an exposure of 3 minutes on the eclipse field at about ten o'clock in the evening; the two plate holders were then left undisturbed in the telescope (closing only the slides) until about 4:15 in the morning, when the same plates were exposed on the check field. In a similar way the procedure at Wallal was to make an exposure on the check field at about 7:45 on the evening preceding the eclipse, to leave the two plates in the telescope, and to use them again the following day for the first pair of eclipse photographs, with a 2 minute exposure, started 10 seconds after the beginning of totality. The plate holders were then changed (the operation taking 50 seconds), and another pair of eclipse plates exposed for 2 minutes 5 seconds, the exposure being stopped 10 seconds before the end of totality. This pair of plates received the impression of the check field on the evening following the eclipse. A visual guiding telescope was attached to the camera, and during all the exposures the observer (Trumpler) corrected the irregularities of the clock work by means of the right ascension slow motion so as to keep the guiding star (β *Virginis* for the eclipse field, 70 *Ophiuchi* for the check field) exactly on the cross wire.

The eclipse photographs show the corona strongly with traces reaching as far as $1\frac{1}{4}°$ from the Sun's center (see outline in Fig. 1). The sky background also is considerably blackened. Nevertheless, thanks to the excellence of the objectives and to the precautions taken with the development, stars as faint as 10.8 photographic magnitude are visible and 92 stars altogether were measured on one or another of the plates. This is probably the first time that such faint stars have been photographed or observed during a solar eclipse. The star images on the eclipse photographs are somewhat fuzzy and diffuse (probably an effect of the fogging of the background) especially for the fainter stars, and different weights were assigned to the measures according to the quality of each star image.

Following the example of the British expeditions of 1919 an intermediate plate was taken at Tahiti with the glass side turned toward the objective. This intermediate plate, having right and left reversed, could be placed film to film with any of the eclipse photographs or the night comparison plates so that corresponding star images on the two plates fell close together. The distance between the two images was then measured with the micrometer of the measuring microscope. Each of the eclipse plates was combined with one of the Tahiti comparison plates, and three of these pairs of plates were measured independently by both observers, the last pair by one observer only.

Subtracting the measures of a comparison plate from those of the corresponding eclipse photograph, the intermediate plate is completely eliminated, and we obtain directly the differences between the two plates to be compared. In these differences the light deflections are mixed up with the effects of the plate constants. Using the stars in the outer portions of the plate (outside of the dotted circle in Fig. 1) as reference stars, the effects of the plate constants were determined and applied without any priori assumptions as to the nature of the light

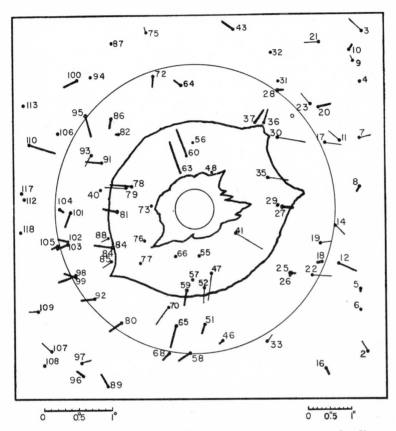

Fig. 1. The deflection of the star light observed during a total eclipse of the sun.

deflections, since, for these stars, the deflections would be very small under any assumption. The resulting star displacements in the mean of the four plates are plotted, 2250 times magnified, in Fig. 1. They are the *relative* displacements of the stars near the Sun as compared with the reference stars near the edge of the plate. For the latter the average displacements were made zero by the way in which the plate constants were determined. Fig. 1 gives a striking illustration of the existence of light deflections near the Sun as predicted by Einstein's theory. If these deflections are not always pointed exactly away from the Sun's center, this is (according to the theory of errors) entirely accounted for by the accidental errors of observation.

Our observations are in good agreement with the prediction of the theory that the deflections are inversely proportional to the star's angular distance from the Sun's center, and on the basis of this law we can calculate from all the stars measured on a plate, what the light deflection of a star at the Sun's limb would be. The results of the least squares solutions made separately for the two observers and the four plates are given, together with their probable errors, in the following table:

Light Deflection at the Sun's Limb

Plates	Campbell	No. of Stars	Trumpler	No. of Stars	Plate mean
CD22—CD15	*1″.72 ± ″.32	62	*1″.88 ± ″.27	69	1″.80
CD23—CD17	1.35 ± .22	77	*1.62 ± .22	81	1.48
AB18—AB12	*1.78 ± .22	80	*1.91 ± .19	84	1.85
AB17—AB10/9			1.76 ± .22	85	1.76 (weight .9)
Mean for each observer	1.60 ± .14		1.78 ± .11		

Mean from four plates..........1″.72 ± ″.11
Einstein's predicted value.......1.745

* These five values, very slightly modified, were the only ones available when the preliminary announcement of our results for the Einstein eclipse problem was made, through the press associations and otherwise, on April 11, 1923. At that date we had also determined, from measures of the check region star images, that any corrections suggested or demanded by existing small plate errors could not operate to diminish any of the five values of the Einstein coefficient.

Our observations agree very closely with Einstein's prediction.

A test was also made to determine whether or not the stars near the Sun's poles show as much deflection as those near the Sun's equator:

Two quadrants near Sun's equator   1″.63 ± ″.15 (53 stars)
Two quadrants near Sun's poles   1.76 ± .18 (39 stars)

The small difference between the two values is within the limits of the errors of observation, and the deflections seem to be essentially the same all around the Sun.

A selected list of 37 stars of the check field was measured intermingled with the eclipse stars. After applying the plate corrections in a similar way as for the eclipse field, the check stars near the center of the plate do not show any displacements away from the plate center;

on the contrary they seem, on the average, rather to be displaced a trifle toward it. This proves that any instrumental errors that might have affected our measures could not have produced the observed light deflections of the stars of the eclipse field; they could only have had the tendency to make our results too small. If corrections based on the residuals of the check stars are applied to the observations of the eclipse stars the light deflection at the Sun's limb derived from the corrected values is 2″.05. It is, however, not the stars nearest to the Sun that differ from the theory; it is only the stars between 1° and 2° from the Sun's center that show, with these corrections, somewhat too large deflections. Perhaps the application of such corrections is not quite justified, or perhaps an effect additional to that of Einstein's theory (like Courvoisier's "yearly refraction") enters for the stars mentioned. The plates of the five foot camera, to be measured shortly, may throw some light upon this question.

61. THE END OF THE WORLD

By A. S. Eddington

The world—or space-time—is a four-dimensional continuum, and consequently offers a choice of a great many directions in which we might start off to look for an end; and it is by no means easy to describe "from the standpoint of mathematical physics" the direction in which I intend to go. I have therefore examined at some length the preliminary question, Which end?

SPHERICAL SPACE

We no longer look for an end to the world in the space dimension. We have reason to believe that so far as its space dimensions are concerned the world is of a spherical type. If we proceed in any direction in space we do not come to an end of space, nor do we continue on to infinity; but, after travelling a certain distance (not inconceivably great), we find ourselves back at our starting-point, having 'gone round the world.' A continuum with this property is said to be finite but unbounded. The surface of a sphere is an example of a finite but unbounded two-dimensional continuum; our actual three-dimensional space is believed to have the same kind of connectivity, but naturally the extra dimension makes it more difficult to picture. If we attempt to picture spherical space, we have to keep in mind that it is the *surface* of the sphere that is the analogue of our three-dimensional space; the inside and outside of the sphere are fictitious elements in the picture which would have no analogue in the actual world.

We have recently learnt, mainly through the work of Prof. Lemaître, that this spherical space is expanding rather rapidly. In fact, if we wish to travel round the world and get back to our starting point, we shall have to move faster than light; because, whilst we are loitering on the way, the track ahead of us is lengthening. It is like trying to run a race in which the finishing-tape is moving ahead faster than the runners. We can picture the stars and galaxies as embedded in the surface of a rubber balloon which is being steadily inflated; so that, apart from their individual motions and the effects of their ordinary gravitational attraction on one another, celestial objects are becoming farther and farther apart simply by the inflation. It is probable that the spiral nebulae are so distant that they are very little affected by mutual gravitation and exhibit the inflation effect in its pure form. It has been known for some years that they are scattering apart rather rapidly, and we accept their measured rate of recession as a determination of the rate of expansion of the world.

From the astronomical data it appears that the original radius of space was 1200 million light years.[1] Remembering that distances of celestial objects up to several million light years have actually been measured, that does not seem overwhelmingly great. At that radius the mutual attraction of the matter in the world was just sufficient to hold it together and check the tendency to expand. But this equilibrium was unstable. An expansion began, slow at first; but the more widely the matter was scattered the less able was the mutual gravitation to check the expansion. We do not know the radius of space today, but I should estimate that it is not less than ten times the original radius.

At present our numerical results depend on astronomical observations of the speed of scattering apart of the spiral nebulae. But I believe that theory is well on the way to obtaining the same results independently of astronomical observation. Out of the recession of the spiral nebulae we can determine not only the original radius of the universe but also the total mass of the universe, and hence the total number of protons in the world. I find this number to be either 7×10^{78} or 14×10^{78}.[2] I believe that this number is very closely connected with the ratio of the electrostatic and the gravitational units of force, and, apart from a numerical coefficient, is equal to the square of the ratio. If F is the ratio of the electrical attraction between a proton and electron to their gravitational attraction, we find $F^2 = 5.3 \times 10^{78}$. There are theoretical reasons for believing that the total number of particles in the world is $a F^2$, where a is a simple geometrical factor (perhaps involving π). It ought to be possible before long to find a theoretical value of a, and so make a complete connexion between the observed rate of expansion of the universe and the ratio of electrical and gravitational forces.

[1] [See Lemaitre, *The Primeval Atom,* B. H. and S. A. Korff, trans. (Van Nostrand, New York, 1950), for a more generally accepted view of the original "size" of the universe.]

[2] This ambiguity is inseparable from the operation of counting the number of particles in finite but unbounded space. It is impossible to tell whether the protons have been counted once or twice over.

Signposts for Time

I must not dally over space any longer but must turn to time. The world is closed in its space dimensions but is open in both directions in its time dimension. Proceeding from 'here' in any direction in space we ultimately come back to 'here'; but proceeding from 'now' towards the future or past we shall never come across 'now' again. There is no bending round of time to bring us back to the moment we started from. In mathematics this difference is provided for by the symbol $\sqrt{-1}$, just as the same symbol crops up in distinguishing a closed ellipse from an open hyperbola.

If, then, we are looking for an end of the world—or, instead of an end, an indefinite continuation for ever and ever—we must start off in one of the two time directions. How shall we decide which of these two directions to take? It is an important question. Imagine yourself in some unfamiliar part of space-time so as not to be biased by· conventional landmarks or traditional standards of reference. There ought to be a signpost with one arm marked 'To the future' and the other arm marked 'To the past.' My first business is to find this signpost, for if I make a mistake and go the wrong way I shall lead you to what is no doubt an 'end of the world,' but it will be that end which is more usually described as the *beginning*.

In ordinary life the signpost is provided by consciousness. Or perhaps it would be truer to say that consciousness does not bother about signposts; but wherever it finds itself it goes off on urgent business in a particular direction, and the physicist meekly accepts its lead and labels the course it takes 'To the future.' It is an important question whether consciousness in selecting its direction is guided by anything in the physical world. If it is guided, we ought to be able to find directly what it is in the physical world which makes it a one-way street for conscious beings. The view is sometimes held that the 'going on of time' does not exist in the physical world at all and is a purely subjective impression. According to that view, the difference between past and future in the material universe has no more significance than the difference between right and left. The fact that experience presents space-time as a cinematograph film which is always unrolled in a particular direction is not a property or peculiarity of the film (that is, the physical world) but of the way it is inserted into the cinematograph (that is, consciousness). In fact, the one-way traffic in time arises from the way our material bodies are geared on to our consciousness:

> "Nature has made our gears in such a way
> That we can never get into reverse."

If this view is right, 'the going on of time' should be dropped out of our picture of the physical universe. Just as we have dropped the old geocentric outlook and other idiosyncracies of our circumstances as observers, so we must drop the dynamic presentation of events which is no part of the universe itself but is introduced in our peculiar mode of apprehending it. In particular, we must be careful not to treat a past-to-future presentation of events as truer or more significant than a future-to-past presentation. [In which case] we must, of course, drop the theory of evolution, or at least set alongside it a theory of anti-evolution as equally significant. . . .

ENTROPY AND DISORGANIZATION

Leaving aside the guidance of consciousness, we have found it possible to discover a kind of signpost for time in the physical world. The signpost is of a rather curious character, and I would scarcely venture to say that the discovery of the signpost amounts to the same thing as the discovery of an objective 'going on of time' in the universe. But at any rate it serves to discriminate past and future, whereas there is no corresponding objective distinction of left and right. The distinction is provided by a certain measurable quantity called entropy. Take an isolated system and measure its entropy S at two instants t_1 and t_2. We want to know whether t_1 is earlier or later than t_2 without employing the intuition of consciousness, which is too disreputable a witness to trust in mathematical physics. The rule is that the instant which corresponds to the greater entropy is the later. In mathematical form

$$dS/dt \text{ is always positive.}$$

This is the famous second law of thermodynamics.

Entropy is a very peculiar conception, quite unlike the conceptions ordinarily employed in the classical scheme of physics. We may most conveniently describe it as the measure of disorganization of a system. Accordingly, our signpost for time resolves itself into the law that disorganization increases from past to future. It is one of the most curious features of the development of physics that the entropy outlook grew up quietly alongside the ordinary analytical outlook for a great many years. Until recently it always 'played second fiddle'; it was convenient for getting practical results, but it did not pretend to convey the most penetrating insight. But now it is making a bid for supremacy, and I think that there is little doubt that it will ultimately drive out its rival.

There are some important points to emphasize. First, there is no other independent signpost for time; so that if we discredit or 'explain away' this property of entropy, the distinction of past and future in the physical world will disappear altogether. Secondly, the test works consistently; isolated systems in different parts of the universe agree in giving the same direction of time. Thirdly, in applying the test we must make certain that our system is strictly isolated. Evolution teaches us that more and more highly organised systems develop as time goes on; but this does not contradict the conclusion that on the whole there is a loss of organisation. It is partly a question of definition of organisation; from the evolutionary point of view it is quality rather than quantity of organisation that is noticed. But, in any case, the high organisation of these systems is obtained by draining organisation from other systems with which they come in contact. A human being as he grows from past to future becomes more and more highly organised— at least, he fondly imagines so. But if we make an isolated system of him, that is to say, if we cut off his supply of food and drink and air, he speedily attains a state which everyone would recognise as 'a state of disorganisation.'

It is possible for the disorganisation of a system to become complete. The state then reached is called thermodynamic equilibrium. The entropy can increase no further, and since the second law of thermodynamics forbids a decrease, it remains constant. Our signpost for time disappears; and so far as that system is concerned, time ceases to go on. That does not mean that time ceases to exist; it exists and extends just as space exists and extends, but there is no longer any one-way property. It is like a one-way street on which there is never any traffic.

Let us return to our signpost. Ahead there is ever-increasing disorganisation. Although the sum total of organisation is diminishing, certain parts of the universe are exhibiting a more and more highly specialised organisation; that is the phenomenon of evolution. But ultimately this must be swallowed up in the advancing tide of chance and chaos, and the whole universe will reach a state of complete disorganisation—a uniform featureless mass in thermodynamic equilibrium. This is the end of the world. Time will *extend* on and on, presumably to infinity. But there will be no definable sense in which it can be said to *go* on. Consciousness will obviously have disappeared from the physical world before thermodynamical equilibrium is reached, and dS/dt having vanished, there will remain nothing to point out a direction in time.

The Beginning of Time

It is more interesting to look in the opposite direction—towards the past. Following time backwards, we find more and more organisation in the world. If we are not stopped earlier, we must come to the time when the matter and energy of the world had the maximum possible organisation. To go back further is impossible. We have come to an abrupt end of space-time—only we generally call it the 'beginning.'

I have no 'philosophical axe to grind' in this discussion. Philosophically, the notion of a beginning of the present order of Nature is repugnant to me. I am simply stating the dilemma to which our present fundamental conception of physical law leads us. I see no way around it; but whether future developments of science will find an escape I cannot predict. The dilemma is this:—Surveying our surroundings, we find them to be far from a 'fortuitous concourse of atoms.' The picture of the world, as drawn in existing physical theories, shows arrangement of the individual elements for which the odds are multillions [3] to 1 against an origin by chance. Some people would like to call this non-random feature of the world purpose or design; but I will call it non-committally anti-chance. We are unwilling to admit in physics that anti-chance plays any part in the reactions between the systems of billions of atoms and quanta that we study; and indeed all our experimental evidence goes to show that these are governed by the laws of chance. Accordingly, we sweep anti-chance out of the laws of physics—out of the differential equations. Naturally, therefore, it reappears in the boundary conditions, for it must be got into the scheme somewhere. By sweeping it far enough away from the sphere of our current physical problems, we fancy we have got rid of it. It is only when some of us are so misguided as to try to get back billions of years into the past that we find the sweepings all piled up like a high wall and forming a boundary—a beginning of time—which we cannot climb over. . . .

I suppose that to justify my title I ought to conclude with a prophecy as to what the end of the world will be like. I confess that I am not very keen on the task. I half thought of taking refuge in the excuse that, having just explained that the future is unpredictable, I ought not to be expected to predict it. But I am afraid that someone would point out that the excuse is a thin one, because all that is required is a computation of averages and that type of prediction is not forbidden by the

---

[3] I use "multillions" as a general term for numbers of order $10^{10^{10}}$ or larger.

principle of indeterminacy. It used to be thought that in the end all the matter of the universe would collect into one rather dense ball at uniform temperature; but the doctrine of spherical space, and more especially the recent results as to the expansion of the universe, have changed that. There are one or two unsettled points which prevent a definite conclusion, so I will content myself with stating one of several possibilities. It is widely thought that matter slowly changes into radiation. If so, it would seem that the universe will ultimately become a ball of radiation growing ever larger, the radiation becoming thinner and passing into longer and longer wave-lengths. About every 1500 million years [4] it will double its radius, and its size will go on expanding in this way in geometric progression for ever.

[4] [Eddington would now at least double this number because of revisions in the distance scale for galaxies.]

62. THE BEGINNING OF THE WORLD

By Georges Lemaître

Sir Arthur Eddington states [1] that, philosophically, the notion of a beginning of the present order of Nature is repugnant to him. I would rather be inclined to think that the present state of quantum theory suggests a beginning of the world very different from the present order of Nature. Thermodynamical principles from the point of view of quantum theory may be stated as follows: (1) Energy of constant total amount is distributed in discrete quanta. (2) The number of distinct quanta is ever increasing. If we go back in the course of time we must find fewer and fewer quanta, until we find all the energy of the universe packed in a few or even in a unique quantum.

Now, in atomic processes, the notions of space and time are no more than statistical notions; they fade out when applied to individual phenomena involving but a small number of quanta. If the world has begun with a single quantum, the notions of space and time would altogether fail to have any meaning at the beginning; they would only begin to have a sensible meaning when the original quantum had been divided into a sufficient number of quanta. If this suggestion is correct, the beginning of the world happened a little before the beginning of space and time. I think that such a beginning of the world is far enough from the present order of Nature to be not at all repugnant.

It may be difficult to follow up the idea in detail as we are not yet able to count the quantum packets in every case. For example, it may be that an atomic nucleus must be counted as a unique quantum, the atomic number acting as a kind of quantum number. If the future development of quantum theory happens to turn in that direction, we could conceive the beginning of the universe in the form of a unique atom, the atomic weight of which is the total mass of the universe. This highly unstable atom would divide in smaller and smaller atoms by a kind of super-radioactive process. Some remnant of the process might, according to Sir James Jeans's idea, foster the heat of the stars until our low atomic number atoms allowed life to be possible.

[1] [See the preceding paper, *Nature 127*, 447ff (1931).]

Clearly the initial quantum could not conceal in itself the whole course of evolution; but according to the principle of indeterminacy, that is not necessary. Our world is now understood to be a world where something really happens; the whole story of the world need not have been written down in the first quantum like a song on the disc of a phonograph. The whole matter of the world must have been present at the beginning, but the story it has to tell may be written step by step.

63. FROM NEWTON TO EINSTEIN

By Willem de Sitter

The theory of relativity may be considered as the logical completion of Newton's theory of gravitation, the direct continuation of the line of thought which dominates the development of the science of mechanics, from Archimides—who may be considered as the first relativist—through Galileo to Newton. Newton's theory had celebrated its greatest triumphs in the eighteenth and nineteenth centuries; one after another all the irregularities in the motions of the planets and the moon had been explained by the mutual gravitational action of these bodies. In the beginning of the nineteenth century Laplace's monumental work completed the application of the theory on the motions of the planets.

The final triumph came in 1846 by the discovery of Neptune, verifying the prediction by Adams and Leverrier, based on the theory of gravitation.

Gradually Newton's law of gravitation had become a model on which physical laws were framed, and all physical phenomena were reduced to laws which were formulated as attractions or repulsions inversely proportional to some power of the distance, such as, e.g., Laplace's theory of capillarity, which was even published as a chapter in his "Mécanique céleste". Gradually, however, during the second half of the nineteenth century, the uncomfortable feeling of dislike of action at a distance, which had been so strong in Huygens and other contemporaries of Newton, but had subsided during the eighteenth century, began to emerge again, and gained strength rapidly.

This was favored by the purely mathematical transformation (which can be compared in a sense with that from the Ptolemaic to the Copernican system), replacing Newton's finite equations by the differential equations, the potential becoming the primary concept, instead of the force, which is only the gradient of the potential. These ideas, of course, arose first in the theory of electricity and magnetism—or perhaps one should say in the brain of Faraday. In electromagnetism also the law of the inverse square had been supreme, but, as a consequence of the work of Faraday and Maxwell, it was superseded by the field.

And the same change took place in the theory of gravitation. By and by the material particles, electrically charged bodies, and magnets—which are the things that we actually observe—come to be looked upon only as "singularities" in the field. So far this transformation from the force to the potential, from the action at a distance to the field, is only a purely mathematical operation. Whether we talk of a "particle of matter" or of a "singularity in the gravitational field" is only a question of a name. But this giving of names is not so innocent as it looks. It has opened the gate for the entrance of hypotheses. Very soon the field is materialized, and is called aether. From the mathematical point of view, of course, "aether" is still just another word for "field," or, perhaps better, for "space"—the absolute space of Newton—in which there may or may not be a field. From the point of view of physical theory (and it is especially in the theory of electromagnetism that this evolution took place), however, the "aether of space," as it used to be called about forty years ago, is not simply space, it is something substantial, it is the carrier of the field, and mechanical models, consisting of racks and pinions and cogwheels, are devised to explain how it does the carrying. These mechanical models have, of course, been given up long ago; they were too crude. But hypotheses have been cropping up on all sides: electrons, atomic nuclei, protons, wave-packets, etc. At first the imagining of mechanical models went on. Fifteen, or even ten, years ago, although an atom was no longer, as the name implies, just a piece of matter that could not be cut into smaller pieces, atoms, electrons, and protons were still thought of as mechanical structures, models of the atom were imagined, having the mechanical properties of ordinary matter. The inconsistency of first explaining matter by atoms and then explaining atoms by matter was only slowly realized, and it is only comparatively recently that we have come to see that there is nothing paradoxical in the fact that an atom or an electron, which are not matter, may have properties different from those of matter, and must be allowed to do things that a material particle could not do.

However, whilst in all other domains of physics hypotheses have been found successful in accounting for the observed facts, and replacing the formal laws, the case of gravitation stands apart. Gravitation has been insusceptible to this general infection. By using this word I do not mean to suggest that the luxuriant growth of hypotheses in physics is a contagious disease—it is not a disease, but a natural development—but it is certainly contagious. Gravitation, however, seems to be immune to it. In the course of history a great number of hypotheses have been proposed in order to "explain" gravitation, but not one of these has ever had the least chance, they have all been failures. Why is that? How does it come about that we have been able to find satisfactory hypotheses to explain electricity and magnetism, light and heat, in short all other physical phenomena, but have been unsuccessful in the case of gravitation? The explanation must be sought in the peculiar position that gravitation occupies amongst the laws of nature. In the case of other physical phenomena there is something to get hold of, there are circumstances on which the action depends. Gravitation is entirely independent of everything that influences other natural phenomena. It is not subject to absorption or refraction, no velocity of propagation has been observed. You can do whatever you please with a body, you can electrify or magnetise it, you can heat it, melt or evaporate it, decompose it chemically, its behaviour with respect to gravitation is not affected. Gravitation acts on all bodies in the same way, everywhere and always we find it in the same rigorous and simple form, which frustrates all our attempts to penetrate into its internal mechanism. Gravitation is, in its generality and rigour, entirely similar to inertia, which has never been considered to require a particular hypothesis for its explanation, as any ordinary special physical law or phenomenon. Inertia has from the beginning been admitted as one of the fundamental facts of nature, which have to be accepted without explanation, like the axioms of geometry.

But gravitation is not only similar to inertia in its generality, it is also measured by the same number, called the mass. The inertial mass is what Newton calls the "quantity of matter": it is a measure for the resistance offered by a body to a force trying to alter its state of motion. It might be called the "passive mass." The gravitational mass, on the other hand, is a measure of the force exerted by the body in attracting other bodies. We might call it the "active" mass. The equality of active and passive, or gravitational and inertial, mass was in Newton's system a most remarkable accidental co-incidence, something like a miracle. Newton himself decidedly felt it as such, and made experiments to verify it, by swinging a pendulum with a hollow bob which could be filled with different materials. The force acting on the pendulum is proportional to its gravitational mass, the inertia to its inertial mass; the period of its swing thus depends on the ratio between these two masses. The fact that the period is always the same therefore proves that the gravitational and inertial masses are equal. Gradually, during the eighteenth century, physicists and philosophers had become so accustomed to Newton's law of gravitation, and to the equality of gravitational and inertial mass, that the miraculousness of it was forgotten and only an acute mind like Bessel's perceived the necessity of repeating those experiments. By the experiments of Bessel about 1830 and of Eötvös in 1909, the equality of gravitational and inertial mass has become one of the best ascertained empirical facts in physics.

In Einstein's general theory of relativity the identity of these two co-efficients, the gravitational and inertial mass, is no longer a miracle, but a necessity, because gravitation and inertia are identical.

There is another side to the theory of relativity. We have pointed out in the beginning how the development of science is in the direction to make it less subjective, to separate more and more in the observed facts that which belongs to the reality behind the phenomena, the absolute, from the subjective element, which is introduced by the observer, the relative. Einstein's theory is a great step in that direction. We can say that the theory of relativity is intended to remove entirely the relative and exhibit the pure absolute.

The physical world has three space dimensions and one time dimension; the position of a material particle at a certain time t is defined by three space coördinates, x, y, z. In Newton's system of mechanics this is unhesitatingly accepted as a property of the outside world: there is an absolute space and an absolute time. In Einstein's theory time and space are interwoven, and the way in which they are interwoven depends on the observer. Instead of three plus one we have four dimensions.

Is the fact that we observe the outside world as a four-dimensional continuum a property of this outside world, or is it a consequence of the particular nature of our consciousness, does it belong to the absolute or to the relative? I do not think the answer to that question can yet be given. For the present we may accept it as an empirically ascertained fact.

The sequence of different positions of the same particle at different times forms a one-dimensional continuum in the four-dimensional space-time, which is called the *world-line* of the particle. All that physical experiments or observations can teach us refers to intersections of world-lines of different material particles, light-pulsations, etc., and how the course of the world-line is, between these points of intersection, is entirely irrelevant and outside the domain of physics. The system of intersecting world-lines can thus be twisted about at will, so long as no points of intersection are destroyed or created, and their order is not changed. It follows that the equations expressing the physical laws must be invariant for arbitrary transformations.

This is the mathematical formulation of the theory of relativity. The metric properties of the four-dimensional continuum are described, as is shown in treatises on differential geometry, by a certain number (ten, in fact) of quantities denoted by $g_{\alpha\beta}$, and commonly called "potentials". The physical status of matter and energy, on the other hand, is described by ten other quantities, denoted by $T_{\alpha\beta}$, the set of which is called the "material tensor." This special tensor has been selected because it has the property which is mathematically expressed by saying that its divergence vanishes, which means that it represents something permanent. The fundamental fact of mechanics is the law of inertia, which can be expressed in its most simple form by saying that it requires the fundamental laws of nature to be differential equations of the second order. Thus the problem was to find a differential equation of the second order giving a relation between the metric tensor $g_{\alpha\beta}$ and the material tensor $T_{\alpha\beta}$. This is a purely mathematical problem, which can be solved without any reference to the physical meaning of the symbols. The simplest possible equation (or rather set of ten equations, because there are ten g's) of that kind, that can be found was adopted by Einstein as the fundamental equation of his theory. It defines the space-time continuum, or the "field." The world-lines of material particles and light quanta are the geodesics in the four-dimensional continuum defined by the solutions $g_{\alpha\beta}$ of these field-equations. The equations of the geodesic thus are equivalent to the equations of motion of mechanics. When we come to solve the field-equations and substitute the solutions in the equations of motion, we find that in the first approximation, i.e. for small material velocities (small as compared with the velocity of light), these equations of motion are the same as those resulting from Newton's theory of gravitation, The distinction between gravitation and inertia has disappeared; the gravitational action between two bodies follows from the same equations, and *is* the same thing, as the inertia of one body. A body, when not subjected to an extraneous force (i.e. a force other than gravitation), describes a geodesic in the continuum, just as it described a geodesic, or straight line, in the absolute space of Newton under the influence of inertia alone.

The field-equations and the equations of the geodesic together contain the whole science of mechanics, including gravitation.

In the first approximation, as has been said just now, the new theory gives the same results as Newton's theory of gravitation. The enormous wealth of experimental verification of Newton's law, which has been accumulated during about two and a half centuries, is therefore at the same time an equally strong verification of the new theory. In the second approximation there are small differences, which have been confirmed by observations, so far as they are large enough for such a confirmation to be possible. Thus especially the anomalous motion of the perihelion of Mercury, which had baffled all attempts at explanation for over half a century, is now entirely accounted for. Further the theory of relativity has predicted some new phenomena, such as the deflection of the rays of light that pass near the sun, which has actually been observed on several occasions during eclipses; and the redshift of spectral lines originating in a strong gravitational field, which is also confirmed by observations, e.g. in the spectrum of the sun, and also in the spectrum of the companion of Sirius, which, being a so-called white dwarf, i.e. a small star with very high density and consequently a strong gravitational field, gives a considerable redshift. We cannot stop to explain these phenomena in detail. It must suffice just to mention them.

Two points should be specially emphasized in connection with the general theory of relativity.

First, that it is a purely *physical* theory, invented to explain empirical physical facts, especially the identity of gravitational and inertial mass, and to coordinate and harmonise different chapters of physical theory, and simplify the enunciation of the fundamental laws. There is nothing metaphysical about its origin. It has, of course, largely attracted the attention of philosophers, and has, on the whole, had a very wholesome influence on metaphysical theories. But that is not what it set out to do, that is only a by-product.

Second, that it is a pure generalization, or abstraction, like Newton's system of mechanics and law of gravitation. It contains *no hypothesis,* as contrasted with other modern physical theories, electron theory, quantum theory, etc., which are full of hypotheses. It is, as has already been said, to be considered as the logical sequence and completion of Newton's Principia.

The great men of science, as well as the great artists, are filled with a spirit of reverence, with a consciousness of the presence of mystery and sublimity in the simplest and smallest as well as the greatest of things and phenomena, and with faith in the order and unity of all things. Only the way in which they try to comprehend this order and penetrate to its deeper meaning is different.

It is a rather common misapprehension that science, by analysing and dissecting nature, by subjecting it to the rigorous rule of mathematical formulae and numerical expression, would lose the sense of its beauty and sublimity. On the contrary, even the purely aesthetical appreciation, say of a landscape, or of a thunderstorm, is, in my opinion, helped rather than impeded by the knowledge, so far as it goes, that the scientific beholder has of the inner structure and the connection of the phenomena. And the measurement and reduction to numbers, "pointer readings," as Sir Arthur Eddington says, is not the ultimate aim of science, but its means to an end. By the use of mathematics, that most nearly perfect and most immaterial tool of the human mind, we try to transcend as much as possible the limitations imposed by our finiteness and materiality, and to penetrate ever nearer to the understanding of the mysterious unity of the Kosmos.

64. REMARKS ON THE GENERAL THEORY OF RELATIVITY [1]

By Albert Einstein

The special theory of relativity, which was simply a systematic development of the electro-dynamics of Clerk Maxwell and Lorentz, pointed beyond itself. Should the independence of physical laws of the state of motion of the co-ordinate system be restricted to the uniform translatory motion of co-ordinate systems in respect to each other? What has nature to do with our co-ordinate systems and their state of motion? If it is necessary for the purpose of describing nature, to make use of a co-ordinate system arbitrarily introduced by us, then the choice of its state of motion ought to be subject to no restriction; the laws ought to be entirely independent of this choice (general principle of relativity).

The establishment of this general principle of relativity is made easier by a fact of experience that has long been known, namely that the weight and the inertia of a body are controlled by the same constant. (Equality of inertial and gravitational mass.) Imagine a co-ordinate system which is rotating uniformly with respect to an inertial system in the Newtonian manner. The centrifugal forces which manifest themselves in relation to this system must, according to Newton's teaching, be regarded as effects of inertia. But these centrifugal forces are, exactly like the forces of gravity, proportional to the masses of the bodies. Ought it not to be possible in this case to regard the co-ordinate system as stationary and the centrifugal forces as gravitational forces? This seems the obvious view, but classical mechanics forbid it.

This hasty consideration suggests that a general theory of relativity must supply the laws of gravitation, and the consistent following up of the idea has justified our hopes.

But the path was thornier than one might suppose, because it demanded the abandonment of Euclidean geometry. This is to say, the

[1] [See selection 58 above for an important deduction from the Special theory of Relativity, and selection 63 for de Sitter's presentation of both the Special and General theories.]

laws according to which fixed bodies may be arranged in space, do not completely accord with the spatial laws attributed to bodies by Euclidean geometry. This is what we mean when we talk of the 'curvature of space.' The fundamental concepts of the 'straight line,' the 'plane,' etc., thereby lose their precise significance in physics.

In the general theory of relativity the doctrine of space and time, or kinematics, no longer figures as a fundamental independent of the rest of physics. The geometrical behaviour of bodies and the motion of clocks rather depend on gravitational fields, which in their turn are produced by matter.

The new theory of gravitation diverges considerably, as regards principles, from Newton's theory. But its practical results agree so nearly with those of Newton's theory that it is difficult to find criteria for distinguishing them which are accessible to experience. Such have been discovered so far:—

(1) In the revolution of the ellipses of the planetary orbits round the sun (confirmed in the case of Mercury).

(2) In the curving of light rays by the action of gravitational fields (confirmed by the English photographs of eclipses).

(3) In a displacement of the spectral lines towards the red end of the spectrum in the case of light transmitted to us from stars of considerable magnitude.

The chief attraction of the theory lies in its logical completeness. If a single one of the conclusions drawn from it proves wrong, it must be given up; to modify it without destroying the whole structure seems to be impossible.

Let no one suppose, however, that the mighty work of Newton can really be superseded by this or any other theory. His great and lucid ideas will retain their unique significance for all time as the foundation of our whole modern conceptual structure in the sphere of natural philosophy.

65. THE PRIMEVAL ATOM

By Georges Lemaître

The primeval atom hypothesis is a cosmogonic hypothesis which pictures the present universe as the result of the radioactive disintegration of an atom.

I was led to formulate this hypothesis, some fifteen years ago, from thermodynamic considerations, while trying to interpret the law of degradation of energy in the frame of quantum theory. Since then, the discovery of the universality of radioactivity shown by artificially provoked disintegration, as well as the establishment of the corpuscular nature of cosmic rays, manifested by the force which the Earth's magnetic field exercises on these rays, made more plausible an hypothesis which assigned a radioactive origin to these rays, as well as to all existing matter.

Therefore, I think that the moment has come to present the theory in deductive form. I shall first show how easily it avoids several major objections which would tend to disqualify it from the start. Then I shall strive to deduce its results far enough to account, not only for cosmic rays, but also for the present structure of the universe, formed of stars and gaseous clouds, organized into spiral or elliptical galaxies, sometimes grouped into large clusters of several thousand galaxies which, more often, are composed of isolated galaxies, receding from one another according to the mechanism known by the name of expanding universe.

For the exposition of my subject, it is indispensable that I recall several elementary geometric conceptions, such as that of the closed space of Riemann, which led to that of space with a variable radius, as well as certain aspects of the theory of relativity, particularly the introduction of the cosmological constant and of the cosmic repulsion which is the result of it.

CLOSED SPACE

All partial space is open space. It is comprised in the interior of a surface, its boundary, beyond which there is an exterior region. Our

habit of thought about such open regions impels us to think that this
is necessarily so, however large the regions being considered may be.
It is to Riemann that we are indebted for having demonstrated that
total space can be closed. To explain this concept of closed space, the
most simple method is to make a small-scale model of it in an open
space. Let us imagine, in such a space, a sphere in the interior of which
we are going to represent the whole of closed space. On the rim surface
of the sphere, each point of closed space will be supposed to be represented twice, by two points, A and A′, which, for example, will be two
antipodal points, that is, two extremities of the same diameter. If we
join these two points A and A′ by a line located in the interior of the
sphere, this line must be considered as a closed line, since the two extremities A and A′ are two distinct representations of the same, single
point. The situation is altogether analogous to that which occurs with
the Mercator projection, where the points on the 180th meridian are
represented twice, at the eastern and western edges of the map. One
can thus circulate indefinitely in this space without ever having to
leave it.

It is important to notice that the points represented by the outer
surface of the sphere, in the interior of which we have represented all
space, are not distinguished by any properties of the other points of
space, any more than is the 180th meridian for the geographic map. In
order to account for that, let us imagine that we displaced the sphere
in such a manner that point A is superposed on B, and the antipodal
point A′ on B′. We shall then suppose that the entire segment AB and
the entire segment A′B′ are two representations of a similar segment
in closed space. Thus we shall have a portion of space which has already been represented in the interior of the initial sphere which is now
represented a second time at the exterior of this sphere. Let us disregard the interior representations as useless; a complete representation
of the space in the interior of the new sphere will remain. In this representation, the closed contours will be soldered into a point which is
twice represented, namely, by the points B and B′, mentioned above,
instead of being welded, as they were formerly, to points A and A′.
Therefore, these latter are not distinguished by an essential property.

Let us notice that when we modify the exterior sphere, it can happen
that a closed contour which intersects the first sphere no longer intersects the second, or, more generally, that a contour no longer intersects
the finite sphere at the same number of points.

Nevertheless, it is evident that the number of points of intersection
can only vary by an even number. Therefore, there are two kinds of

closed contours which cannot be continuously distorted within one another. Those of the first kind can be reduced to a point. They do not intersect the outer sphere or they intersect it at an even number of points. The others cannot be reduced to one point; we call them the *odd contours* since they intersect the sphere at an odd number of points.

If, in a closed space, we leave a surface which we can suppose to be horizontal, in going toward the top we can, by going along an odd contour, return to our point of departure from the opposite direction without having deviated to the right or left, backward or forward, without having traversed the horizontal plane passing through the point of departure.

Elliptical Space

That is the essential of the topology of closed space. It is possible to complete these topological ideas by introducing, as is done in a geographical map, scales which vary from one point to another and from one direction to another. That can be done in such a manner that all the points of space and all the directions in it may be perfectly equivalent. Thus, Riemann's homogeneous space, or elliptical space, is obtained. The straight line is an odd contour of minimum length. Any two points divide it into two segments, the sum of which has a length which is the same for all straight lines and which is called the tour of space.

All elliptical spaces are similar to one another. They can be described by comparison with one among them. The one in which the tour of the straight line is equal to $\pi = 3.1416$ is chosen as the standard elliptical space. In every elliptical space, the distances between two points are equal to the corresponding distances in standard space, multiplied by the number R which is called the radius of elliptical space under consideration. The distances in standard space, called space of unit radius, are termed angular distances. Therefore, the true distances, or linear distances, are the product of the radius of space times the angular distances.

Space of Variable Radius

When the radius of space varies with time, space of variable radius is obtained. One can imagine that material points are distributed evenly in it, and that spatio-temporal observations are made on these points.

The angular distance of the various observers remains invariant, therefore the linear distances vary proportionally to the radius of space. All the points in space are perfectly equivalent. A displacement can bring any point into the center of the representation. The measurements made by the observers are thus also equivalent, each one of them makes the same map of the universe.

If the radius increases with time, each observer sees all points which surround him receding from him, and that occurs at velocities which become greater as they recede further. It is this which has been observed for the galaxies that surround us. . . .

THE PRIMEVAL ATOM

These are the geometric concepts that are indispensable to us. We are now going to imagine that the entire universe existed in the form of an atomic nucleus which filled elliptical space of convenient radius in a uniform manner.

Anticipating that which is to follow, we shall admit that, when the universe had a density of 10^{-27} gram per cubic centimeter, the radius of space was about a billion light-years, that is, 10^{27} centimeters. Thus the mass of the universe is 10^{54} grams. If the universe formerly had a density equal to that of water, its radius was then reduced to 10^{18} centimeters, say, one light-year. In it, each proton occupied a sphere of one angstrom, say, 10^{-8} centimeter. In an atomic nucleus, the protons are contiguous and their radius is 10^{-13}, thus about 100,000 times smaller. Therefore, the radius of the corresponding universe is 10^{13} centimeters, that is to say, an astronomical unit.

Naturally, too much importance must not be attached to this description of the primeval atom, a description which will have to be modified, perhaps, when our knowledge of atomic nuclei is more perfect.

Cosmogonic theories propose to seek out initial conditions which are ideally simple, from which the present world, in all its complexity, might have resulted, through the natural interplay of known forces. It seems difficult to conceive of conditions which are simpler than those which obtained when all matter was unified in an atomic nucleus. The future of atomic theories will perhaps tell us, some day, how far the atomic nucleus must be considered as a system in which associated particles still retain some individuality of their own. The fact that particles can issue from a nucleus, during radioactive transformations, certainly does not prove that these particles pre-existed as such. Protons issue from an atom of which they were not constituent parts, electrons appear there, where they were not previously, and the theoreticians deny them an individual existence in the nucleus. Still more protons or alpha particles exist there, without doubt. When they issue forth, their existence becomes more independent, nevertheless, and their degrees of freedom more numerous. Also, their existence, in the course of radioactive transformations, is a typical example of the degradation of energy, with an increase in the number of independent quanta or increase in entropy.

That entropy increases with the number of quanta is evident in the case of electromagnetic radiation in thermodynamical equilibrium. In fact, in black body radiation, the entropy and the total number of photons are both proportional to the third power of the temperature. Therefore, when one mixes radiations of different temperatures and one allows a new statistical equilibrium to be established, the total number of photons has increased. The degradation of energy is manifested as a pulverization of energy. The total quantity of energy is maintained, but it is distributed in an ever larger number of quanta, it becomes broken into fragments which are ever more numerous.

If, therefore, by means of thought, one wishes to attempt to retrace the course of time, one must search in the past for energy concentrated in a lesser number of quanta. The initial condition must be a state of maximum concentration. It was in trying to formulate this condition that the idea of the primeval atom was germinated. Who knows if the evolution of theories of the nucleus will not, some day, permit the consideration of the primeval atom as a single quantum?

FORMATION OF CLOUDS

We picture the primeval atom as filling space which has a very small radius (astronomically speaking). Therefore, there is no place for superficial electrons, the primeval atom being nearly an *isotope of a neutron*. This atom is conceived as having existed for an instant only; in fact it was unstable and, as soon as it came into being, it was broken into pieces which were again broken, in their turn; among these pieces electrons, protons, alpha particles, etc., rushed out. An increase in volume resulted. The disintegration of the atom was thus accompanied by a rapid increase in the radius of space which the fragments of the primeval atom filled, always uniformly. When these pieces became too small, they ceased to break up; certain ones, like uranium, are slowly disintegrating now, with an average life of four billion years, leaving us a meager sample of the universal disintegration of the past.

In this first phase of the expansion of space, starting asymptotically with a radius practically zero, we have particles of enormous velocities (as a result of recoil at the time of the emission of rays) which are immersed in radiation, the total energy of which is, without doubt, a notable fraction of the mass energy of the atoms.

The effect of the rapid expansion of space is the attenuation of this radiation and also the diminution of the relative velocities of the atoms. This latter point requires some explanation. Let us imagine that an atom has, along the radius of the sphere in which we are representing closed space, a radial velocity which is greater than the velocity normal to the region in which it is found. Then this atom will depart faster from the center than the ideal material particle which has normal velocity. Thus the atom will reach, progressively, regions where its velocity is less abnormal, and its proper velocity, that is, its excess over normal velocity, will diminish. Calculation shows that proper velocity varies in this way in reverse ratio to the radius of space. We must therefore look for a notable attenuation of the relative velocities of atoms in the first period of expansion. From time to time, at least, it will happen that, as a result of favorable chances, the collisions between atoms will become sufficiently moderate so as to give rise to atomic transformations or emissions of radiation, but that these collisions will be elastic collisions, controlled by superficial electrons, so considered in the theory of gases. Thus we shall obtain, at least locally, a beginning of statistical equilibrium, that is, the formation of gaseous clouds. These gaseous clouds will still have considerable velocities, in relation to one another, and they will be mixed with radiations that are themselves attenuated by expansion.

It is these radiations which will endure until our time in the form of cosmic rays, while the gaseous clouds will have given place to stars and to nebulae by a process which remains to be explained. . . .

The theory of relativity has thus unified the theory of Newton. In Newton's theory, there were two principles posed independently of one another: universal attraction and the conservation of mass. In the theory of relativity, these principles take a slightly modified form, while being practically identical to those of Newton in the case where these have been confronted with the facts. But universal attraction is now a result of the conservation of mass. The size of the force, the constant of gravitation, is determined experimentally.

The theory again indicates that the constancy of mass has, as a result, besides the Newtonian force of gravitation, a repulsion proportional to the distance of which the size and even the sign can only be determined by observation and by observation requiring great distances. . . .

COSMIC RAYS

Finally, we said in the beginning that the radiation produced during the disintegrations, during the first period of expansion, could explain cosmic rays. These rays are endowed with an energy of several billion electron-volts. We know no other phenomenon currently taking place which may be capable of such effects. That which these rays resemble most is the radiation produced during present radioactive disintegrations, but the individual energies brought into play are enormously greater. All that agrees with rays of superradioactive origin. But it is not only by their quality that these rays are remarkable, it is also by their total quantity. In fact, it is easy, from their observed density which is given in ergs per centimeter, to deduce their density of energy by dividing by c, then their density in grams per cubic centimeter by dividing by c^2. Thus one finds 10^{-34} grams per cubic centimeter, about one ten-thousandth the present density of the matter existing in the form of stars. It seems impossible to explain such an energy which represents one part in ten thousand of all existing energy, if these rays had not been produced by a process which brought into play all existing matter. In fact, this energy, at the moment of its formation, must have been at least ten times greater, since a part of it was able to be absorbed and the remainder has been reduced as a result of the expansion of space. The total intensity observed for cosmic rays is therefore just about that which might be expected.

CONCLUSION

The purpose of any cosmogonic theory is to seek out ideally simple conditions which could have initiated the world and from which, by the play of recognized physical forces, that world, in all its complexity, may have resulted.

I believe that I have shown that the hypothesis of the primeval atom satisfies the rules of the game. It does not appeal to any force which is not already known. It accounts for the actual world in all its complexity. By a single hypothesis it explains stars arranged in galaxies within an expanding universe as well as those local exceptions, the clusters of galaxies. Finally, it accounts for that mighty phenomenon, the ultra-penetrating rays. They are truly cosmic; they testify to the primeval activity of the cosmos. In their course through wonderfully empty space, during billions of years, they have brought us evidence of the superradioactive age, indeed they are a sort of fossil rays which tell us what happened when the stars first appeared.

I shall certainly not pretend that this hypothesis of the primeval atom is yet proved, and I would be very happy if it has not appeared to you to be either absurd or unlikely. When the consequences which result from it, especially that which concerns the law of the distribution of densities in the galaxies, are available in sufficient detail, it will doubtless be possible to declare oneself definitely for or against.

66. THE PERFECT COSMOLOGICAL PRINCIPLE [1]

By H. Bondi and T. Gold

The unrestricted repeatability of all experiments is the fundamental axiom of physical science. This implies that the outcome of an experiment is not affected by the position and the time at which it is carried out. A system of cosmology must be principally concerned with this fundamental assumption and, in turn, a suitable cosmology is required for its justification. In laboratory physics we have become accustomed to distinguish between conditions which can be varied at will and the inherent laws which are immutable.

Such a distinction between the "accidental" conditions and the "inherent" laws and constants of nature is justifiable so long as we have control over the "accidental," and can test the validity of the distinction by a further experiment. In astronomical observations we do not have this control, and we can hence never prove which is "accidental" and which "inherent." This difficulty, though logically a very real one, need not concern us in an interpretation of the dynamics of the solar system. We may be satisfied when we discover that the solar system with all its numerous orbits is accurately one of the many systems permitted by our "inherent" laws.

But when we wish to consider the behaviour of the entire universe, then the logical basis for a distinction between "inherent" laws and "accidental" conditions disappears. Any observation of the structure of the universe will give as unique a result as, for instance, a determination of the velocity of light or the constant of gravitation. And yet, if we were to contemplate a changing universe we should have to assume some such observations to represent "accidental" and other "inherent" laws.

Such assumptions were in fact implied in all theories of evolution of the universe; they were necessary to specify the problem. Without them, there would be no rules and hence unlimited freedom in any extrapolation into the future or into the past. Some such sets of assump-

[1] [This excerpt serves as an introduction to the authors' work on the steady-state universe. Compare F. Hoyle's somewhat similar steady-state hypothesis, *Monthly Notices of the Royal Astronomical Society 108*, 372ff (1948).]

tions may be intellectually much more satisfying than others, and accordingly they are adopted. We may place much reliance in such aesthetic judgments, but we cannot claim any logical foundation for them.

Let us take an example to demonstrate the point of this criticism. Present observations indicate that the universe is expanding. This suggests that the mean density in the past has been greater than it is now. If we are now to make any statement regarding the behaviour of such a denser universe, and the way in which it is supposed to have reached its present condition, then we have to know the physical laws and constants applicable in denser universes. But we have no determinations for those. An assumption which is commonly made is to say that those physical laws which we have learnt to regard as "intrinsic" because they are unaffected by any changes of conditions, which we may produce, are in fact not capable of any change. It is admitted that such a change of the density of the universe would have slight local effects; there would be more or less light arriving here from distant galaxies. But, it is assumed, there would be no change in the physical laws which an observer would deduce in a laboratory.

Such a philosophy may be intellectually very agreeable: it gives permanence to the abstract things, the physical laws, whilst it regards the present condition of the universe as merely a particular demonstration of the consequences of such laws. There are, however, grave difficulties inherent in such a view. The most striking of those is concerned with the absolute state of rotation of a body. Mach examined this problem very thoroughly and all the advances in theory which have been made have not weakened the force of his argument. According to "Mach's Principle" inertia is an influence exerted by the aggregate of distant matter which determines the state of motion of the local frame of reference by means of which rotation or acceleration is measured. A particular rotational state which is described as "non-rotating" can be found by a purely local experiment (a Foucault pendulum, for instance). This rotational state is always found to coincide accurately with the rotational state of the distant stars around the observer. Mach's principle is the recognition that this coincidence must in fact be due to a causal relation. If, as is now widely agreed, we adopt Mach's principle, then we imply that the nature of any local dynamical experiment is fundamentally affected by distant matter. We can hence not contemplate a laboratory which is shielded so as to exclude all influence from outside; and for the same reason we cannot have any logical basis for choosing physical laws and constants and assigning to them an existence independent of the structure of the universe.

Any interdependence of physical laws and large-scale structure of the universe might lead to a fundamental difficulty in interpreting observations of light emitted by distant objects. For if the universe, as seen from those objects, presented a different appearance, then we should not be justified in assuming familiar processes to be responsible for the emission of the light which we analyze. This difficulty is partly removed by the "cosmological principle." According to this principle all large-scale averages of quantities derived from astronomical observations (i.e. determination of the mean density of space, average size of galaxies, ratio of condensed to uncondensed matter, etc.) would tend statistically to a similar value independent of the positions of the observer, as the range of the observation is increased; provided only that the observations from different places are carried out at equivalent times. This principle would mean that there is nothing outstanding about any place in the universe, and that those differences which do exist are of only local significance; that seen on a large scale the universe is homogeneous.

This principle is widely recognized, and the observations of distant nebulae have contributed much evidence in its favor. An analysis of these observations indicates that the region surveyed is large enough to show us a fair sample of the universe, and that this sample is homogeneous.

But in the sense in which the principle came to be adopted there is the qualification regarding the time of the observation. The universe is still presumed to be capable of altering its large-scale structure, but only in such a way as not to upset its homogeneity. The result of a large-scale observation may hence be a measure of a universal time.

We might have looked to the cosmological principle for a justification of the assumption of the general validity of physical laws; but whilst the principle supplies the justification with respect to changes of place, it still leaves the possibility of a change of physical laws with universal time. Any system of cosmology must still involve a speculation about this dependence as one of its basic assumptions. Indeed, we are not even in a position to interpret observations of very distant objects without such an assumption, for the light which we receive from them was emitted at a different instant in this scale of universal time, and accordingly the processes responsible for its emission may be unfamiliar to us.

The systems of cosmology may well be classified according to the assumption made or implied at this stage of the argument. While one school of thought considers all the results of laboratory physics to be always applicable, without regard to the state of the universe, another starts from the narrow cosmological principle and with the aid of a number of intellectually agreeable assumptions arrives at the conclusion that laboratory physics is qualitatively permanently valid, though some of its "constants" are changing. There are yet other schools of thought which attempt to distinguish the changeable from the permanent constants by their magnitudes.

We shall proceed quite differently at this point. As the physical laws cannot be assumed to be independent of the structure of the universe, and as conversely the structure of the universe depends upon the physical laws, it follows that there may be a stable position. We shall pursue the possibility that the universe is in such a stable, self-perpetuating state, without making any assumptions regarding the particular features which lead to this stability. We regard the reasons for pursuing this possibility as very compelling, for it is only in such a universe that there is any basis for the assumption that the laws of physics are constant; and without such an assumption our knowledge, derived virtually at one instant of time, must be quite inadequate for an interpretation of the universe and the dependence of its laws on its structure, and hence inadequate for any extrapolation into the future or the past.

Our course is therefore defined not only by the usual cosmological principle but by that extension of it which is obtained on assuming the universe to be not only homogeneous but also unchanging on the large scale. This combination of the usual cosmological principle and the stationary postulate we shall call the *perfect cosmological principle,* and all our arguments will be based on it. The universe is postulated to be homogeneous and stationary in its large-scale appearance as well as in its physical laws.

We do not claim that this principle must be true, but we say that if it does not hold, one's choice of the variability of the physical laws becomes so wide that cosmology is no longer a science. One can then no longer use laboratory physics without relying on some arbitrary principle for their extrapolation.

But if the perfect cosmological principle is satisfied in our universe then we can base ourselves confidently on the permanent validity of all our experiments and observations and explore the consequences of the principle. Unless and until any disagreement appears we therefore accept the principle, since this is the only assumption on the basis of which progress is possible without further hypothesis.

XIII

SURVEYS OF ASTROPHYSICAL PROGRESS

An examination of the 66 preceding selections reveals that at the end of the fifth decade of this century the wide-ranging science of astronomy had largely recovered from the interruptions of war and was increasing in activity along many lines—in observational techniques and astrophysical theory, in particular. Will the next five decades see further acceleration? In many ways the essentially new subscience of radio astronomy will underwrite the growth, for it enters many fields; it brings new light on meteors, moon, sun, planetary surfaces, exploded stars, the hydrogen substratum, and the remotest galaxies. The two large American national observatories, Green Bank for radio, Kitt Peak for optical research, will soon be in full production. Russian, South African, and Australian optical instruments will supplement the new radio telescopes in a dozen countries.

Before we ponder and analyze the hoped for contributions of future years, it will be well to give at least a part of the past a summarizing review. That is accomplished to some extent by the three concluding papers of this volume. The boldness and charm of modern astrophysical theory is set before us by Svein Rosseland, a pioneer in the field. In the final paper Otto Struve briefly surveys the accomplishments with spectroscope and photographic plate. And, in his setting forth the science of cosmochemistry, Rupert Wildt leads us all over the universe from little dust particles in the air to the turbulent stars in galaxies on the "rim" of the universe.

Perhaps the most appealing phase of our concern with what the cosmos is made of, and how it is constructed, is the knowledge that however objective we attempt to be we are all the time concerned with ourselves, for the chemistry of a galaxy is also the chemistry of man and his biological environment.

67. INTRODUCTION TO THEORETICAL ASTROPHYSICS

By Svein Rosseland

Among natural sciences astronomy is unique as the mother of all other sciences, the oldest by far in years, and probably still the one which is dearest to the heart of man. It was originally inextricably linked up with religious thought and practice, and it will always have something to say about the general outlook of man on the universe, which must take place, so to speak, through the astronomer's telescope.

It is the most poetical of sciences, and many a beautiful human dream was woven into a celestial language. But with oncoming age astronomy has taken on a sterner countenance. Poetry may still be there, but it has been forced into a different garb. The machine age has revolutionized astronomy like everything else. There is the ever-insistent demand for the mass production of observations, for increased precision in measurements, for further detail in arithmetical reductions, and for more satisfying theoretical interpretations. As a consequence an astronomical observatory of to-day looks more like a factory plant than an abode for philosophers. The poetry of constellations has given way to the lure of plate libraries, and the angel of cosmogonic speculation has been caught in a cobweb of facts insistently clamouring for explanations.

Who has not experienced the mysterious thrill of springtime in a forest, with sunbeams flickering through the foliage, and the low humming of insect life? It is the feeling of unity with nature, which is the counterpart of the attitude of the scientist analysing the sunbeams into light quanta and the soft rustling of a dragon-fly into condensations and rarefactions of the air. But what is lost in fleeting sentiment is more than regained in the feeling of intellectual security afforded by the scientific attitude, which may grow into a trusting devotion, challenging the peace of a religious mystic. For in the majestic growth of science, analytical in its experimental groping for detail, synthetic in its sweeping generalizations, we are watching at least one aspect of the human mind, which may be believed to have a future of dizzy heights and a nearly unlimited perfectibility.

The analytic nature of science has revealed itself most clearly in the constantly growing importance of the atomic conceptions; and astronomy is the science which, after physics and chemistry, has benefited most from this development. It is at present the avowed goal of these sciences to build up a complete theoretical self-contained structure, based on atomic theory, sufficient for the adequate description of all physical and chemical processes in the universe. . . .

It is true, of course, that the idea that matter is ultimately atomistic in structure goes back to the very beginning of science; but the saturation of nearly all physical and chemical theories with atomic conceptions is of more recent origin. Chemistry has always been the stronghold of atomism, and rightly, because the limited number of chemical elements, and the formation of chemical combinations out of pure elements mixed in constant multiple proportions, should suffice as an irrefutable proof that the ultimate laws of nature are atomistic, and hence not reconcilable with a description of nature in purely continuous terms. . . .

Just as many scientists of the last century believed it possible to describe the universe adequately in purely mechanical terms, so we are now expecting the physical universe to yield its secrets when attacked with the artillery of quantum-theory conceptions. In pursuing this course we are liable to make some of the errors of our forefathers all over again. But there is nevertheless ground for some confidence, since there is a much greater margin of safety this time, the quantum theory having already been subjected to many severe tests, which show that its range of application extends all over the field of terrestrial physics and chemistry. Moreover, during the last fifteen years astronomy has become permeated by atomic theory, and it is by now abundantly clear that further progress in theoretical astrophysics will depend intimately on the astronomers understanding how to express their observed facts in the language of atomic theory.

The beginning of this development goes back to the foundation of astronomical spectroscopy about 1860. Already Secchi's first survey of the stellar spectra showed that the stars are made from the same chemical elements as the earth and the sun, indicating very strongly that terrestrial physics and chemistry, properly administered, must apply throughout the starry heaven. This result has been further fortified by every new improvement in laboratory technique, leading to a match of previously unidentified lines in stellar spectra by lines in terrestrial sources.

The first really sensational case of this sort was the discovery by Lockyer and Ramsay (1893) that the yellow doublet discovered by Lockyer in the chromospheric spectrum of the sun (1868) was really due to helium, which was previously unknown on the earth. Through the work of Bohr and Fowler (1915) it became clear that the so-called 'secondary hydrogen lines' first discovered by E. Pickering in the spectrum of Zeta Puppis were also due to helium, this time in its singly ionized state. This was the first strict proof that the elements in stellar atmospheres are in a state of ionization. It is true that Lockyer many years before (cf. *Chemistry of the Sun,* or *Sun's Place in Nature*) had urged the idea that the change in the spectrum of the elements on passing from the arc to the spark corresponds to a progressive change in the atoms themselves due to the increased temperature. But the notions of atomic structure entertained at those times were too crude for Lockyer's ideas to assume a more concrete form. Yet it must be admitted that he was right in his contentions in this particular respect, although their value is slightly diminished by the fact that in his writings the finer points are sometimes obscured by controversies.

Although Bohr must thus be considered the pioneer in the field, it was the Indian physicist Megd Nad Saha who (1920) first attempted to develop a consistent theory of the spectral sequence of the stars from the point of view of atomic theory. Saha's work is, in fact, the theoretical formulation of Lockyer's view along modern lines, and from that time the idea that the spectral sequence indicates a progressive transmutation of the elements has been definitely abandoned. From that time dates the hope that a thorough analysis of stellar spectra will afford complete information about the state of the stellar atmospheres, not only as regards the chemical composition but also as regards the temperature and various deviations from a state of thermal equilibrium, the density distribution of the various elements, the value of gravity in the atmosphere and its state of motion. The impetus given to astrophysics by Saha's work can scarcely be overestimated, as nearly all later progress in this field has been influenced by it, and much of the subsequent work has the character of refinements of Saha's ideas.

Fowler and Milne showed how the analysis could be made much more precise by focusing attention on the maximum intensity of the lines, instead of considering the marginal appearance or disappearance as Saha preferred to do. In this way it was possible to show conclusively that the density in the reversing layer of the stars is several thousand times smaller than the density of ordinary air.

It was further shown by Fowler that the conspicuous contrast between the persistence of hydrogen lines with increasing temperature beyond their maximum and the very rapid decline of metallic arc lines beyond their maximum is a simple consequence of the different atomic constitution of these elements. Milne, on the other hand, showed how varying ionization with height in the atmosphere, and the dependence of general opacity on temperature and density, are reflected in the behaviour of the spectral lines, and are responsible for some of the more striking spectral differences between giant and dwarf stars. . . .

First Unsöld on one hand and Stewart on the other showed how the coefficient of scattering predicted by classical dispersion theory would account for the wide wings of strong solar and stellar lines. Unsöld worked out in this way a programme for analysing the state of an atmosphere from the profiles of the spectral lines. It is true that the applicability of the method was originally, and still is, very limited; partly because so many things yet remain to be worked out in atomic theory; still more because the theory of transmission of radiation through a stellar atmosphere is only very crudely developed; most of all because the observational determination of line profiles is one of the most difficult tasks in astronomy, which in many cases will demand a wholly new instrumental equipment and technique.

The cruder procedure of observing only the total intensity of a line must therefore remain for a long time to come the principal way of approach. Here conditions are much less uncertain, as was shown by Minnaert and his collaborators. In the case of strong lines the intensity appears to be a simple function of the product of the number of atoms above the photosphere, the oscillatory strength of the line, and, finally, its damping constant. When the intensity of the line diminishes, the intensity becomes gradually more and more independent of the argument, passes through a minimum of dependence, and when for still smaller concentrations of atoms the intensity begins to fall more rapidly, the damping constant no longer plays any part. In its place the intensity is influenced by the intrinsic width of the line which is due to thermal Doppler effect.

This description refers to an ideal case of a quiescent atmosphere. If the atmosphere is in some state of motion, conditions will be different, and it is here that the study of the lines reveals essentially new information. Rotating stars are thus recognized from their broad and diffuse lines with dish-shaped profiles. Expanding stars, as we observe them in novae, Wolf-Rayet stars, and *O* stars have very broad emission lines which are easy to recognize. Recently Struve and Elvey believe they have found what may be called 'boiling' stars, where the apparent thermal Doppler effect exceeds the one to be expected by a factor of the order ten, and this fact they have interpreted to mean that the atmospheres are in a state of irregular motion. The study of the Doppler effect in interstellar lines, so important for the study of galactic rotation, also belongs to this field. It is in fact not merely a question of a simple bodily displacement of the lines, as in the case of routine radial velocity measurements, but of an accumulated Doppler effect, in which the width of the line increases proportionately to its displacement from the zero position.

Another line of study in which the finer details of atomic theory play a considerable part was opened up by Bowen in 1927, when he removed the veil which up to that time had hidden the secrets of nebular radiation by showing conclusively that the previously unknown nebular lines must be due to forbidden transitions in the ionized atoms of oxygen and nitrogen and some other elements. It goes without saying that this discovery at once stimulated theoretical research into the quantitative theory of forbidden transitions, and astrophysicists became on the alert as regards other possibilities where such transitions might play a part.

Forbidden transitions entered the field of cosmical physics for the first time when McLennan discovered that the green auroral line is really due to a transition from a metastable state in the neutral oxygen atom. In our own atmosphere we have also other instances of forbidden transitions in the well-known red absorption bands of oxygen. Recently Adams and Dunham found forbidden bands of carbon dioxide in the spectrum of Venus. Still more sensational was the discovery that the absorption bands in the spectra of outer planets could be interpreted as forbidden bands of methane and ammonia, which was brought out by the theoretical work of Wildt and the observations of Dunham.

The discovery by Menzel, Payne, and Boyce, that many lines in novae and O stars of previously unknown origin are due to forbidden lines in atoms of ordinary elements, also serves to show that astrophysics in just the field where forbidden transitions may be studied with advantage.

The story of the role of atomic theory in astrophysics has not been told in full, but enough has been said to indicate how it intervenes at strategic points, and how it promises in due time to reveal a full picture of the constitution of stellar atmospheres. Already we begin to perceive in outline the possibilities which are lying ahead. We know enough now to maintain with considerable confidence that the study of the atmospheres is not an isolated field. On the contrary, the structure of the atmospheres is so intimately connected with the internal structure of the stars, that it is possible to derive from the spectra such typical properties of the star in bulk as the mass and the luminosity, which again leads to a knowledge of the distance of the star in space. In fact, stellar spectroscopy has in this way already threatened to supplant many methods of classical astronomy, and is likely to do so much more in the future.

68. THE CHEMISTRY OF THE COSMOS

By Rupert Wildt

Astronomy and chemistry both arose in the elusive dawn of human contemplation of nature and its mysterious forces, but the birth of the age of modern science was marked by the transition from a kinematic to a dynamic astronomy. Chemistry could not keep pace with the achievements of theoretical and practical astronomy, though an enormous wealth of empirical knowledge had been piled up by generations of chemists from the days of Newton. Newton himself had been profoundly concerned with the experimental science of matter. His paper "De Natura Acidorum" shows him pondering over the problem of chemical affinity, the full comprehension of which was not reached until late in the 19th century. The keenest observers were baffled by the variety and diffuseness of chemical processes, in contrast to the striking regularity of many a celestial phenomenon inviting interpretation by geometrical and mechanical models.

There is scarcely a better example of the persistent growth and fruition of rather ill-defined pre-scientific ideas than the historic development of the theoretical conceptions of chemistry. Affinity, valency and many other notions likewise indispensable to the chemist's vocabulary not only for a long time lacked a precise meaning, but preserved anthropomorphic ideas about the nature of the chemical forces. Yet the last decades have justified the slowly increasing confidence that all chemical phenomena are ultimately reducible to physical principles. The palpable substance of the ancient philosophers has given place to atomic and molecular models, whereby the diversity of seemingly disparate qualities of matter is conceived as a functional unity, and the *Mysterium Cosmographicum,* which Kepler sought in vain among the geometrical relations of the Platonic bodies and the proportions of the planetary system, has truly been proclaimed anew in the mathematical formalism of quantum theory.

The first rational convergence of astronomy and chemistry was brought about by the perception of the cosmic origin of the meteorites and by the invention of spectroscopy. There is no need to repeat here the story of the advancing spectroscopic exploration of the celestial bodies. A final triumph was Humason's securing of the spectra of faint extragalactic nebulae and his measuring of the apparent Doppler shift of the ionized calcium lines discovered in these objects which are many millions of light years away. The philosophic significance of this latest accomplishment has hardly been emphasized enough. Thus at the very outermost boundaries of the observable universe we find refuted once more Comte's famous thesis that the physical and chemical constitution of the celestial bodies would forever defy human curiosity. On this note of an unqualified statement there came to a close the strain of metaphysical speculation about the substance of the heavens. Peripatetic philosophy had held that the stars and the planets consisted of a matter different from the four elements of Aristotle and, therefore, called *quinta essentia*. This proposition was still widely discussed among the scholastics. Then the great cardinal Nicholas of Cusa took a new stand and taught the identity of terrestrial and cosmic matter. Similar ideas are said to be found in the scientific memoranda of Leonardo da Vinci. A comprehensive elucidation of these early trends of cosmochemical speculation is much to be desired, as they must have influenced the development of the exact sciences. Writings like Boyle's *Sceptical Chemist* and Huygens' *Cosmotheoros* reveal the familiarity of the fathers of modern science with these philosophical ideas. Huygens was deeply convinced of the uniformity of matter, inanimate and alive, throughout the cosmos, and so was Newton, but at their time such beliefs were without any strictly scientific foundation.

The rapid advance of analytical chemistry since the end of the 18th century has been accompanied by a careful examination of all matter accessible to the chemist's scrutiny. This first circumnavigation of the chemical globe led to the discovery of numerous elements and opened vistas of the systematic chemistry to come. Yet its most remarkable result was perhaps the realization of how small the number of chemical elements is which in innumerable combinations make up our terrestrial environment. In 1802 Howard proved that the meteorites contained only the well-known elements, a conclusion substantiated by all later investigators, and Count de Bournon pointed out the mineralogical peculiarities of these celestial fragments.

This direct proof of the fundamental unity of terrestrial and cosmic matter fell into a period when manifold cosmogonic and geological speculations flourished. Gradually novel notions began to take shape in the minds of such scientists as J. H. Davy and A. von Humboldt, but it was Chr. F. Schönbein who in 1838 coined the term geochemistry and mapped out the following programme: "We have to investigate in greatest detail the properties of every single geognostic formation; we have to find out as precisely as possible the relations in which these products stand to one another as regards their chemical nature, their physical properties and chronological succession, and likewise we have to compare carefully the products of the chemical forces still active today with those of the inorganic primeval world. In a word, a comparative geochemistry ought to be launched, before geognosy can become geology, and before the mystery of the genesis of our planet and its inorganic matter may be revealed." [1]

Although much detail remains to be cleared up, and the problem of the influence of living matter on the geochemical processes is only vaguely grasped, this proposal today may be regarded as fulfilled in broad outlines. Furthermore the idea now gains ground that the chemistry of the Earth is only a special chapter in the universal chemistry of planetary bodies. Moreover, contemporary astrophysical research has led to unexpected insight into the constitution and composition of stellar and nebular matter. Indeed, the spectroscopic analysis of the cosmic light sources has been carried to the point where it can confidently be asserted that matter is qualitatively identical and, to a large extent, similar in quantitative composition throughout the entire universe. Whereas the older astronomy mainly treated of the magnitude, the spatial arrangement and the motions of the celestial bodies, the substance of which was preponderatingly considered as mechanical mass, a great deal of research has recently been concentrated upon the material substrate of the cosmic phenomena, which has gained so much distinct individuality. Even the interstellar space, once thought of merely as an immense void affording freedom of motion for the pointlike stars, is now known to harbour huge amounts of disperse matter, both gaseous and solid.

On observing how the material aspect of cosmic phenomena has come into the focus of interest, the question may be raised whether it would not be a timely expansion of the notion of geochemistry to adopt the term cosmochemistry to designate the science which shall deal with matter under all cosmic conditions. No cogency can be claimed for such a generalization. Cosmochemistry has no specific methods, but geochemistry did not have such either. Another objection suggests itself, namely that astrophysics may well take care of the problems of cosmic matter. If the purely observational side, comprising photometry and

---

[1] *Poggendorf's Annalen der Physik und Chemie 45,* 263 (1838).

spectroscopy, be disregarded, the aim of astrophysics is a rather narrow one. An outstanding authority in the field of theoretical astrophysics, Prof. E. A. Milne, has maintained that theoretical astrophysics is as exact a science as geometry, and that to science as a whole it contributes theorems dealing with idealized models, which are then available for the ultimate amalgamation of theory and observation.

Descriptive astronomy has for its subject matter the celestial bodies, regarded as individuals, and in interpreting their features it is guided by those models furnished by theoretical astrophysics. Scientific cosmogony aims at the elucidation of the history of the celestial bodies, and this endeavor entails the intimate study of the behaviour of matter under all possible conditions of state realized in the cosmic expanse. It may be useful, therefore, to denote this partial aspect of the general cosmogonic problem by a name of its own—cosmochemistry offers itself as the most convenient one. Here the component 'chemistry' should be understood in its broadcast implications, meaning the science of matter in all its manifestations. Traditionally, formation and decomposition of the substances, affinity and reaction-velocity were the central conceptions of chemistry. Further, the physical properties of chemical substances occupied a modest place in the textbooks, and nobody felt quite certain whether they belonged properly to the realm of chemistry or to that of physics. In classical physics they appeared as the so-called material constants. Compared to the universal constants of nature, the legitimate possession of the physicists, they seemed like strangers. Only the new quantum mechanics has made them indigenous to physics and, moreover, has revealed their close relation to the chemical qualities proper. So we may look forward to a unified theory of matter, in which traditional chemistry will be merged with the physics of matter, and which may be called chemistry in the metaphorical sense alluded to before.

If it be understood this way, the scope of cosmochemistry is very wide. The aim of geochemistry has been to evaluate the quantitative composition of the Earth and to elucidate the genesis of the distribution of the elements and their compounds throughout the globe. Cosmochemistry has to comply with more far reaching demands. By virtue of its methods of research it has to handle a large amount of spectroscopic material, usually claimed by astrophysics. In fact, the first rational theory of stellar spectra, as given by Saha, was based on what was then considered a strictly chemical theorem, namely the theory of the reaction isochore (van t'Hoff, Nernst), and so had been Eggert's first treatment of the ionization equilibrium in a stellar interior. Meanwhile the entire theory of chemical equilibria and reaction velocities has been refounded on the solid ground of statistical mechanics, and it is significant that R. H. Fowler has deemed it appropriate to include a sketch of the theory of stellar atmospheres in his famous textbook of statistical mechanics. New triumphs for this science may be foreseen in the field of stellar constitution, the cardinal problem of which is now the liberation of energy by subatomic processes continually going on in the deep interior of the stars.

The experimental study of atomic transmutations and of the ensuing energy changes has reached the stage where the results of this 'nuclear chemistry' have a direct bearing on our conception of the constitution of stellar cores. There is a curious parallelism between the relation of the physics of stellar atmospheres to the structure of the electronic fringes of the atoms, on the one hand, and that of the constitution of the stellar cores to the structure of the atomic nuclei, on the other hand. It is to be hoped that again statistical mechanics will provide, in conjunction with the laboratory data of nuclear physics, the basis for an understanding of the constitution of a star in bulk and, indirectly, of stellar evolution. Fowler's interpretation of the white dwarf stars as bodies composed of degenerate matter has inaugurated a new branch of cosmochemical speculation, which concerns itself with the analysis of the properties of matter under extremely high pressures and temperatures. Consequently, during recent years a great many theoretical studies have centered around the hypothetical existence of dense stellar cores, supposed to consist of degenerate electron gas or pure neutron gas.

It is too early to judge the astrophysical implications of the discovery of new elementary particles, like the neutron, the positron, and the meson. But while laboratory research continues to present us with such surprises as the fission of heavy atomic nuclei under neutron bombardment, it may well be that many traits of our current notions of stellar constitution and evolution will unexpectedly and rapidly become obsolete. Still, there can be no doubt about the pivotal role the atomic transmutations play in the evolution of the stars. More than a hundred years ago Prout advanced the hypothesis that hydrogen atoms were the building stones of all the heavier atoms, for no better reason than the fact that the atomic weights are approximately integers if measured by that of hydrogen as unity. In consequence of the discovery of the neutron as a regular constituent of atomic nuclei the original version of this hypothesis, as far as it was meant to reveal the constitutive principle of the chemical elements, has been superseded. However, it may be said that Prout's idea survives in modified form, as our contemporary cosmogonists try to devise a scheme of stellar evolution according to which in a star consisting in the beginning of pure hydrogen all the heavier elements evolve by spontaneous synthesis. The recent researches of von Weizsäcker, Gamov, and Bethe, illuminating as they are, have not yet led to a consistent picture of the evolution of the atoms and the stars. However, they seem to account satisfactorily for the energy production of the normal stars, which is now believed to be derived from the transmutation of hydrogen into helium. A rather intriguing problem remains unsolved, namely how under stellar conditions the nuclei of the elements heavier than oxygen could have been built up by starting from hydrogen alone. Since no such process can be visualized at present, we are left with the stimulating hypothesis that the heavy elements may have existed before the agglomeration of matter into the stars observed now.

In the range of extremely high temperatures, the transformation of the elements and the liberation of energy, maintaining the radiation flux emanating from the surface of the self-luminous stars, form an integral part of cosmochemistry. Among the dark bodies, at the opposite end of the temperature scale, nature displays an overwhelming variety of features. With the condensation of matter from shapeless gases into liquids and solids there appear a multiplicity of forms which reach their zenith in the living organisms. In studying the planetary bodies as prospective abodes of life, cosmochemistry even embraces the material basis of the physiological processes. The material constituents of these low-temperature phenomena are the molecules, varying in size between the diatomic compounds, as encountered in the solar atmosphere, and the giant molecules forming the colloidal systems of the terrestrial biosphere.[2] Passing from the sun to the cooler fixed stars, we find their atmospheres populated by increasing numbers of diatomic molecules, the dissociation equilibrium of which is ruled by the same principles of statistical mechanics as the thermal ionization in stellar atmospheres.

Polyatomic molecules like water vapor, carbon dioxide, ammonia and methane abound in the atmospheres of different planets. While these molecules are thermally stable at the comparatively low temperatures now prevailing at the planetary surfaces, they are subject to manifold photochemical reactions induced by the ultraviolet radiation of the sun, most of which are not yet fully understood for lack of adequate laboratory data. Quite recently it has been suggested, by E. Desguin and A. Dauvillier, that the photochemical formation of complicated organic

[2] [For Wildt's report on "Molecules in Stars," see *Mémoires Roy. Soc. Sci. Liege* (4) *18*, 319–331 (1957).]

compounds under the influence of ultra-violet sunlight may have been the *primum agens* responsible for the spontaneous generation of living matter in the primordial terrestrial ocean. The gaseous envelopes of the solid matter constituting the heads of the comets are also the seat of important photochemical reactions, which produce those diatomic molecules which are the carriers of the characteristic band spectrum of the comets. Solid inorganic matter is found throughout the universe in aggregates of all sizes between the finest interstellar dust and the bulky planets. In order to deal with the intricate problems of condensed cosmic matter and to reconstruct the hypothetical evolution of the planetary bodies, cosmochemistry has to resort to the thermodynamical theory of phase equilibria and heterogeneous chemical equilibria. Here geochemistry has its place in the wider frame of planetary chemistry. Especially the internal constitution of the giant planets involves exceedingly interesting chemical problems, on account of the high pressures realized in the inner zones of Jupiter and Saturn.

The chemical nature of the solid cometary matter is still somewhat obscure, though it is generally supposed to resemble closely and to be genetically related to the meteors. Their origin however is itself uncertain. It may well be that the genesis of the meteors which are members of the solar system is fundamentally different from that of the interstellar ones. The petrographic structure of the majority of meteorite specimens gathered at the earth's surface proves them to be products of the crystallization of a liquid refractory phase. If meteors are formed by slow condensation from the cold dilute interstellar gas, a hypothesis now seriously entertained, such particles may conceivably be expected to have a peculiar structure, which should be predictable from the experiments and theories on the formation of condensations from a vapor phase. The comparison of such a prediction, which does not seem to have been undertaken so far, with the more unusual types of meteorites of predominantly amorphous structure would be of great interest.

The problem of the condensation of solid particles from a gas phase also arises in studying the atmospheres of the coolest stars, many of which appear to have surface temperatures so low as to permit the precipitation of the most refractory substances. On this account the irregular variations in brightness of certain types of variable stars have tentatively been attributed to the transitory and recurrent formation of veils of solid particles either in their atmospheres or, after ejection of matter from the stars, in their immediate surroundings.[3]

[3] [Compare P. W. Merrill's "veils" as possible factors in long-period and irregular variation *The Nature of Variable Stars* (Macmillan, New York, 1938), chap. 8.]

These examples may suffice to outline the scope of cosmochemistry. It is obvious how far the analysis of the structure and history of cosmic matter is from completion. There are many sciences which contribute to this end, viz. the statistical and thermodynamical theory of matter, nuclear physics, the theory of spectra and that old-fashioned chemistry, which Sir Arthur Eddington once has delineated by jestingly upholding the notion "that atoms are physics, but molecules are chemistry." To integrate these diverse subjects into the conception of cosmochemistry may seem a sweeping generalization. Yet it will serve a heuristic purpose by emphasizing certain aspects of the cosmic phenomena, the intrinsic coherence of which can perhaps be characterized best by contrasting them to what is now called cosmology. This term has come to be used in the restricted sense of denoting the study of the geometrical and mechanical side of the universe at large, which are indeed, as the metrics of the universe, conceived as a unity by the theory of general relativity. By a somewhat liberal interpretation of this definition, cosmology also would include what goes under the name of the statistical and dynamical astronomy of the stars and nebulae of our galaxy. Just as general relativity and the quantum theory of matter are still coordinate in modern physics, the conception of cosmochemistry may be regarded as complemental to that of cosmology. If it be permitted to close with a paraphrase, cosmology and cosmochemistry might be said to treat of the universe *more geometrico* and *sub specie materiae,* respectively.

69. FIFTY YEARS OF PROGRESS IN ASTROPHYSICS

By Otto Struve

Fifty years ago the science of astrophysics had just been recognized as a separate branch of astronomy, and when George E. Hale of Chicago founded the Yerkes Observatory at Williams Bay, Wisconsin, he decided to devote the new 40-inch refracting telescope primarily to the study of the physical properties of the sun and the stars. The most important problem which confronted him and his associates was to determine, if possible, the chemical constitution of the various celestial bodies.

Astronomers already knew from the work of the early spectroscopists that the spectrum of the sun consists of a continuous band of radiation whose intensity distribution resembles that of a black body at a temperature of approximately 6000°, and of a series of narrow absorption lines scattered throughout the continuous spectrum whose origin could be attributed to the vapors of the common elements known on the earth.

The work of H. A. Rowland at Johns Hopkins University, which was published in the *Astrophysical Journal* between 1895 and 1897, had led to the identification of thousands of solar absorption lines with the laboratory lines of such common elements as iron, calcium, chromium, manganese, nickel, titanium, hydrogen, etc. It was also known that among the fixed stars there were many whose spectra closely resembled that of the sun, but there were many more that were entirely different. Not only were the continuous spectra of certain stars richer in blue light or in red light than the spectrum of the sun but there were also larger differences in the absorption lines. The blue stars, for example, possessed strong lines of hydrogen and weak lines of helium. Many of them had no lines of the metals, such as iron. On the other hand, the yellow stars had relatively weak lines of hydrogen, no helium, but very strong lines of iron and other metals. The red stars were even known to have absorption bands corresponding to some of the well-known diatomic molecules, such as CH, CN, TiO, and others. It was not surprising that astronomers confronted with these remarkable spectroscopic differences concluded that there were corresponding differences

Fig. 1. The 40-inch refractor of the Yerkes Observatory.

in the chemical compositions of the stars. The blue stars were designated as hydrogen or helium stars. The yellow ones as metallic stars, etc.

There must have been a trace of doubt in the minds of the most active astrophysicists, even as long ago as in 1897. The fact that the apparent chemical composition was so closely related to the color of the star and hence to its temperature must have caused them some concern: could it not be possible that the high temperature of the blue stars somehow tended to suppress the absorption lines of the ordinary metals and enhance those of the light gases, and was it not reasonable to suppose that the low temperatures of the red stars might favor the origin of molecules?

It took a long time to find the answers to these questions. The work of Sir Norman Lockyer in England was perhaps the greatest single contribution in this field. He discovered that a chemical element, such as iron, produces different line spectra at different temperatures. When the temperatures are barely sufficient to vaporize the iron terminals of an electric arc, certain lines appear in the spectrum which are characteristic of the element but which can be observed only if the temperature does not exceed a very definite value. Above that critical temperature the so-called "low-level" lines are weak or absent and a new pattern of lines makes its appearance. These "enhanced" lines again give way to a pattern of lines of still greater enhancement which appear at temperatures greatly in excess of those required to produce either the first or the second sets. Lockyer correctly suggested that the spectra of stars of different temperatures should show the different groups of lines and that it should be possible to obtain an estimate of the temperature from the particular pattern of lines of each element observed in a star.

The final solution of this problem came in 1919 when the Indian physicist Saha announced his famous theory of ionization. As long as an atom of iron possesses its entire retinue of twenty-six electrons, the pattern of spectral lines is that observed at low temperatures in the laboratory or in relatively cool stars like the sun. In the atmospheres of such stars the outermost electron among the twenty-six jumps from one orbit to another and thereby produces the characteristic "low-level" lines. When the temperature rises above a certain value the energy of the collisions between the various particles in the gas causes certain iron atoms to lose their outermost electrons so that only twenty-five are left. It is then the twenty-fifth electron which is essentially responsible for the energy changes within the atom and the corresponding "enhanced" emission of lines. These lines are quite different from those produced by the normal atom. The ionized atoms and the liberated free electrons move about at random so that occasionally a free electron may approach an ionized atom and be captured by the latter. There will be established after a sufficiently long interval of time an equilibrium between the number of processes of ionization and the number of captures. This equilibrium will, of course, depend greatly upon the pressure of the gas. When the pressure is low the distances between the individual particles are great. Processes of collision will occur at relatively infrequent intervals while processes of ionization, depending only upon the intensity of the light from the star, are independent of the pressure. Hence ionization will be favored at low pressures and will be retarded when the pressures are great.

The theory of Saha led to an interesting prediction. Suppose we observe the spectra of two stars whose temperatures, as determined from their continuous spectra, are the same. If the pressures in the atmospheres of the two stars are also the same, then according to Saha, the spectra should be identical—provided of course that there are no real differences of chemical composition. If, however, the pressure in one star is greatly in excess of that in the other the normal spectrum of iron should appear in the one of higher pressure and the enhanced spectrum of iron should be observed in the star of lower pressure. This prediction has been amply confirmed by a long series of investigations in which astronomers of many countries have taken an active part. Perhaps the greatest progress was made by W. S. Adams and his associates at the Mount Wilson Observatory. They discovered, even before the theory of Saha had been announced and fully accepted, that it was possible, from the relative intensities of the enhanced and normal lines, to estimate the intrinsic brightness of the stars. It became at once evident that a very luminous star is one of large volume and consequently low pressure and density, while a star of low luminosity is one of relatively small size and therefore of high pressure and density. The work at Mount Wilson Observatory on the luminosity effects in the spectra of the stars was one of the greatest advancements of science in the early 1920's. It not only gave a complete confirmation of the theory of ionization, but it also provided a tool of great power for the determination of the distances of very remote objects. It was possible to conclude that if two stars have the same apparent brightness—for example, if both appear to the eye as objects of the sixth magnitude—and if from the spectroscopic criteria it is found that one star is 10,000 times more luminous than the other, then their distances must be very different. Hence, if the distance of the nearer star could be measured by ordinary trigonometric methods, the distance of the other would be derived by multiplying the observed value by a factor of $\sqrt{10,000} = 100$. It is not an exaggeration to say that almost all our knowledge of the structure of the Milky Way which has developed during the past quarter of a century has come from the Mount Wilson discovery of spectroscopic luminosity criteria.

The theory of ionization dominated astrophysics for a period of ten or fifteen years. Great progress was made in the development of this theory by H. N. Russell at Princeton and E. A. Milne at Oxford. Equally significant advances were made at several observatories, including Mount Wilson, Lick, Harvard, Yerkes and many others. It seemed for a while that the large differences in the spectra of the stars could be entirely accounted for by differences in the pressures and temperatures of their outer layers.

It was certainly one of the most important conclusions of all physical science that there exist in the universe no chemical elements which have not been recognized on the earth and that, in turn, all those elements whose abundances on earth are large enough, whose spectra are excited under conditions of temperature and pressure similar to those of the stars and whose lines, furthermore, fall within the region of the spectrum which can be observed with astronomical instruments, were actually found in the sun and stars. For a number of years progress in astrophysics depended upon the assumption that the abundances of the elements were strictly the same in all stars and nebulae and that any remaining differences in their spectra must be accounted for by differences in the physical conditions.

During the early 1930's various observations were made which registered appreciable spectroscopic differences among the stars that could not be explained by means of the theory of ionization. It was found, for example, that the relative intensities of the members of a single multiplet of neutral iron could be quite different in different stars. In many simple multiplets the theoretical relative intensities or transition probabilities were already known from the sum rules of Burger and Dorgelo and their various extensions. These relative intensities had been verified by means of laboratory measurements. It was found that they would also apply in the case of absorption lines observed in certain stars. However, in other stars the relative intensities of the multiplet members were proportional to the square roots of the theoretical values, and in still others the range between the strongest and the weakest lines would be even less than would correspond to these square roots. Significantly, no cases were found in which the stellar absorption-line intensities had a greater range than that predicted by the multiplet rules. At first this phenomenon seemed quite obscure. It was not directly related to the temperature because the energy levels within a single multiplet are nearly identical. Hence differences in the degree of excitation could produce only negligible differences in the relative intensities within a multiplet.

Further investigations led to the discovery of an entirely new and unexpected phenomenon. When the observed intensities of the stellar absorption lines are plotted against the theoretical transition probabilities a curve is obtained which has received the name "the curve of growth." This curve has a somewhat complicated appearance. For the smaller transition probabilities the intensities increase, falling upon an approximately straight line whose slope is approximately 45° when the conventional units are used for the coordinates. The strongest absorption lines, which correspond to large transition probabilities, again fall upon a nearly straight line, but its slope is only 22.5°. The intermediate portion of the curve represents the transition from one straight line to the other, and this section is approximately horizontal with a slight amount of curvature at the two ends where it joins the two straight lines. This peculiar looking curve of growth is different for different stars. In some, the intermediate, horizontal section occurs for relatively weak absorption lines, while for other stars it occurs when the absorption lines are fairly strong. It has been shown that this difference is one that depends upon the degree of turbulence or mass motion within the atmospheres of the stars. There are currents of gas, some moving outward, others inward, which can be compared with the thermal turbulence on a sunny day in the atmosphere of the earth.

We have not yet reached a complete understanding of the turbulent motions in stellar atmospheres. For example, we do not definitely know whether the turbulent cells are large or are very small compared to the thickness of the entire atmosphere. It was at first suggested that the turbulent cells were relatively small in size so that a great many of them having different motions, could exist simultaneously one above the other. The combined absorptions of all such cells would then reproduce with great faithfulness the best determined curves. However, in recent months several curves of growth have been published, especially by K. O. Wright of the Dominion Astrophysical Observatory at Victoria, B. C., which throw some doubt upon this assumption. Perhaps we should rather suppose that the turbulent cells are not small but are more like a field of dense prominences similar in structure to those of the sun, which surround the stars. Some of these prominences would be moving outward; others would be moving inward. Some might move tangentially to the surface of the star or at different inclinations. However, there is reason to believe that the forces which activate the prominences of the stars are directed mainly along the radii so that the predominant motions would also be inward or outward along each radius. This, if confirmed by further investigations, would indicate that the turbulent motions are not random ones, as had been assumed in the earlier investigations. There might well be a tendency to produce a distribution with two separate maxima, one corresponding to an outward motion, the other to an inward motion. Observations by W. S. Adams at Mount Wilson have shown that such a tendency is indeed observed in several bright stars. For example, in the supergiant Betelgeuse many lines are not simple symmetrical structures as they are in the sun. Instead they have two unequal depressions, one of which corresponds to motions away from the star, while the other corresponds to motions inward.

All of this work would have been relatively unimportant if it were not for the fact that we have finally reached the stage where we can begin to talk intelligently about the abundances of the different elements in the atmospheres of the stars. It has been known for some time that hydrogen is enormously more abundant in the sun and in most other stars than it is on the earth. Roughly speaking there are 10,000 hydrogen atoms for a single atom of a metal. A similar abundance of hydrogen has been established in the case of most ordinary stars, as well as in the nebulae and in the exceedingly rarified gas of interstellar space. Because of this prevalence of hydrogen in the Universe, it is of special interest to record those objects in which hydrogen is absent or, at any rate, of small abundance. Foremost among these objects is the remarkable variable star R Coronae Borealis. Its spectrum abounds in absorption lines of all the usual metals, but the lines of hydrogen are amazingly weak—so weak in fact that we must conclude that the abundance of hydrogen is one hundred or one thousand times less than it is in the sun. Another equally interesting star is Upsilon Sagittarii, an object which has been investigated in great detail by J. L. Greenstein at the Yerkes Observatory. Its spectrum resembles that of Alpha Cygni, or Deneb. But in the latter the hydrogen lines are very strong, while in Upsilon Sagittarii they are exceedingly weak. Greenstein concluded that in this star hydrogen is only one hundred times as abundant as iron, and that helium is nearly one hundred times as abundant as hydrogen. In a list of chemical elements arranged according to abundance the relative places of hydrogen and helium are therefore interchanged. It is tempting in this connection to think of the nuclear transmutations which take place in the interiors of the stars and thereby give rise to the light and heat radiated by them into space. According to H. A. Bethe of Cornell University, the predominant nuclear process in the stellar interiors consists of the gradual conversion of hydrogen into helium. This should slowly change not only the luminosity of a star but also its spectrum. However, before we can make safe deductions from our observations it is necessary to have an understanding of the process which leads to the formation of the stars and their gradual evolution. Might it not be possible that some stars were originally formed from a cloud of interstellar matter that was rich in hydrogen while other stars originated in a region of low hydrogen content? The interpretation of the observed abundances of the elements must therefore be supplemented by a careful study of possible regional effects. There is little theoretical information to lead us. We know, or we think we know, that atoms in interstellar space may collide and form molecules. These in turn may start to grow through the process of combining with other molecules so that we may gradually have a medium consisting of small particles, grains of matter too large to be called molecules. Through the work of L. Spitzer at Princeton and F. Whipple at Harvard, we believe that clouds of such particles will tend to form condensations which will gradually attract material from the surroundings and grow into stars. Such embryonic stars must be peculiar in their properties. They will be small in mass and luminosity and there is reason to believe that their spectra will be unusual.

A few years ago A. H. Joy at the Mount Wilson Observatory announced the discovery of some 40 stars with unusual spectra located in and near the famous obscuring dust clouds in the constellation Taurus. More recently a star of the fifteenth apparent magnitude was found at the McDonald Observatory in Western Texas to possess an especially interesting spectrum. Its lines of hydrogen, helium and calcium are not in absorption, as they are in all normal stars, but appear as strong bright lines on top of a blue continuous spectrum. The star is seen projected upon the very center of an exceedingly dark and opaque cloud of cosmic dust. It is located on the nearer side of the dust cloud and produces something like a halo of diffuse light in the material of the dust cloud, just as a street light produces a visible glow around it on a foggy night. Since the distance of the dust cloud is known, and since the star must be quite close to it, we know the distance of the star itself. We can therefore compute the intrinsic candlepower of the star. It turns out to be approximately 100,000 times fainter than a normal "full-grown" star of the same color. Hence, it must be very small in size. But we have as yet no knowledge of its mass and cannot compute the density of the matter within it.

It is possible that the star which we have described is a newly formed object produced as a condensation in the great dust cloud of Taurus. It is as yet much smaller and much less luminous than an ordinary star and it is probably only part way through the process of gathering up the dust around it. The impact of the particles of the cloud with the newly formed star could well explain the bright lines of hydrogen, helium, and calcium. As long as the star remains intrinsically faint, the surrounding grains of dust in the great cloud of Taurus will presumably continue falling into the newly-formed star, evaporating on the way and adding their mass and kinetic energy to the star. The latter will thus grow in mass and in size, and probably also in intrinsic luminosity. Gradually, the increasing candle power of the star will presumably tend to reduce the infall of the dust-particles through the action of radiation pressure. This pressure is always present when a beam of light falls on a particle, and it was measured in the laboratory, many years ago, by the Russian physicist Lebedeff. In the case of an ordinary star the light pressure exerted upon a very small particle is many times greater, in amount, than the force of gravity. Hence, all small particles are driven away from the stars, and are not falling into their surfaces. But an embryonic star, like the one in Taurus, has relatively little light, and the light pressure which it exerts must be quite small.

What happens in this case is the following. A particle of dust located on our side of the great cloud in Taurus will be subjected to the pressure of radiation from all the stars of the Milky Way, except those which are hidden by the obscuring cloud. The pressure will therefore be unsymmetrical, and the particle will be accelerated towards the cloud and the embryonic star associated with it. This motion will be resisted by the opposite pressure of the light of the embryonic star. Only when this star has grown in size will its own radiation pressure be sufficient to prevent the infall of dust particles.

APPENDIX

IDENTIFICATION AND SOURCES

Selection 1. George Ellery Hale, 1868–1938, Mount Wilson Observatory; founder of Kenwood, Yerkes, Mount Wilson, California Institute of Technology, and Palomar Observatories. From "The Astrophysical Observatory of the California Institute of Technology," *Astrophysical Journal 82*, 111–139 (1935).

Selection 2. Bernhard Schmidt, 1879–1935, Hamburg Observatory, Bergedorf, Germany. From "Ein lichtstarkes komafreies Spiegelsystem," *Mitteilungen der Hamburger Sternwarte in Bergedorf 7*, 15–17, (1932). Translated by Dr. Nicholas Ulrich Mayall, *Publications of the Astronomical Society of the Pacific 58*, 285–289 (1946).

Selection 3. Bernard Lyot, 1897–1952, Pic du Midi, France. From "A Study of the Solar Corona and Prominences without Eclipses," *Monthly Notices of the Royal Astronomical Society 99*, 580–594 (1939).

Selection 4. Meghnad Saha, 1893–1956, Universities of Alahabad and Calcutta, India; Visiting Professor, Harvard Observatory, 1937. From "A Stratosphere Solar Observatory," *Harvard Observatory Bulletin*, No. 905, 1–7 (1937).

Selection 5. Grote Reber, 1911–   , private observatory, Wheaton, Illinois; later with the U. S. National Bureau of Standards. From "Cosmic Static," *Astrophysical Journal 91*, 621–624 (1940).

Selection 6. George Ellery Hale, 1868–1938, Mount Wilson Observatory. From "The Earth and Sun as Magnets," *Smithsonian Report for 1913*, 145–158.

Selection 7. Edison Pettit, 1889–   , Mount Wilson Observatory. "Solar Prominences," *Astronomical Society of the Pacific*, Leaflet 137 (1940).

Selection 8. Bengt Edlén, 1906–   , Uppsala Observatory, Sweden; now at Lund Observatory. From "An Attempt to Identify the Emission Lines in the Spectrum of the Solar Corona," *Arkiv för Matematik, Astronomi och Fysik, 28B*, No. 1 (1941), and from the George Darwin Lecture, *Monthly Notices of the Royal Astronomical Society 105*, 332–333 (1945).

Selection 9. Walter Orr Roberts, 1915–   , High Altitude Observatory, University of Colorado, Boulder, Colorado. From "A Preliminary Report on Chromospheric Spicules of Extremely Short Lifetime," *Astrophysical Journal 101*, 136–140 (1945).

Selection 10. Ernest William Brown, 1866–1938, Haverford College, Pennsylvania; later at Yale University. From "On the Completion of the Solution of the Main Problem in the New Lunar Theory," *Monthly Notices of the Royal Astronomical Society 65*, 104–108 (1904).

Selection 11. Charles Pollard Olivier, 1884–   , Leander McCormick Observatory, University of Virginia; later at the University of Pennsylvania. From Olivier's "Meteors," Williams and Wilkins, Baltimore (1925), Chapter XIII.

Selection 12. Hisashi Kimura, 1870–1943, Mizusawa, Japan. From "The International Latitude Service," *Vierteljahrschrift der Astronomische Gesellshaft 63*, 359–361 (1928). Translated by Kimura, with verbal changes by Shapley.

Selection 13. Edison Pettit, 1889–   , and Seth Barnes Nicholson, 1891–   , Mount Wilson Observatory. From "Lunar Radiation and Temperatures," *Astrophysical Journal 71*, 102–103 (1930).

Selection 14. Clyde William Tombaugh, 1906–   , Lowell Observatory, Flagstaff, Arizona; now at Las Cruces, New Mexico. "The Search for the Ninth Planet, Pluto," *Astronomical Society of the Pacific*, Leaflet 209 (1946).

Selection 15. Leonid Alekseevich Kulik, 1883–1942, Committee on Meteorites of the Academy of Sciences, Moscow, USSR. A paper read at the fifth annual meeting of the Society for Research on Meteorites, Denver, Colorado, June 1937. Translated by E. L. Krinov.

E. L. Krinov, 1906–   , Committee on Meteorites of the Academy of Sciences, Moscow, USSR. The comments on Kulik's paper are taken by permission from a personal letter of November 1958.

Selection 16. Fred Lawrence Whipple, 1906–   , Harvard Observatory; later also director of the Smithsonian Astrophysical Observatory. From "Upper Atmosphere Densities and Temperatures from Meteor Observations," *Popular Astronomy 47*, 419–425 (1939).

Selection 17. Harold Spencer-Jones, 1890–   , Astronomer Royal (retired), Royal Observatory, Greenwich and Herstmonceux, England; formerly His Majesty's Astronomer at the Cape of Good Hope. From a report to the Royal Astronomical Society on June 13, 1941; *The Observatory 64*, 99–104 (1941).

Selection 18. Gerald Maurice Clemence, 1908–   , U. S. Naval Observatory, Washington, D. C.; Dirk Brouwer, 1902–   , Yale University Observatory, New Haven, Connecticut; Wallace John Eckert, 1902– , Watson Scientific Computing Laboratory, New York City. "The Motions of the Five Outer Planets," *Research Reviews* (Office of Naval Research), December pp. 6–13 (1950).

Selection 19. Jacobus Cornelius Kapteyn, 1851–1922, University of Groningen, Holland; Visiting Scientist, Mount Wilson Observatory. From a paper presented at the South African meeting of the British Association in 1905, and printed in the report of that meeting, Section A, 257ff (1905).

Selection 20. Frank Schlesinger, 1871–1943, Yale University Observatory. From the first George Darwin Lecture, "Some Aspects of Astronomical Photography of Precision," *Monthly Notices of the Royal Astronomical Society 87,* 506–523 (1927).

Selection 21. Grigory Abramovich Shajn, 1892–1956, Simeis Observatory, USSR, and Otto Struve, 1897–   , Yerkes Observatory; now director of the National Radio Astronomy Observatory, Green Bank, West Virginia. From "On the Rotation of the Stars," *Monthly Notices of the Royal Astronomical Society 89,* 222–239 (1929).

Selection 22. Ralph Elmer Wilson, 1886–   , Dudley and Mount Wilson Observatories, Carnegie Institution of Washington. "The General Catalogue of Stellar Positions and Proper Motions," *Astronomical Society of the Pacific,* Leaflet 128 (1939).

Selection 23. Viktor Amazaspovich Ambartsumian, 1908–   , Erevan Observatory, USSR. From "Zvezdnye Assotsiatsii," *USSR Astronomical Journal 26,* 1–9 (1949). Translated for this volume by Mrs. Zdenka Kadla-Michailov of Poulkova.

Selection 24. William Hammond Wright, 1871–1959, Lick Observatory. From "On a Proposal to use the Extragalactic Nebulae in Measuring the Proper Motions of Stars and in Evaluating the Precessional Constant," *Proceedings of the American Philosophical Society 94,* 1–12 (1950).

Selection 25. Annie Jump Cannon, 1863–1941, Harvard Observatory. From "The Henry Draper Memorial," *Journal of the Royal Astronomical Society of Canada 9,* 203–215 (1915).

Selection 26. Walter Sydney Adams, 1876–1956, Mount Wilson Observatory, and Arnold Kohlschütter, 1883–   , Mount Wilson Observatory, later at Bonn University, Germany. From "Some Spectral Criteria for the Determination of Absolute Stellar Magnitudes," *Astrophysical Journal 40,* 385–398 (1914).

Selection 27. Walter Sydney Adams, 1876–1956, Mount Wilson Observatory. From "Stellar Distances and Stellar Motions," *Scientia, 32,* 289–300 (1922). *Cf.* 68 below.

Selection 28. Ira Sprague Bowen, 1898–   , California Institute of Technology. From "The Origin of the Nebular Lines and the Structure of the Planetary Nebulae," *Astrophysical Journal 67,* 1–15 (1928).

Selection 29. William Wilson Morgan, 1906–   , Philip Childs Keenan, 1908–   , and Edith Kellman, 1911–   , Yerkes Observatory. From "An Atlas of Stellar Spectra, with an Outline of Spectral Classification," *Astronomical Monographs,* University of Chicago Press (1943), pp. 1–6.

Selection 30. Willem Jacob Luyten, 1899–   , University of Minnesota. "The White Dwarfs," *Astronomical Society of the Pacific,* Leaflet 202 (1945).

Selection 31. Paul Willard Merrill, 1887–   , Mount Wilson Observatory. "Iron in the Stars." *Astronomical Society of the Pacific,* Leaflet 233 (1948).

Selection 32. Henrietta Swan Leavitt, 1868–1921, Harvard College Observatory. From "Periods of 25 Variable Stars in the Small Magellanic Cloud," *Harvard Circular,* No. 173 (1912).

Selection 33. Harlow Shapley, 1885–   , Mount Wilson Observatory; later at Harvard Observatory. From "On the Nature and Cause of Cepheid Variation," *Astrophyical Journal 40,* 105–122 (1914).

Selection 34. Alfred Harrison Joy, 1882–   , Mount Wilson Observatory. From "Radial Velocities of Cepheid Variable Stars," *Astrophysical Journal 86,* 363–436 (1937).

Selection 35. Joel Stebbins, 1878–   , Lick Observatory; formerly at the Washburn Observatory of the University of Wisconsin. From "Six-Color Photometry of Stars," *Astrophysical Journal 101,* 47–55 (1945), and *103,* 108–113 (1946).

Selection 36. Zdenek Kopal, 1914–   , Harvard Observatory and the Massachusetts Institute of Technology; now at Manchester University, England. From a communication to *Centennial Symposia* (Harvard Observatory Monographs No. 7; Harvard College Observatory, Cambridge, 1946), pp. 261–262.

Selection 37. Albert Abraham Michelson, 1852–1931, University of Chicago and Mount Wilson Observatory, and Francis Gladheim Pease, 1881–1938, Mount Wilson Observatory. From "Measurement of the Diameter of α Orionis with the Interferometer," *Astrophysical Journal 53,* 249–259 (1921).

Selection 38. Arthur Stanley Eddington, 1882–1944, Cambridge, England. From *Stars and Atoms* (Yale University Press, New Haven, 1927), pp. 9–28.

Selection 39. Gerard Peter Kuiper, 1905– , Yerkes Observatory. From "The Empirical Mass-Luminosity Relation," *Astrophysical Journal 88*, 472–476 (1938).

Selection 40. Henry Norris Russell, 1877–1957, Princeton University. From "Stellar Energy," *Proceedings of the American Philosophical Society 81*, 295–307 (1939). See also Hans Albrecht Bethe, "Energy Production in Stars," *Physical Review 55*, 434–456 (1939).

Selection 41. Subrahmanyan Chandrasekhar, 1910– , Yerkes Observatory. From the Henry Norris Russell lecture, "Turbulence—A Physical Theory of Astrophysical Interest," *Astrophysical Journal 110*, 329–339 (1949).

Selection 42. Ejnar Hertzsprung, 1873– , Copenhagen; formerly at Potsdam and Leiden Observatories. From "Zur Strahlung der Sterne," *Zeitschrift für wissenschaftliche Photographie 3, Abt. 2* (1905), pp. 429–442. Translated by the editor.

Selection 43. Henry Norris Russell, 1877–1957, Princeton University. From "Relations between the Spectra and other Characteristics of the Stars," *Popular Astronomy 22*, 275–294, 331–351 (1914).

Selection 44. Robert Julius Trumpler, 1886–1956, Lick and Leuschner Observatories. From "Spectral Types in Open Clusters," *Publications of the Astronomical Society of the Pacific 37*, 307–318 (1925).

Selection 45. Walter Baade, 1893–1960, Mount Wilson and Palomar Observatories. From "The Resolution of M 32, NGC 205, and the Central Region of the Andromeda Nebula," *Astrophysical Journal 100*, 137–146 (1944).

Selection 46. Johannes Franz Hartmann, 1865–1936, Potsdam Observatory, Germany. From "Investigations on the Spectrum and Orbit of δ Orionis," *Astrophysical Journal 19*, 268–286 (1904).

Selection 47. Bart Jan Bok, 1906– , Harvard Observatory; later at Mount Stromlo, Canberra, Australia. From *The Distribution of the Stars in Space* (University of Chicago Press, Chicago, 1937), pp. 40–43.

Selection 48. Hendrik Christoffel van de Hulst, 1918– , Leiden Observatory, Holland. From "Origin of the Radio Waves," *Nederlandsch Tijdschrift voor Natuurkunde 2*, 201–221 (1945); translation by Elsa van Dien for Cornell University lectures.

Selection 49. William Albert Hiltner, 1914–   , McDonald Observatory, and John Scoville Hall, 1908–   , Amherst College, later at U. S. Naval Observatory. From "Polarization of Light from Distant Stars by Interstellar Medium," and "Observations of the Polarized Light from Stars," *Science 109*, 165–167 (1949).

Selection 50. ₐBart Jan Bok, 1906–   , Harvard Observatory; later at Mount Stromlo, Canberra, Australia. From "Dimensions and Masses of Dark Nebulae," *Centennial Symposia* (Harvard Observatory Monographs No. 7; Harvard College Observatory, Cambridge, 1948), pp. 55–72.

Selection 51. Lyman Spitzer, Jr., 1914–   , Princeton Observatory. From "The Formation of Stars," *Physics Today 1*, No. 5, 7–11 (1948).

Selection 52. Harlow Shapley, 1885–   , Harvard Observatory; formerly at the Mount Wilson Observatory. From *Star Clusters* (Harvard Observatory Monographs No. 2; McGraw-Hill, New York, 1930), pp. 171–178.

Selection 53. Jan Hendrik Oort, 1900–   , Leiden Observatory, Holland. From "Observational Evidence Confirming Lindblad's Hypothesis of the Galactic System," *Bulletin of the Astronomical Institutes of the Netherlands 3*, 275–282, (1927).

Selection 54. Edwin Powell Hubble, 1889–1953, Mount Wilson Observatory. From "A Relation between Distance and Radial Velocity among Extra-galactic Nebulae," *Proceedings of the National Academy of Sciences 15*, 168–173 (1929).

Selection 55. Edwin Powell Hubble, 1889–1953, and Milton Lasell Humason, 1891–   , Mount Wilson Observatory. From "The Velocity-Distance Relation among Extra-galactic Nebulae," *Astrophysical Journal 74*, 43–80 (1931).

Selection 56. Milton Lasell Humanson, 1891–   , Mount Wilson Observatory. From "The Large Apparent Velocities of Extra-galactic Nebulae," *Astronomical Society of the Pacific*, Leaflet 37 (1931).

Selection 57. James Hopwood Jeans, 1877–1946, Trinity College, Cambridge, England. "A Suggested Explanation of Radio-activity," *Nature 70*, 101 (1904).

Selection 58. Albert Einstein, 1879–1955. "Does the Inertia of a Body Depend upon its Energy Content?" Translated by the Editor from *Annalen der Physik 18*, 639–641 (1905).

Selection 59. Henri Poincaré, 1854–1912, Académie Française. From the preface of "Leçons sur les Hypothèses Cosmogoniques," Hermann and Co., Paris (1913).

Selection 60. William Wallace Campbell, 1862–1938, and Robert Julius Trumpler, 1886–1956, Lick Observatory. From "Observations on the Deflection of Light in Passing through the Sun's Gravitational Field, Made during the Total Solar Eclipse of September 21, 1922," *Publications of the Astronomical Society of the Pacific 35,* 158–163 (1923).

Selection 61. Arthur Stanley Eddington, 1882–1944, Cambridge, England. From the Presidential address to the Mathematical Association, delivered on January 5, 1931, *Nature 127,* 447–453 (1931).

Selection 62. Georges Lemaître, 1894–   , Louvain University, Belgium. "The Beginning of the World from the Point of View of the Quantum Theory," *Nature 127,* 706 (1931).

Selection 63. Willem de Sitter, 1872–1934, Leiden Observatory, Holland. From the chapter on "Relativity and Modern Theories of the Universe," in *Kosmos* (Harvard University Press, Cambridge, Mass., 1932), pp. 103–112.

Selection 64. Albert Einstein, 1879–1955. From "The World as I See It," New York (1934), pp. 78–81. Copyright held by Crown Publishers, New York.

Selection 65. Georges Lemaître, 1894–   , Louvain University, Belgium. From "L'Hypothèse de l'Atome Primitif: Essai de Cosmogonie," a lecture given at Freibourg in 1945; English edition: "The Primeval Atom: An Essay on Cosmogony," Van Nostrand, New York (1950), pp. 134–138, 139–144, 147–148, 161–163; translation by Betty H. and Serge A. Korff.

Selection 66. Hermann Bondi, 1919–   , Trinity College, Cambridge, England, now at Kings College, London, and Thomas Gold, 1920–   , Trinity College, now at Cornell University. From "The Steady-state Theory of the Expanding Universe," *Monthly Notices of the Royal Astronomical Society 108,* 252–270 (1948).

Selection 67. Svein Rosseland, 1894–   , Oslo, Norway. From *Theoretical Astrophysics,* Clarendon Press, Oxford (1936), pp. xi–xix.

Selection 68. Rupert Wildt, 1905–   , Princeton Observatory, now at Yale Observatory. From "Cosmochemistry," *Scientia 67,* 85–90 (1940); reprinted with the permission of *"Scientia,* International Review of Scientific Synthesis."

Selection 69. Otto Struve, 1897–   , Yerkes Observatory; later at University of California, Berkeley, now Director of National Radio Observatory, Green Bank, West Virginia. "Fifty Years of Progress in Astrophysics," *The Science Counselor 11,* 4–6, 26–27 (1948).

INDEX